Springer Series in Surface Sciences

Volume 60

Series editors

Roberto Car, Princeton, USA
Gerhard Ertl, Berlin, Germany
Hans-Joachim Freund, Berlin, Germany
Hans Lüth, Jülich, Germany
Mario Agostino Rocca, Genova, Italy

AF212774

This series covers the whole spectrum of surface sciences, including structure and dynamics of clean and adsorbate-covered surfaces, thin films, basic surface effects, analytical methods and also the physics and chemistry of interfaces. Written by leading researchers in the field, the books are intended primarily for researchers in academia and industry and for graduate students.

More information about this series at http://www.springer.com/series/409

Nong Moon Hwang

Non-Classical Crystallization of Thin Films and Nanostructures in CVD and PVD Processes

Springer

Nong Moon Hwang
Department of Material Science
and Engineering
Seoul National University
Seoul
Korea (Republic of)

ISSN 0931-5195 ISSN 2198-4743 (electronic)
Springer Series in Surface Sciences
ISBN 978-94-024-1395-3 ISBN 978-94-017-7616-5 (eBook)
DOI 10.1007/978-94-017-7616-5

© Springer Science+Business Media Dordrecht 2016
Softcover reprint of the hardcover 1st edition 2016
This work is subject to copyright. All rights are reserved by the Publisher, whether the whole or part of the material is concerned, specifically the rights of translation, reprinting, reuse of illustrations, recitation, broadcasting, reproduction on microfilms or in any other physical way, and transmission or information storage and retrieval, electronic adaptation, computer software, or by similar or dissimilar methodology now known or hereafter developed.
The use of general descriptive names, registered names, trademarks, service marks, etc. in this publication does not imply, even in the absence of a specific statement, that such names are exempt from the relevant protective laws and regulations and therefore free for general use.
The publisher, the authors and the editors are safe to assume that the advice and information in this book are believed to be true and accurate at the date of publication. Neither the publisher nor the authors or the editors give a warranty, express or implied, with respect to the material contained herein or for any errors or omissions that may have been made.

Printed on acid-free paper

This Springer imprint is published by Springer Nature
The registered company is Springer Science+Business Media B.V. Dordrecht

Preface

Thin films and nanostructures have a wide spectrum of applications such as microelectronics, optoelectronics, solar cells, flat panel displays, wear-resistant coatings, data storage media, sensors, micromechanical system (MEMS), and biological implants. Considering the huge number of publications in this field, it is not difficult to imagine how important this field is. Nevertheless, an understanding on the growth mechanism of thin films and nanostructures surprisingly lags behind. There are many puzzling phenomena regarding the growth of thin films and nanostructures, which cannot be explained by the textbook knowledge. In order to make a scientific, systematic, and efficient approach to this important area, the highest priority would be to understand the underlying principle of the microstructure evolution of thin films and nanostructures as well as the puzzling phenomena.

This effort should start with asking 'Why?' followed by the effort to answer the question. This is a typical way scientific progresses are made. By doing so, 'How?' can be approached most systematically and efficiently with trials-and-errors being minimized. This book will focus on 'Why' as to the puzzling microstructure evolutions of thin films and nanostructures.

The first question was "Why can diamond be synthesized at low pressure where graphite is more stable?" This question is answered based on the thermodynamic and kinetic concepts in Chap. 5. The next question was "Why do diamond crystals and porous skeletal soot deposit respectively on Si and Fe substrates under the same processing conditions?" The related figure is Fig. 6.1 of Chap. 6. Another question was "How is it possible that deposition of less stable diamond and etching of stable graphite occur simultaneously?" This puzzling phenomenon, which was experimentally observed, appears to violate the second law of thermodynamics. It turned out that these questions could not be answered with any conventional theory of crystal growth, where diamond films grow by individual atoms, in the textbook. Instead, these questions could be answered by a new growth mechanism, where diamond films grow by charged nanoparticles generated in the gas phase. It is a paradigm-shifting discovery about the growth mechanism of diamond films. Besides, this mechanism turns out to be very general in other chemical vapor deposition processes.

In other words, many thin films and nanostructures synthesized in the gas phase grow by non-classical crystallization, which has been studied recently and extensively in solution especially in biomineralization. Although non-classical crystallization is a relatively new and revolutionary concept in crystal growth, it has now become quite established that a few related books have been published and its tutorial and technical sessions have been included respectively in the spring meetings of Materials Research Society and European Materials Research Society in 2014. With the establishment of non-classical crystallization, many crystals that were believed to grow by atomic, molecular, or ionic entities turn out to grow actually by nanoparticles.

Chapter 1 is the general introduction of non-classical crystallization and its relation with the growth mechanism of thin films and nanostructures. Chapters 2–4 treat respectively thermodynamics of CVD, nucleation and growth, which are some fundamental concepts for the classical theory of crystal growth. For educational purposes, there are some questions for undergraduate seniors or graduate students to think upon before reading any further. Nevertheless Chaps. 2–4 contain some essential knowledge to understand Chaps. 5–7, linking the basic knowledge to non-classical crystallization of thin films and nanostructures. Chapter 5 covers the thermodynamic and kinetic concepts to explain the diamond synthesis at low pressure. Chapter 6 covers the diamond growth mechanism based on the theory of charged nanoparticles. Chapters 7–12 cover the extension of the theory to other systems. Chapter 13 deals with the charge-enhanced kinetics. Chapter 14 deals with the implications and applications of this new understanding.

I like to thank my former and present collaborators and students for their efforts to study the growth mechanism of thin films and nanostructures. I am indebted to Profs. Miyoung Kim, Seungwu Han, Heungnam Han, Seonghyeon Hong, Kibum Kim, and Jungjoong Lee in my department for their fruitful discussion and comments. I express my sincere gratitude to Dr. J. Cahn formerly at the National Institute of Standards and Technology, Prof. M. Hillert formerly at the Royal Institute of Technology, Prof. S. Iijima at Meijo University, Prof. J.K. Lee formerly at Michigan Technological University, Prof. I. Sunagawa formerly at Tohoku University, Prof. P. Bennema formerly at Nijmegen University, Prof. T. Yoshida formerly at the University of Tokyo, Prof. P. Roca i Cabarrocas at Ecole Polytechnique, Profs. K. Ostrikov and A.B. Murphy at CSIRO for fruitful advices, discussions and comments. I thank my graduate students, Jihye Kwon, Daseul Kim, Taeyoung Kim, Jonghwan Park, and Jungwoo Yang for their help in preparation of the manuscript related to drawing figures and references. Finally, financial support for the recent work from the Global Frontier R&D Program (2013M3A6B1078874) in Center for Hybrid Interface Materials, the National Research Foundation of Korea (NRF) Grant funded by the Government (MSIP) (No. 2015R1A5A1037627), and the Asian Office of Aerospace Research and Development (AOARD) of AFOSR is greatly appreciated.

Seoul, Korea (Republic of) Nong Moon Hwang

Contents

Chapter 1
Non-classical Crystallization

Crystals consist of atoms in a lattice, which are so regularly arrayed as to be diffracted by X-ray. After such an extraordinary regularity of atoms was discovered in crystals by X-ray diffraction, it is quite natural to believe that the building block of crystals should be individual ions, atoms or molecules. The theory of classical crystal growth was established based on this belief. However, a few scientists found out some puzzling phenomena that could not be explained by this classical mechanism. They claimed that in order to explain such puzzling phenomena, crystals should grow by the building block of nanoparticles. However, due to the limited experimental tools available at the time, they could not provide direct evidences to prove their claims and failed to convince those with strong beliefs in classical crystallization. Thanks to the development of advanced experimental tools such as transmission electron microscopy (TEM), now the time is ripe for providing convincing evidences for non-classical crystallization. Nanoparticle-mediated crystal growth is a new paradigm, which was actually discovered long time ago but rediscovered recently.

1.1 Classical Versus Non-classical Crystallization

In classical crystallization, the way of atomic incorporation into a crystal depends on whether the interfacial structure is atomically smooth or rough. On the smooth interface, atoms first land on the terrace, diffuse to the ledge and finally become incorporated into a crystal at the kink. On the rough interface, which consists of lots of kinks, there exists no barrier for atomic incorporation, which typically leads to diffusion-controlled growth, which again typically leads to dendritic growth.

Recently, however, increasing evidences indicating that crystals grow by the building block of nanoparticles were reported. The crystal growth by the building block of nanoparticles is called non-classical crystallization, which has been extensively studied over the past decade (Banfield et al. 2000; Qi et al. 2000;

© Springer Science+Business Media Dordrecht 2016

N.M. Hwang, *Non-Classical Crystallization of Thin Films and Nanostructures in CVD and PVD Processes*, Springer Series in Surface Sciences 60,
DOI 10.1007/978-94-017-7616-5_1

Alivisatos 2000; Cölfen and Mann 2003; Leite et al. 2003; Yu and Cölfen 2004; Zhang et al. 2005; Niederberger and Cölfen 2006; Gebauer et al. 2008, 2010; Gebauer and Cölfen 2011; Demichelis et al. 2011; Li et al. 2012; Teng 2013). This new concept is a paradigm changer, which is drastically different from the classical understanding of crystal growth. Although numerous evidences indicated that some crystallization including biomineralization is mediated by nanoparticles, they are indirect in the sense that the nanoparticle-based growth is not observed directly. Recently, the direct observation was made in solution using a liquid-cell in situ transmission electron microscopy (TEM) technique (Zheng et al. 2009; Liao et al. 2012; Yuk et al. 2012).

Figure 1.1 illustrates schematically classical versus non-classical crystallization pathways. The classical crystallization model (Fig. 1.1a) starts from primary building blocks like atoms, ions or molecules, forming clusters (Wohlrab et al. 2005; Niederberger and Cölfen 2006). These clusters may grow or shrink, depending on the relative magnitude of surface and bulk energies as described in the classical nucleation theory. Once clusters reach the size of the critical nucleus, they can continue to grow into macro crystals by the attachment of an individual atom or molecule.

In contrast to the classical crystallization pathway, non-classical crystallization events are particle mediated and involve a mesoscopic transformation process (Cölfen and Mann 2003). The main pathways of non-classical crystallization are

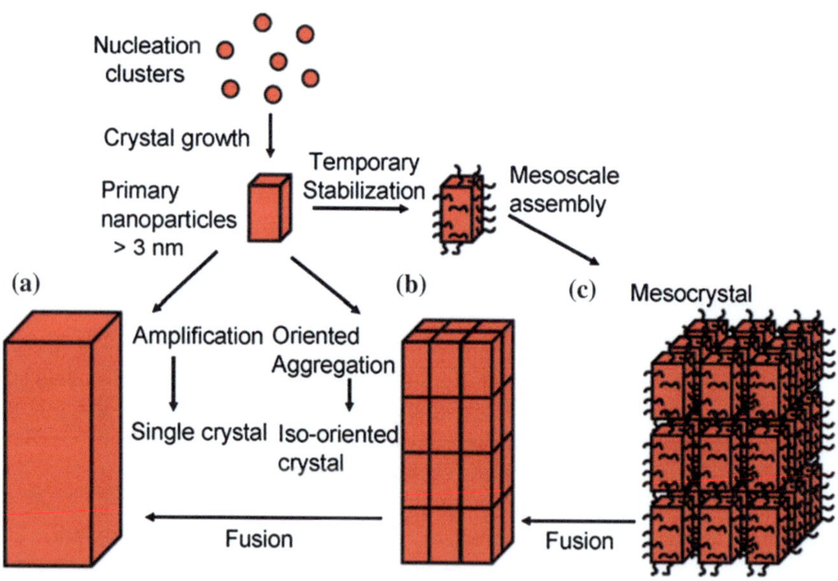

Fig. 1.1 Schematic representation of classical and non-classical crystallization. **a** Classical crystallization pathway, **b** oriented attachment of primary nanoparticles forming an iso-oriented crystal upon fusing, **c** mesocrystal formation via self-assembly of primary nanoparticles covered with organics (Wohlrab et al. 2005)

summarized in Fig. 1.1b, where an iso-oriented crystal grow by oriented attachment of primary nanoparticles, which can form a single crystal upon fusion of the nanoparticles. If the nanoparticles are coated by some organic components, they can form a mesocrystal via mesoscale assembly (path (c)).

Cölfen and Antonietti (2005) defined a mesocrystal as a superstructure of crystalline nanoparticles with external crystal faces on the scale of some hundred nanometers to a few micrometers. Mesocrystal intermediates can also lead by fusion to the formation of single crystals with included organic additives, sometimes allowing observation of highly oriented nanoparticle-based intermediates as shown in Fig. 1.1c. Mesocrystal intermediates clearly reveal the building block of nanoparticles. If these mesocrystal intermediates are heated at sufficiently high temperature, they can transform to a single crystal.

Because of organic components in Fig. 1.1c, the mesocrystal intermediates clearly reveal the kinetic path of crystallization. Therefore, the mesocrystal intermediates play an important role in discovering the mechanism of non-classical crystallization. However, when the kinetics follows the path of Fig. 1.1b, it would be difficult to distinguish from a final morphology of the crystal whether it grows by an individual atom or nanoparticle. This is why the crystal growth by nanoparticles had a great resistance in the crystal growth community in early years.

For example, more than 40 years ago, a concept similar to non-classical crystallization was suggested by Glasner et al. (Glasner and Kenat 1968; Glasner and Skurnik 1968; Glasner and Tassa 1974a, b) in their study of the crystal growth of KBr and KCl in the presence of Pb^{2+} in aqueous solution. In this case, nanometre-sized nuclei were formed in the solution and became the building blocks of the crystal. They confirmed the formation of these invisible nuclei in the solution indirectly by a thermal method, where the heat generated during cluster precipitation from the solution is measured.

They demonstrated that a nearly perfect crystal could grow by orderly packing or self-assembly of the nanosized block nuclei. The cluster size decreased with increasing concentration of Pb^{+2}. And with decreasing cluster size, the crystalline quality increased even though the kinetics of crystallization decreased significantly. Small clusters crystallized into a transparent crystal, while large clusters crystallized into an opaque crystal.

At that time, their suggestion was so revolutionary that they received severe criticism (Botsaris and Reid 1967) and subsequently did not receive any attention from the crystal growth community. Later, Sunagawa, who allegedly had a sabbatical leave in Glasner's group earlier, made a similar suggestion that the growth unit of synthetic diamonds is not an atom but a much larger unit based on their morphological features (Sunagawa 1987, 1990). Also Sunagawa's suggestion has not been taken seriously in the community of diamond synthesis.

In relation with such a historical background, Cölfen and Antonietti (2008) made a meaningful remark that, reanalyzing the literature, non-classical crystallization mechanism turns out to be a "rediscovery", as it seems that many important original observations are meanwhile forgotten and hidden in the past literature, as they simply did not comply with the classical crystallization model.

1.2 Theory of Charged Nanoparticles

In 1996, Hwang et al. (1996a, b) and Hwang and Yoon (1996) published 3 papers suggesting that diamonds synthesized at low pressure by chemical vapour deposition (CVD) grow by clusters formed in the gas phase. In other words, CVD diamonds grow by non-classical crystallization. Actually, two of those three papers had been submitted since 1992, but rejected for the 4 following years. This indicated how strong the resistance to this new paradigm was. Since it was difficult to provide a direct experimental evidence for the non-classical crystallization of diamonds at that time, a new manuscript was prepared to claim that the diamond deposition with simultaneous graphite etching, a well-known experimental observation in the CVD diamond community, violates the second law of thermodynamics unless diamonds grow by gas-phase nucleated diamond nanoparticles. The original writing of a reviewer's comment is shown in Fig. 1.2.

This is how the first paper (Hwang and Yoon 1996) on non-classical crystallization of diamonds was published. At that time, the new mechanism was called 'charged cluster model'. Later, it turned out that although in the case of CVD diamond, the size of nanoparticles, which is 1–2 nm, is small enough to be called clusters but the size in other CVD systems is so large that charged nanoparticles sounded more appropriate. Therefore, the new mechanism was called 'theory of charged nanoparticles (TCN) (Hwang and Lee 2010), where the role of electric charges in the nanoparticles is emphasized because charges play a critical role in non-classical crystallization.

According to the TCN, charged nanoparticles (CNPs) become the building block of diamond films when they deposit on the Si substrate; when CNPs deposit on the Fe substrate, they lose charge to the Fe substrate through tunnelling and the resulting neutral nanoparticles (NNPs) deposit as a porous skeletal structure of soot.

In colloidal solution with so many ions, it is difficult to compare the crystal growth behaviour between CNPs and NNPs because it is difficult to separate NNPs

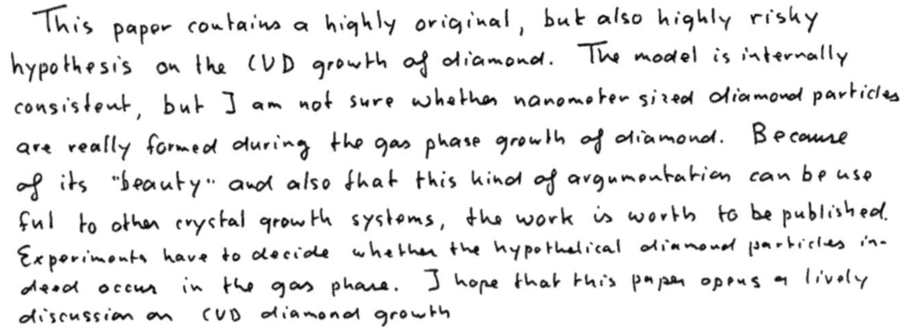
This paper contains a highly original, but also highly risky hypothesis on the CVD growth of diamond. The model is internally consistent, but I am not sure whether nanometer sized diamond particles are really formed during the gas phase growth of diamond. Because of its "beauty" and also that this kind of argumentation can be useful to other crystal growth systems, the work is worth to be published. Experiments have to decide whether the hypothetical diamond particles indeed occur in the gas phase. I hope that this paper opens a lively discussion on CVD diamond growth

Fig. 1.2 Reviewer's comment on the first paper suggesting that CVD diamonds grow by charged nanoparticles formed in the gas phase from CNPs. In the gas phase, however, it is easy to compare the crystal growth behaviour between CNPs and NNPs because CNPs loose charge just before landing on the Fe substrate whereas they retain charge on the Si substrate. (The details will be treated in Chap. 6.) In this sense, studying the non-classical crystallization in the gas phase has some advantage over that in solution. The evolution of diamond crystals by the building block of CNPs requires that CNPs should be liquid-like as well as undergo self-assembly and oriented attachment. The evolution of soot by the building block of NNPs implies that NNPs should undergo random Brownian coagulation, which is in accordance with the well-known behaviour of NNPs.

The liquid-like property of CNPs or the concept of charge-enhanced diffusion was derived from the deposition behaviour of CNPs in comparison with that of NNPs. Many experimental results indicate that charges not only enhance atomic diffusion but also enhance chemical kinetics. Thus, charge-enhanced diffusion is one of the phenomena that belongs to charge-enhanced kinetics. To explain the charge-enhanced kinetics, it must be assumed that the charge should weaken the bond strength. Therefore, the most important assumption of the TCN that CNPs are liquid-like would be related to the weakening of the bond strength by charge. Clare et al. (1994) published ab initio calculations most relevant to this assumption.

1.3 Effect of Charge on the Bond Strength

Clare et al. (1994) studied the effect of charge on bond strength in hydrogenated amorphous silicon to assess whether charge trapping in a solid-state lattice might promote the photo-degradation of amorphous silicon alloys (the Staebler-Wronski effect). They did ab initio calculations of the effect of a single negative or positive charge on the strength of a silicon-silicon and silicon-hydrogen bond in the molecules SiH_4 and Si_2H_6. Because the atoms in a lattice are less free to relax than those in a simple gaseous silane, they performed the calculations both with the atoms completely free to relax and under complete geometrical constraint.

To determine the difference in the energy needed to break a single Si–H bond in SiH_4, SiH_4^+, and SiH_4^-, calculations were done on six species: SiH_3, SiH_4, SiH_3^-, SiH_4^-, SiH_3^+, and SiH_4^+ and the required energies were found by subtraction. Similar calculations were done with the species Si_2H_6, Si_2H_5, $Si_2H_6^-$, $Si_2H_5^-$, $Si_2H_6^+$, and $Si_2H_5^+$ to observe the effect of a lower charge/size ratio and to examine the effect of charge on the Si–Si bond energy.

The Si–H bond strengths are obtained from the relations $E(SiH_4^c) - E(SiH_3^c) - E(H^-)$ for monosilane and $E(SI_2H_6^c) - E(Si_2H_5^c) - f(H^-)$ for disilane, where c is the charge, and may be 0, −1, or +1. The silicon-silicon bond strength is obtained from the formula $E(SI_2H_6^c) - E(SiH_3^c) - E(SiH_3)$, where c is again the charge. The calculations were done at a level for which all required energies were available, and subject to this, which gave the lowest energy.

Table 1.1 Summary of Si–H and Si–Si bond strengths (Clare et al. 1994)

Compound	Si–H (eV)	Si–Si (eV)
SiH_4 (optimized)	3.9	–
SiH_4^- (optimized)	0.98	–
SiH_4^+ (optimized)	0.30	–
Si_2H_6 (optimized)	3.5	3.2
$Si_2H_6^-$ (optimized)	1.02	1.11
$Si_2H_6^+$ (optimized)	1.59	1.6
SiH_4^- (unoptimized)	1.35	–
SiH_4^+ (unoptimized)	0.09	–
$Si_2H_6^-$ (unoptimized)	1.34	1.3
$Si_2H_6^+$ (unoptimized)	1.49	1.6

It is clear from Table 1.1 that both positive and negative charges on the clusters significantly weaken both the silicon-silicon and silicon-hydrogen bonds. Positive charge is more effective than negative in weakening the bond strength. When the atoms are embedded in a lattice, they will not be free to attain geometries resembling the optimized ion geometry. They will, however, be able to relax to some degree. Thus the actual effects of charge on bond strength will be between those indicated by the unoptimized (adiabatic) and optimized (vertical) rows of Table 1.1. They are likely to be closer to those for the unoptimized rows.

The effect of charge on the bond strength can be approached by bond order. A bond order, which represents the strength or stability of bond, is the number of bonding electron pairs shared by two atoms in a molecule. In molecular orbital theory, bond order is defined as half the difference between the number of bonding electrons and the number of antibonding electrons as expressed by the following equation,

$$\mathit{Bond\ order} = \frac{\mathit{no.\ of\ bonding\ electrons} - \mathit{no.\ of\ antibonding\ electrons}}{2} \tag{1.1}$$

Electrons are either added to antibonding orbitals (negative charge) or removed from bonding orbitals (positive charge). Therefore, both positive and negative charge would decrease the bond order and thereby weaken the bond strength.

We could not find literatures on the effect of charge on the bond strength of nanoparticles. However, similar to the weakening effect of charge on the bond strength in molecules, the charge is expected to weaken the bond strength in nanoparticles. Compared with a NNP, a CNP has extra electrostatic energy, which, for a spherical conductor with uniformly distributed charge, is given by

$$E = \frac{3}{5}\frac{1}{4\pi\varepsilon_o}\frac{q^2}{r}, \tag{1.2}$$

where q is the charge, r the radius and ε_o the vacuum permittivity (Feynmann et al. 1970). Therefore, a CNP is in a higher energy state than a NNP although they are at the same temperature. This extra energy would roughly contribute to weakening the bond strength. In order to understand what it means by the higher energy state at the same temperature, let's consider a liquid and a solid in equilibrium at the melting temperature. The liquid has the higher energy than the solid by the latent heat. And the bonding in the liquid is much weaker than that in the solid and the atoms in the former vibrate much more actively than those in the latter. The higher energy in the liquid is reflected in the active atomic vibration, which would enhance atomic diffusion and any chemical kinetics.

Equation (1.2) shows that the electrostatic energy is inversely proportional to the radius, which is in agreement with the experimental observation that the liquid-like behaviour of CNPs is diminished with increasing size. The electrostatic energy is proportional to the square of the charge, which means that if CNPs are multiply charged, they would have a liquid-like property even at lower temperature. Multiple charging would be favoured when the high density plasma such as microwave plasma or inductively coupled plasma (ICP) is used.

A few examples would be introduced to show how the liquid-like property of CNPs is related to the non-classical crystallization (Sects. 1.4 and 1.5) and how the charge enhances diffusion (Sect. 1.6).

1.4 Growth of Nanowires by Non-classical Crystallization

Tang et al. (2002) observed that CdTe nanoparticles were spontaneously reorganized into crystalline nanowires upon controlled removal of the protective shell of the organic stabilizer as shown in Fig. 1.3. This result shows that the mechanism of non-classical crystallization also applies to the nanowire growth.

Such chain aggregation is initiated by the overall decrease of the stabilizer concentration. Nanowires in Fig. 1.3a, b grow respectively by the building blocks of 3.4 and 5.4 nm CdTe nanoparticles. The diameters of the nanowires were virtually identical to the diameters of precursor nanoparticles. The produced

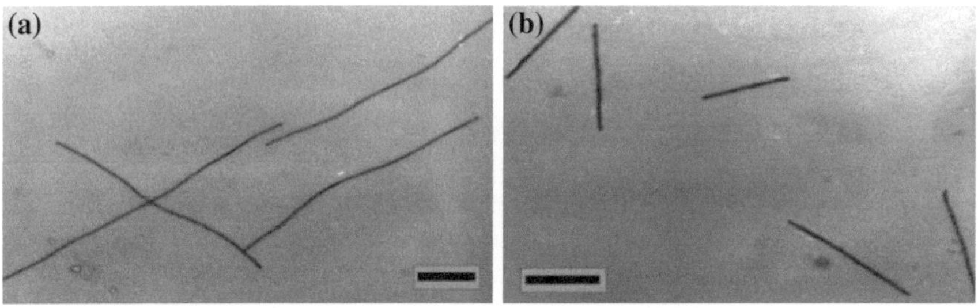

Fig. 1.3 Transmission electron microscopy (TEM) images of CdTe nanowires made from 3.4- (**a**) and 5.4-nm (**b**) nanoparticles. Bars, 100 nm (Tang et al. 2002)

nanowires were single-crystalline with high luminescence quantum yields indicating the high degree of crystallinity.

Then some questions arise, "Which interaction is responsible for the self-assembly of nanoparticles into nanowires and how is it possible for single crystalline nanowires to grow by the building block of nanoparticles at room temperature?" Tang et al. (2002) suggested that strong dipole-dipole interaction would be the driving force for nanoparticle self-organization.

To reveal the intermediate steps in the nanoparticle to nanowire transition, they examined the aliquots of the dispersions in the early stages of the nanowire formation by TEM. In these samples, "pearl-necklace" agglomerates as shown in Fig. 1.4 were observed in great abundance. The "pearl-necklace" agglomerates provide important information as to how the nanoparticles arrange themselves into

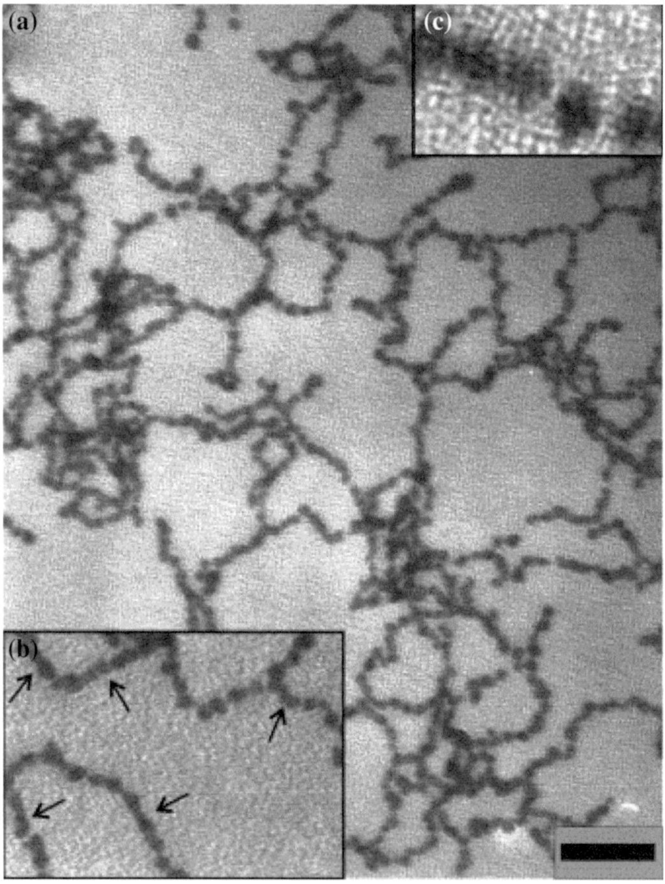

Fig. 1.4 **a** TEM image of intermediate state of nanoparticle-nanowire transition for 5.4-nm nanoparticles. **b** The enlarged portion of the chain, with short rods marked by *arrows*. **c** The high resolution TEM (HRTEM) image of the adjacent nanoparticles in the chain (Tang et al. 2002)

a linear chain and clearly reveal that the nanoparticles are the building block of nanowires shown in Fig. 1.3. The enlarged TEM image of Fig. 1.4b shows how the short nanorods are formed by coalescence of nanoparticles in the initial stage of nanowire formation. Therefore, the size of nanoparticles determines the diameter of nanowires.

The formation of nanowires or necklace nanocrystal chains via the oriented attachment of nanoparticles has been reported for various materials such as Au, Ag, TiO_2, ZnS, CdTe, PbSe, ZnO and SnO_2 (Zhang et al. 2009). The formation mechanism would be similar to more complex nanostructures such as nanodendrites, nanoleaves and nanomultipods because these were also suggested to be synthesized by the oriented attachment of nanoparticles (Zitoun et al. 2005; Wen et al. 2006; Liu et al. 2006).

The oriented attachment is not sufficient to explain the growth of single crystalline nanowires by nanoparticles because it can explain neither the linear self-assembly nor the neck growth kinetics at room temperature. The effect of charge should be considered. The electrostatic interaction and the liquid-like property of CNPs can explain respectively the linear self-assembly and the rapid neck growth kinetics.

The evolution of the chain-like structure in Fig. 1.4 should be noted. Tang et al. (2002) mentioned that the "pearl necklace" aggregates were not observed in the standard dispersions of CdTe. The chain-like structure in Fig. 1.4 is characterized by the presence of the neck between two nanoparticles in contact, which indicates the relatively slow neck growth. In other words, the atomic diffusion from the neck of the high chemical potential to the low curvature area of the low chemical potential is rather slow.

From the viewpoint of TCN, it is believed that the "pearl necklace" aggregates in Fig. 1.4 can be formed when the discharge takes place and CNPs lose their charge and the nanoparticles become neutral after their contact for some reason. The chain-like structure is highly in contrast with the nanowires with a smooth surface in Fig. 1.3, where the neck growth occurred completely. According to the TCN, this drastic difference is attributed to the presence and absence of charge.

1.5 Direct Observation of Non-classical Crystallization

1.5.1 Growth of Pt_3Fe Nanorods

Recently, crystal growth by nanoparticles in solution was directly observed by TEM using a liquid cell (Liao et al. 2012; Yuk et al. 2012), which provided direct evidences for non-classical crystallization. Using a silicon nitride liquid cell for in situ TEM observation, Liao et al. (2012) carried out detailed real-time imaging to show how Pt_3Fe nanorods grow by nanoparticles in solution.

Figure 1.5 shows sequential images depicting the growth trajectory of a twisted Pt_3Fe nanowire. In the first stage, many small nanoparticles are formed when the Pt and Fe precursors are reduced by electron beam illumination. Some of them grow by monomer attachment; others undergo coalescence. The nanoparticles were

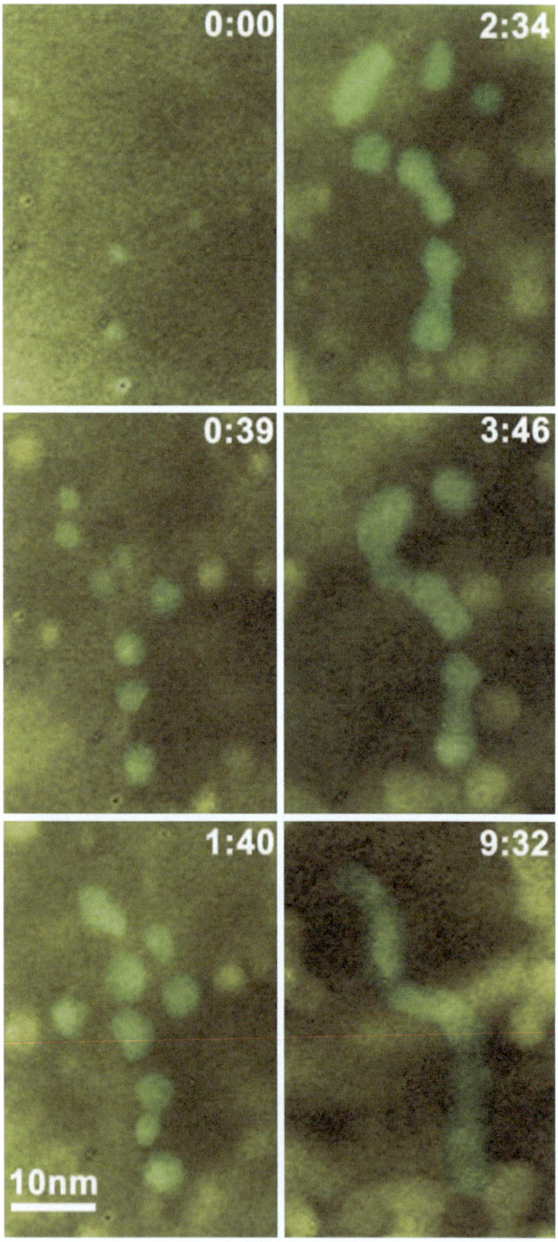

Fig. 1.5 Sequential TEM images showing the initial nucleation and growth of Pt_3Fe nanowires in the molecular precursor solution. Time is displayed as minutes: seconds. The initial time is arbitrary (Liao et al. 2012)

merged by coalescence and then relax into spherical nanoparticles. Eventually, these nanoparticles reach an average size of 5.3 ± 0.9 nm.

In the second stage, nanoparticles interact with each other to form nanoparticle chains. The nanoparticle chain is formed by shape-directed nanoparticle attachment with successive structural relaxation into straight Pt_3Fe nanorods and reorientation, revealing critical mechanisms of the growth into nanorods from nanoparticle building blocks. Therefore, even when nanoparticles attach without orientation, single-crystalline nanorods are formed eventually. Initially, one nanoparticle meets another, forming a dimer. Unlike in the first stage of growth, the dimer does not relax into a sphere. Instead, a trimer is formed when another particle connects to the dimer end, and additional end-to-end attachments generate a nanoparticle chain.

Straight nanorods are produced when additional surfactant (10 % oleic acid) is added to the growth solution. After a nanorod straightens, it takes some time to form a single-crystal structure. It also takes longer to form a long single-crystal nanorod than a short one. A Pt_3Fe nanorod imaged using high-angle annular dark-field (HAADF) scanning TEM (STEM) image shows that although the rod is straight with a smooth surface, polycrystalline features remain as shown in Fig. 1.6a.

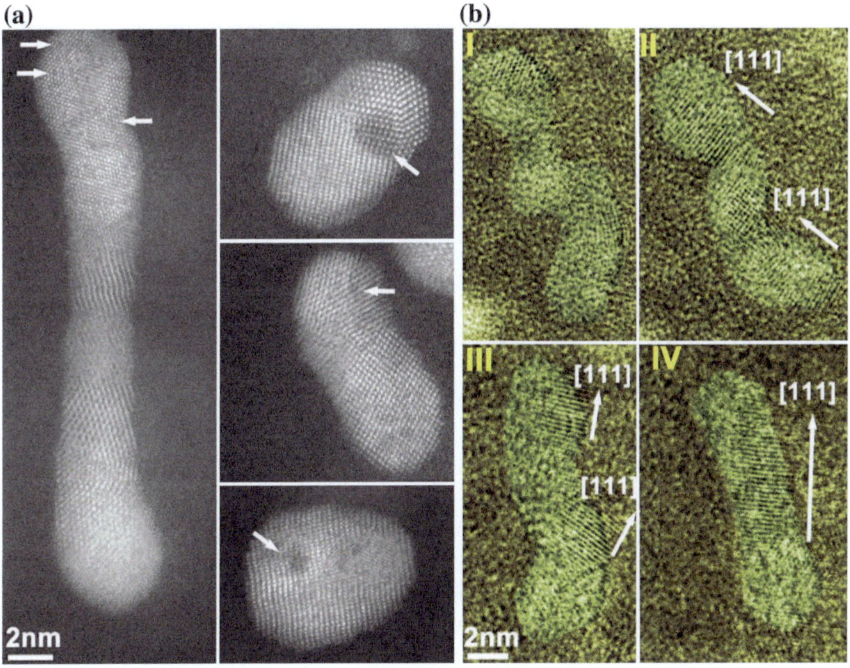

Fig. 1.6 **a** HAADF STEM image of a polycrystalline Pt_3Fe nanorod, dimers, and nanoparticles obtained in a liquid cell. The *dark spots* (examples highlighted by *arrows*) indicate the iron-rich regions. **b** Sequential HRTEM images (I–IV) show both crystal orientation and shape changes during the straightening of a twisted nanoparticle chain (Liao et al. 2012)

The sizes of iron-rich regions are smaller at the later stage as structure relaxation proceeds, which suggests that diffusion and mass redistribution are not limited to the nanocrystal surface during the structural relaxation. Notably, a twisted polycrystalline Pt_3Fe nanoparticle chain assembled by imperfect nanoparticle attachment can straighten and correct the orientation to yield a single-crystal nanorod with a perfectly straight shape. This aspect is shown in Fig. 1.6b, where a chain of three connected nanoparticles possesses different orientations from imperfect attachment. Lattice rotation accompanied by mass redistribution and straightening is observed afterward. A single crystal nanorod is formed shortly after the chain straightens.

The author did not mention the role of charge although charging of nanoparticles is unavoidable during TEM observation. In enhancing the kinetics of the nanorod growth by nanoparticles, the charge plays a critical role.

1.5.2 *Growth of Pt Crystals*

By in situ TEM observation using graphene liquid cells, Yuk et al. (2012) carried out direct atomic-resolution imaging to show how Pt crystals grow in solution. The microscope is operated at 80 kV with a beam intensity of 10^3–10^4 A/m^2 maintained during nanocrystal growth. Upon locating a liquid pocket on the TEM grid, the beam intensity is optimized, which reduces the Pt precursor and initiates nanocrystal growth (Zheng et al. 2009). The use of graphene liquid cells made it possible to discern colloidal Pt nanoparticles with radii as small as 0.1 nm and to track their motion, which was not possible by previous cells with silicon nitride windows (Zheng et al. 2009).

The image clearly revealed that Pt nanocrystals grow not only by monomers but also by the coalescence of nanoparticles. Once nanocrystals collide at {111} planes, they merge quickly, within 0.26 s (the limit of their acquisition time). Among their observation of twenty coalescence events between nanocrystals with diameters greater than 1 nm, most coalescence events distinguished by lattice planes occur along the {111} planes of either one or both nanocrystals with only two coalescence events merged on different planes of the contacting nanocrystals: {110} plane for one and off-zone axis for the other. The reason why most coalescence events proceed along the same crystallographic direction may be attributed to the fact that {111} planes of a fcc crystal have the lowest surface energy and, therefore, perhaps the lowest ligand coverage. Coalescence on the {111} planes often yields nanocrystals with a twin boundary.

Before coalescing along the {111} orientation, the nanocrystals exhibit a prolonged period of correlated motion that facilitates lattice alignment and unification. The correlated motion is presumably attributed to interparticle attractive forces. The authors suggested that the attractive forces would be van der Waals interactions, steric repulsions, and depletion forces arising from surface ligands (Bishop et al. 2009). Also the authors did not mention the charge although charging of nanoparticles is inevitable under e-beam irradiation during TEM observation. In the interparticle attractive and repulsive forces, the electrostatic interaction should play a very important role. It should be noted that the attractive force arising from the electrostatic interaction between two charged particles described by (7.5) and (7.6) in Chap. 7 is much stronger than the van der Waals force.

According to Yuk et al. (2012), as the freely moving nanocrystals draw close, the center-to-center distance rapidly decreases. This initial event occurs within 40 s, and the center-to-center separation fluctuates between 4–6 nm for the next 25 s. While this dynamic event proceeds, the nanocrystals also grow in size from ~0.75 to 1 nm in diameter due to monomer addition. The sustained correlated motion is suggested to occur with weakly touching surface ligand layers (Schapotschnikow et al. 2008). The center-to-center distance shows even more confined fluctuation of ~1 nm in the time interval from 80 to 130 s. In this time range, the correlated motion of the two nanocrystals exhibits 3-dimensional behaviour of rolling and relative sliding of nanocrystals over each other. Throughout this period, the nanocrystals continue to grow by monomer addition. After 130 s, the center-to-center distance approaches the sum of the radii of the two nanocrystals, and the nanocrystal lattices align, leading to coalescence at 160 s.

In addition to the surface ligand layers, the electrostatic interaction between two charged nanocrystals, which depends on the inter-distance, the size and the amount of charge, should play an important role in their correlated motion. The epitaxial alignment before coalescence is similar to the TEM observation of Au nanoparticles by Iijima and Ajayan (1991), who described that the lattice fringes of both particles were seen to align even before the particles actually made contact. The rotation of particles indicates that there exists a torque, which is given by the derivative of the interaction energy with respect to orientation. This means again that the interaction between two charged nanocrystals being apart depends not only on the distance but also on the misorientation, which would be minimized by the epitaxial alignment.

Figure 1.7 shows the corresponding TEM images. The nanocrystals are connected by a neck at the initial stage of coalescence. Neck growth is accompanied by a decrease in length (l) and thickness (t), which indicates that the atoms migrate to the neck region, presumably by surface diffusion (Fang and Wang 2008). After coalescence, the nanocrystal structure also gradually reorganizes, evolving truncated surfaces.

Figures 1.5, 1.6 and 1.7 show how crystals can grow by the building block of nanocrystals, providing the direct evidence for non-classical crystallization. It should be noted that in Fig. 1.7 it took only 2.6 s for the complete neck growth of two nanocrystals. The authors mentioned in the supporting information that all images were collected under ambient conditions at 23 °C. It should be noted that such enhanced kinetics of liquid-like coalescence at such a low temperature can never be expected from neutral nanocrystals and should be explained by considering the liquid-like property of CNPs. It should be also noted that the growth of platinum nanocrystals was initiated by electron beam irradiation. This result indicates that the precursor decomposition is also triggered by charge, which implies that charge enhances kinetics of chemical reactions.

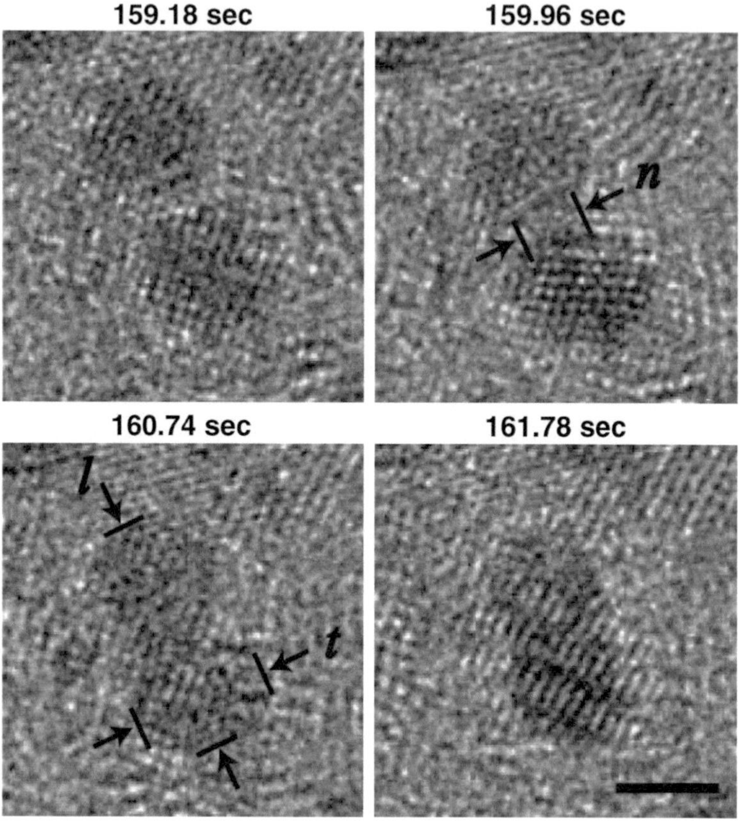

Fig. 1.7 Pt nanocrystal dynamics of coalescence. *l*, *t* and *n* in the figure represent respectively the length along the center-to-center direction, the thickness in vertical direction to the length and the neck diameter. Scale bar, 2 nm (Yuk et al. 2012)

1.6 Healing of Mg Nanopores by Electron Beam Irradiation

Zheng et al. (2013) fabricated nanopores in a Mg substrate using a focused e-beam inside the TEM and subsequently, observed that the nanopores were shrinking and eventually disappeared by defocused e-beam irradiation with the intensity of 10^1–10^2 A cm^{-2}. Figure 1.8a–c schematically illustrates the healing of a nanopore under defocused electron beam irradiation. Figure 1.8d–f shows the time evolution of TEM images revealing that the nanopore with an original size of 3.3 nm shrank continuously (Fig. 1.8d, e) and finally disappeared (Fig. 1.8f) under e-beam

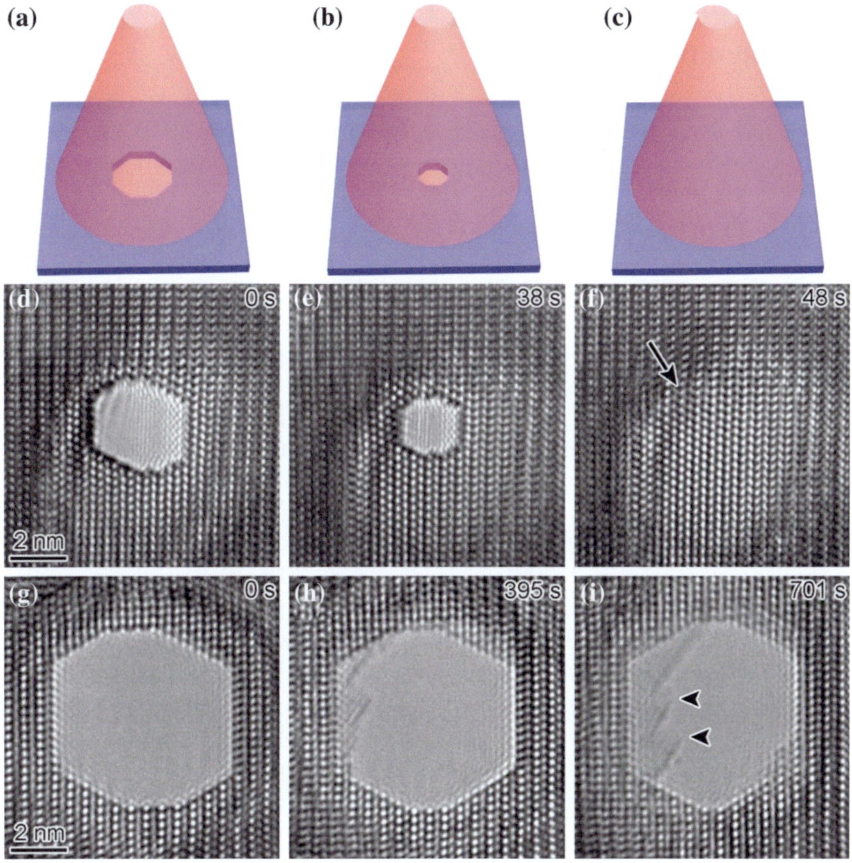

Fig. 1.8 **a–c** A schematic illustration showing the healing of a nanopore under wide-field electron beam irradiation. **d–f** Time-lapsed images illustrating the experimental observations of the healing process of a nanopore with an original size of 3.3 nm. **g–i** Sequential images showing a nanopore with an original size of 6.0 nm under e-beam irradiation. The *arrow heads* indicate the existence of a MgO phase (Zheng et al. 2013)

irradiation. The entire procedure took only 48 s. When the e-beam was turned off, the nanopore retained its shape, suggesting an e-beam-assisted healing mechanism.

The authors attributed this pore elimination solely to electron beam irradiation effect. From the viewpoint of TCN, however, this is the direct evidence of charge-enhanced diffusion. It is believed that if electrically floating and grounded Mg substrates are compared, the pore elimination rate of the former would be much faster than that of the latter.

1.7 Particle Formation in the Chemical Vapour Deposition Process

Particle formation has been identified as a problem in the silicon CVD, as well as in other microelectronic processing systems (Eversteijn 1971; Murthy et al. 1976; Herrick and Woodruff 1984; Wu and Flagan 1987; Selwyn et al. 1990). If particles created in the gas phase adsorb on the surface of the growing film, they can lead to failure of the device in various ways. For this reason, lots of efforts have been made to find out the processing condition under which no gas phase nucleation occurs.

Although the detection of invisible nanometer-sized gas phase nuclei is not easy, recently appropriate instruments have been developed. In the case of atmospheric CVD, the nano-differential mobility analyser (DMA) (Seto et al. 1997; Seol et al. 2001; Kim et al. 2004) can measure the size distribution of nanoparticles in the range of 1–100 nm. In low pressure CVD, low pressure DMA and particle beam mass spectroscopy (PBMS) (Ziemann et al. 1995; Hong et al. 2013) can be used. In the case of plasma CVD, Shiratani and his colleagues (Shiratani et al. 1996; Koga et al. 2000; Watanabe et al. 2001; Nunomura et al. 2006) made extensive studies using the time evolution of a laser light scattering (LLS) intensity. Using the double-pulse-discharge method, Koga et al. (2000) could measure the size and the number density of clusters in a size range of 0.5–4 nm. Based on the thermal coagulation of particles and the time evolution of the LLS intensity, Nunomura et al. (2006) developed a simple in situ method for measuring the size and number density of nanoparticles with a minimum detectable size down to ~ 1 nm.

The extensive study on detecting the generation of clusters or nanoparticles in the plasma or non-plasma CVD process concludes that the generation of clusters or nanoparticles in the gas phase is almost unavoidable at least at the moderate growth rate required in the commercial fabrication of thin films. To inhibit the generation of clusters or nanoparticles, the film growth rate should be made impractically low (typically $< \sim$ nm/h). This means that in order to get the moderate film growth rate, films must be grown under the condition where nanoparticles are generated in the gas phase. It turns out that in most CVD processes with optimized conditions, these gas-phase generated nanoparticles are not so harmful to the growing films and rather contribute to the deposition of thin films. In other words, gas-phase generated nanoparticles make it possible to produce films of relatively high quality at a high growth rate. This means that most CVD films grow by non-classical crystallization.

If gas-phase generated nanoparticles are neutral, they are harmful to the growing film and may lead to failure of the device by a variety of mechanisms. If they are charged and small enough and the substrate temperature is high enough, they are liquid-like and produce high quality films and even epitaxial ones like individual atoms do. The exclusive atomic deposition would produce the best quality film like the film prepared by atomic layer deposition (ALD), but the growth rate would be very low. The slight increase of the growth rate induces gas-phase generated clusters, which would be normally neutral. Then, the film quality is degraded significantly with lots of voids.

Small clusters tend to be neutral whereas large clusters tend to be charged because ionization potential and electron affinity of small clusters are respectively much higher and lower than those of large clusters. It should be noted that both ionization potential and electron affinity of large clusters approach the work function value of their bulk (Persson et al. 1997; Seidl et al. 1998). Large charged clusters are more liquid-like than small neutral clusters as long as the size difference is not extreme. Therefore, under certain conditions, the higher deposition rate would produce the better quality films than the lower deposition rate, which is often reported in the thin film growth (Levlin et al. 1997; Hwang and Dubson 1992; Baski and Fuchs 1994).

Realizing that the generation of nanoparticles in the gas phase is almost unavoidable in many thin film processes, all we have to do is to make the most use of this fact. For this, we need to understand the growth, charging and deposition behaviour of nanoparticles. CNPs in the gas phase behave like colloids. Their deposition behaviour as thin films can be understood best by considering the colloidal crystallization. CNPs undergo colloidal crystallization in the thin film reactor but the size of CNPs is usually much smaller than that of normal colloidal particles and the temperature of film growth, which can be the reactor temperature or the substrate temperature, is much higher than that of normal colloidal crystallization. Because of the small size of CNPs and the high growth temperature, CNPs are liquid-like. Because of this liquid-like property, CNPs can deposit epitaxially on the growing surface, producing dense micron-sized single crystalline or polycrystalline films, leaving no void behind. The final morphology is so perfect (for example, see Figs. 1.3 and 5.2) that it would be very difficult to distinguish whether nanowires in Fig. 1.3 or films in Fig. 5.2 grow by individual atoms or by CNPs.

Therefore, although the effort to avoid the gas phase nucleation during thin film growth is limited by the decrease of the film growth rate, the effort to control the size, the number density and the charging behaviour of gas phase generated nanoparticles would produce the high quality films at a moderate growth rate. The new concept of charge-enhanced diffusion and charge-enhanced kinetics would help in understanding many puzzling phenomena and in solving many technical problems in relation to the growth of thin films and nanostructures.

References

Alivisatos A (2000) Naturally aligned nanocrystals. Science 289(5480):736–737

Banfield JF, Welch SA, Zhang H, Ebert TT, Penn RL (2000) Aggregation-based crystal growth and microstructure development in natural iron oxyhydroxide biomineralization products. Science 289(5480):751–754

Baski A, Fuchs H (1994) Epitaxial growth of silver on mica as studied by AFM and STM. Surf Sci 313(3):275–288

Bishop KJ, Wilmer CE, Soh S, Grzybowski BA (2009) Nanoscale forces and their uses in self-assembly. Small 5(14):1600–1630
Botsaris G, Reid R (1967) Comments on the letter by Glasner and Skurnik entitled "Growth of potassium chloride crystals from aqueous solutions. I. The effect of lead chloride". J Chem Phys 47(9):3689–3690
Cölfen H, Antonietti M (2005) Mesocrystals: Inorganic superstructures made by highly parallel crystallization and controlled alignment. Angew Chem Int Edit 44(35):5576–5591
Cölfen H, Antonietti M (2008) Mesocrystals and Nonclassical Crystallization. Wiley, New York, p 73–101
Cölfen H, Mann S (2003) Higher-order organization by mesoscale self-assembly and transformation of hybrid nanostructures. Angew Chem Int Edit 42(21):2350–2365
Clare BW, Talukder G, Jennings PJ, Cornish JCL, Hefter GT (1994) Effect of charge on bond strength in hydrogenated amorphous silicon. J Comput Chem 15(6):644–652. doi:10.1002/jcc.540150608
Demichelis R, Raiteri P, Gale JD, Quigley D, Gebauer D (2011) Stable prenucleation mineral clusters are liquid-like ionic polymers. Nat Commu 2:590
Eversteijn F (1971) Gas-phase decomposition of silane in a horizontal epitaxial reactor. Philips Res Rep 26(2):134–144
Fang Z, Wang H (2008) Densification and grain growth during sintering of nanosized particles. Int Mater Rev 53(6):326–352
Feynmann R, Leighton R, Sands M (1970) The Feynman lectures on physics, vol 3. Reading Mass., Addison Wesley, Boston
Gebauer D, Cölfen H (2011) Prenucleation clusters and non-classical nucleation. Nano Today 6(6):564–584
Gebauer D, Völkel A, Cölfen H (2008) Stable prenucleation calcium carbonate clusters. Science 322(5909):1819–1822
Gebauer D, Gunawidjaja PN, Ko JYP, Bacsik Z, Aziz B, Liu L, Hu Y, Bergström L, Tai C-W, Sham T-K, Edén M, Hedin N (2010) Proto-calcite and proto-vaterite in amorphous calcium carbonates. Angew Chem Int Edit 49(47):8889–8891. doi:10.1002/anie.201003220
Glasner A, Kenat J (1968) The crystallization of KCl from aqueous solutions in the presence of lead ions. I. A calorimetric study. J Cryst Growth 2(3):119–127
Glasner A, Skurnik S (1968) A new mechanism for the crystallization and growth of ionic crystals, with special reference to KCl in the presence of Pb2 + ions. Isr J Chem 6(1):69–72
Glasner A, Tassa M (1974a) The thermal effects of nucleation and crystallization of KBr and KCl solutions. II. The heat of nucleation and the supersaturated solution. Isr J Chem 12(4):799–816
Glasner A, Tassa M (1974b) The thermal effects of nucleation and crystallization of KBr and KCL solutions. III. The heat of crystallization and the co-precipitation of lead ions. Isr J Chem 12(4):817–826
Herrick CS, Woodruff DW (1984) The homogeneous nucleation of condensed silicon in the gaseous Si-H-Cl system. J Electrochem Soc 131(10):2417–2422
Hong J-S, Kim C-S, Yoo S-W, Park S-H, Hwang N-M, Choi H-M, Kim D-B, Kim T-S (2013) In-situ measurements of charged nanoparticles generated during hot wire chemical vapor deposition of silicon using particle beam mass spectrometer. Aerosol Sci Tech 47(1):46–51. doi:10.1080/02786826.2012.725959
Hwang J, Dubson M (1992) Atomically flat gold films grown on hot glass. J Appl Phys 72(5):1852–1857
Hwang N-M, Lee D-K (2010) Charged nanoparticles in thin film and nanostructure growth by chemical vapour deposition. J Phys D: Appl Phys 43(48):483001
Hwang NM, Yoon DY (1996) Thermodynamic approach to the paradox of diamond formation with simultaneous graphite etching in the low pressure synthesis of diamond. J Cryst Growth 160(1–2):98–103. doi:http://dx.doi.org/10.1016/0022-0248(95)00549-8
Hwang NM, Hahn JH, Yoon DY (1996a) Charged cluster model in the low pressure synthesis of diamond. J Cryst Growth 162(1):55–68

Hwang NM, Hahn JH, Yoon DY (1996b) Chemical potential of carbon in the low pressure synthesis of diamond. J of Cryst Growth 160(1–2):87–97. doi:http://dx.doi.org/10.1016/0022-0248(95)00548-X

Iijima S, Ajayan PM (1991) Substrate and size effects on the coalescence of small particles. J Appl Phys 70(9):5138–5140. doi:http://dx.doi.org/10.1063/1.348990

Kim J, Seto T, Sakiyama K, Kim D (2004) Characterization of low pressure DMA system for the size selection of magnetic nano-particles. Appl Phys A 79(4–6):1497–1499. doi:10.1007/s00339-004-2829-1

Koga K, Matsuoka Y, Tanaka K, Shiratani M, Watanabe Y (2000) In situ observation of nucleation and subsequent growth of clusters in silane radio frequency discharges. Appl Phys Lett 77(2):196–198

Leite E, Giraldi T, Pontes F, Longo E, Beltran A, Andres J (2003) Crystal growth in colloidal tin oxide nanocrystals induced by coalescence at room temperature. Appl Phys Lett 83(8):1566–1568

Levlin M, Laakso A, Niemi H-M, Hautojärvi P (1997) Evaporation of gold thin films on mica: effect of evaporation parameters. Appl Surf Sci 115(1):31–38

Li D, Nielsen MH, Lee JR, Frandsen C, Banfield JF, De Yoreo JJ (2012) Direction-specific interactions control crystal growth by oriented attachment. Science 336(6084):1014–1018

Liao H-G, Cui L, Whitelam S, Zheng H (2012) Real-time imaging of Pt_3Fe nanorod growth in solution. Science 336(6084):1011–1014

Liu J, Huang X, Li Y, Sulieman K, He X, Sun F (2006) Self-assembled CuO monocrystalline nanoarchitectures with controlled dimensionality and morphology. Cryst Growth Des 6(7):1690–1696

Murthy T, Miyamoto N, Shimbo M, Nishizawa J (1976) Gas-phase nucleation during the thermal decomposition of silane in hydrogen. J Cryst Growth 33(1):1–7

Niederberger M, Cölfen H (2006) Oriented attachment and mesocrystals: non-classical crystallization mechanisms based on nanoparticle assembly. Phys Chem Chem Phys 8(28):3271–3287

Nunomura S, Kita M, Koga K, Shiratani M, Watanabe Y (2006) In situ simple method for measuring size and density of nanoparticles in reactive plasmas. J Appl Physs 99(8):083302

Persson JL, Andersson M, Holmgren L, Åklint T, Rosén A (1997) Ionization potentials of oxidized copper clusters. Chem Phys Lett 271(1):61–66

Qi L, Coelfen H, Antonietti M (2000) Crystal design of barium sulfate using double-hydrophilic block copolymers. Angew Chem Int Edit 39(3):604–607

Schapotschnikow P, Pool R, Vlugt TJ (2008) Molecular simulations of interacting nanocrystals. Nano Lett 8(9):2930–2934

Seidl M, Perdew JP, Brajczewska M, Fiolhais C (1998) Ionization energy and electron affinity of a metal cluster in the stabilized jellium model: Size effect and charging limit. J Chem Phys 108(19):8182–8189

Selwyn GS, Mckillop J, Haller KL, Wu J (1990) Insitu plasma contamination measurements by HeNe laser light scattering: A case study. J Vac Sci Technol A 8(3):1726–1731

Seol KS, Tsutatani Y, Fujimoto T, Okada Y, Takeuchi K, Nagamoto H (2001) New in situ measurement method for nanoparticles formed in a radio frequency plasma-enhanced chemical vapor deposition reactor. J Vac Sci Technol B 19(5):1998–2000. doi:10.1116/1.1404979

Seto T, Nakamoto T, Okuyama K, Adachi M, Kuga Y, Takeuchi K (1997) Size distribution measurement of nanometer-sized aerosol particles using dma under low-pressure conditions. J Aerosol Sci 28(2):193–206. doi:http://dx.doi.org/10.1016/S0021-8502(96)00071-7

Shiratani M, Kawasaki H, Fukuzawa T, Yoshioka T, Ueda Y, Singh S, Watanabe Y (1996) Simultaneous in situ measurements of properties of particulates in rf silane plasmas using a polarization-sensitive laser-light-scattering method. J Appl Phys 79(1):104–109

Sunagawa I (1987) Morphology of minerals. Morphology of crystals, part B. Terra Scientific, Tokyo, p 511–587

Sunagawa I (1990) Growth and morphology of diamond crystals under stable and metastable conditions. J Cryst Growth 99(1):1156–1161

Tang Z, Kotov NA, Giersig M (2002) Spontaneous organization of single CdTe nanoparticles into luminescent nanowires. Science 297(5579):237–240. doi:10.1126/science.1072086

Teng HH (2013) How ions and molecules organize to form crystals. Elements 9(3):189–194

Watanabe Y, Shiratani M, Koga K (2001) Formation kinetics and control of dust particles in capacitively-coupled reactive plasmas. Phys Scripta 2001(T89):29

Wen X, Xie Y-T, Mak WC, Cheung KY, Li X-Y, Renneberg R, Yang S (2006) Dendritic nanostructures of silver: facile synthesis, structural characterizations, and sensing applications. Langmuir 22(10):4836–4842

Wohlrab S, Pinna N, Antonietti M, Cölfen H (2005) Polymer-induced alignment of dl-alanine nanocrystals to crystalline mesostructures. Chem Eur J 11(10):2903–2913

Wu JJ, Flagan RC (1987) Onset of runaway nucleation in aerosol reactors. J Appl Phys 61(4):1365–1371

Yu S-H, Cölfen H (2004) Bio-inspired crystal morphogenesis by hydrophilic polymers. J Mater Chem 14(14):2124–2147

Yuk JM, Park J, Ercius P, Kim K, Hellebusch DJ, Crommie MF, Lee JY, Zettl A, Alivisatos AP (2012) High-resolution EM of colloidal nanocrystal growth using graphene liquid cells. Science 336(6077):61–64. doi:10.1126/science.1217654

Zhang Q, Liu S-J, Yu S-H (2009) Recent advances in oriented attachment growth and synthesis of functional materials: concept, evidence, mechanism, and future. J Mater Chem 19(2):191–207

Zhang Z-p, Sun H-p, Shao X-q, Li D, Yu H, Han M (2005) Three-dimensionally oriented aggregation of a few hundred nanoparticles into monocrystalline architectures. Adv Mater 17(1):42–47

Zheng H, Smith RK, Jun Y-W, Kisielowski C, Dahmen U, Alivisatos AP (2009) Observation of single colloidal platinum nanocrystal growth trajectories. Science 324(5932):1309–1312

Zheng H, Liu Y, Cao F, Wu S, Jia S, Cao A, Zhao D, Wang J (2013) Electron beam-assisted healing of nanopores in magnesium alloys. Sci Rep 3:1920. doi:10.1038/srep01920. http://www.nature.com/articles/srep01920#supplementary-information

Ziemann PJ, Liu P, Rao NP, Kittelson DB, McMurry PH (1995) Particle beam mass spectrometry of submicron particles charged to saturation in an electron beam. J Aerosol Sci 26(5):745–756. doi:http://dx.doi.org/10.1016/0021-8502(95)00009-2

Zitoun D, Pinna N, Frolet N, Belin C (2005) Single crystal manganese oxide multipods by oriented attachment. J Am Chem Soc 127(43):15034–15035

Chapter 2
Thermodynamics of Physical and Chemical Vapour Deposition

The ultimate goal of this chapter is to formulate the driving force for deposition of diamond in the chemical vapour deposition. Before reaching that goal, we will study the basics of the related topics in several steps. First, we will study how to formulate the thermodynamics of ideal gases. Then we will study how to formulate the driving force for evaporation and condensation of water, which we experience during our everyday lives, such as drying wet clothes and mist formation. Then, we will study how we formulate the driving force for deposition in the physical vapour deposition process. In the next step, we will study how to formulate the irreversibility when chemical reactions are involved. Finally, we will study how to formulate the driving force for deposition in the CVD process. At the end of the chapter, some kinetic concept will be briefly described. The motivation for formulating the driving force for deposition was to compare the nucleation curves between diamond and graphite. This allows us to compare not only the stability but also the activation energy for nucleation between diamond and graphite.

2.1 Ideal Gases

For a closed system of fixed composition at constant temperature, dG becomes (Gaskell 1995)

$$dG = VdP \tag{2.1}$$

Since V = RT/P for one mole of ideal gas,

$$dG = \frac{RT}{P} dP = RTd \ln P \tag{2.2}$$

For an isothermal change of pressure from P_1 to P_2 at T,

© Springer Science+Business Media Dordrecht 2016

N.M. Hwang, *Non-Classical Crystallization of Thin Films and Nanostructures in CVD and PVD Processes*, Springer Series in Surface Sciences 60,
DOI 10.1007/978-94-017-7616-5_2

$$G(P_2, T) - G(P_1, T) = RT \ln \frac{P_2}{P_1} \tag{2.3}$$

$$G(P, T) = G(P_o, T) + RT \ln \frac{P}{P_o} \tag{2.4}$$

where $G(P_o,T)$ is a reference state. Since the choice of a reference state is arbitrary, it would be good if we make a convenient choice. If the reference state is chosen as $P_o = 1$ atm at the given temperature, (2.4) is simplified to

$$G = G^o + RT \ln P \tag{2.5}$$

Due to the simplicity of (2.5), this special reference state is called the standard state, which is defined by the state of 1 mol of pure gas at 1 atm pressure and the temperature of interest.

The chemical potential of pure gas is equal to the molar free energy. Therefore, (2.5) can also be written as

$$\mu = \mu^o + RT \ln P \tag{2.6}$$

However, the chemical potential of a component A in a mixture gas is equal to the partial molar Gibbs free energy, which is written as

$$\bar{G}_A = G_A^o + RT \ln p_A \tag{2.7}$$

where p_A is the partial pressure of the component A in the mixture gas.

And the corresponding chemical potential is

$$\mu_A = \mu_A^o + RT \ln p_A \tag{2.8}$$

(Q) What would be the driving force for wet clothes to dry?

2.2 Driving Force for Evaporation

The evaporation process is the irreversible transfer of atoms or molecules from the liquid to the gas phase. Water will evaporate when its chemical potential of the liquid is higher than that of the gas phase. This can be expressed as

$$\Delta\mu_{H_2O}^{liquid \to gas} = \mu_{H_2O}^{gas} - \mu_{H_2O}^{liquid} \tag{2.9}$$

If it is assumed that the water vapour follows the ideal gas behaviour, the chemical potential of water in the gas phase is expressed as

$$\mu_{\mathrm{H_2O,gas}} = \mu^{o}_{\mathrm{H_2O,gas}} + RT\ln p_{\mathrm{H_2O}} \tag{2.10}$$

where p_{H_2O} is the vapour pressure of water in the gas phase. Then, what remains to do is to determine the chemical potential of water.

As in (2.1), the molar free energy or the chemical potential of liquid water can be expressed at the given temperature as

$$d\mu = VdP \tag{2.11}$$

The molar volume of liquid is insensitive to the change of pressure, (2.11) can be integrated with respect to P.

$$\mu(P,T) = \mu(P_o,T) + V(P - P_o) \tag{2.12}$$

If the reference state is defined as the pure water at 1 atm pressure and the temperature of interest, which is identical to the standard state, (2.12) can be written as,

$$\mu_{\mathrm{H_2O,liquid}} = \mu^{o}_{\mathrm{H_2O,liquid}} + V(P - 1) \tag{2.13}$$

Since the molar volume of liquid is much smaller than that of gas, the term V (P − 1) is negligibly small compared with the chemical potential of the standard state. So (2.13) can be safely approximated as

$$\mu_{\mathrm{H_2O,liquid}} = \mu^{o}_{\mathrm{H_2O,liquid}} \tag{2.14}$$

The approximation in (2.14) is valid for most condensed phases unless the pressure is very high.

The chemical potential in (2.14) cannot be directly compared with that in the gas phase in (2.10) because the reference state is different. For comparison, the same reference state should be used. If the standard state for gas is used as a reference state, (2.14) can be written as

$$\mu_{\mathrm{H_2O,liquid}} = \mu^{o}_{\mathrm{H_2O,gas}} + RT\ln p^{eq}_{\mathrm{H_2O}} \tag{2.15}$$

where $P^{eq}_{H_2O}$ is the equilibrium vapour pressure of water at the given temperature.

If (2.10) and (2.15) are substituted in (2.9),

$$\Delta\mu^{liquid\rightarrow gas}_{H_2O} = \mu^{o}_{\mathrm{H_2O,gas}} + RT\ln p_{\mathrm{H_2O}} - \mu^{o}_{\mathrm{H_2O,gas}} - RT\ln p^{eq}_{\mathrm{H_2O}} = RT\ln\left(\frac{p_{\mathrm{H_2O}}}{p^{eq}_{\mathrm{H_2O}}}\right) \tag{2.16}$$

which can also be written as

$$\Delta\mu_{\mathrm{H_2O}}^{liquid\rightarrow gas} = -RT\ln\left(\frac{p_{\mathrm{H_2O}}^{eq}}{p_{\mathrm{H_2O}}}\right) \tag{2.17}$$

If the relative humidity is 50 %, the value in the parenthesis is 2 and the driving force for evaporation would be

$$\Delta\mu_{\mathrm{H_2O}}^{liquid\rightarrow gas} = -RT\ln 2 \tag{2.18}$$

The ratio in the parenthesis in (2.17) is defined as the supersaturation ratio for evaporation.

(Q) What would be the driving force for precipitation of water when the water vapour, which was saturated at 300 K, is suddenly supercooled to 290 K?

2.3 Driving Force for Condensation

In this process, the water molecule is transferred from the vapour to the liquid state. Using the similar scheme above, the driving force can be written as

$$\Delta\mu_{\mathrm{H_2O}}^{gas,300\mathrm{K}\rightarrow \mathrm{liquid},290\mathrm{K}} = \mu_{\mathrm{H_2O}}^{liquid,290\mathrm{K}} - \mu_{\mathrm{H_2O}}^{gas,300\mathrm{K}} \tag{2.19}$$

$$= \mu_{\mathrm{H_2O,gas}}^{o} + RT\ln p_{\mathrm{H_2O}}^{290\mathrm{K}} - \left(\mu_{\mathrm{H_2O,gas}}^{o} + RT\ln p_{\mathrm{H_2O}}^{300\mathrm{K}}\right) \tag{2.20}$$

where $\mathrm{P_{H_2O}^{290K}}$ and $\mathrm{P_{H_2O}^{300K}}$ represent the equilibrium vapour pressure of water respectively at 290 and 300 K.

The driving force for condensation becomes

$$\Delta\mu^{300K\rightarrow 290K} = -RT\ln\left(\frac{P_{H_2O}^{300K}}{P_{H_2O}^{290K}}\right) \tag{2.21}$$

The ratio of the parenthesis is called the supersaturation ratio for condensation.

2.4 Driving Force for Deposition in the Physical Vapour Deposition Process

In the physical vapour deposition (PVD) process, the materials to be deposited are evaporated thermally or by sputtering. The species in the vapour state are supersaturated for condensation or precipitation. Therefore, the PVD process consists of evaporation and condensation. When the metal evaporates at equilibrium vapour pressure of P^{eq} at temperature T with the ambient hydrostatic pressure P of the evaporant vapour, the evaporation flux is expressed by the Hertz-Knudsen equation as

$$J = \frac{\alpha_e(P^{eq} - P)}{\sqrt{2\pi mkT}} \quad \text{mols/s cm}^2 \tag{2.22}$$

where α_e is the evaporation coefficient for vapour molecules onto the surface. The evaporation coefficient is usually much less than unity because the evaporant vapour molecules impinging upon the condensed phase surface are reflected back.

Since the mass of the evaporated flux is conserved, the number density of the evaporated molecules decrease along with the distance from the evaporation source. If the evaporation source is a point, the number density decreases with distance, being inversely proportional to the square distance. If the evaporation source is an infinite line, the number density decreases, being inversely proportional to the distance. In the steady state, a certain pressure P will be maintained in the vicinity of the growing surface. This pressure will be in equilibrium with the adatoms on the growing surface or substrate. If this pressure is denoted as P^* and the equilibrium vapour pressure of the evaporant at the substrate temperature is P^{eq}_{Tsub}, the supersaturation ratio for deposition is given as

$$\alpha = \frac{P^*}{P^{eq}_{Tsub}} \tag{2.23}$$

And the driving force for deposition is given as

$$\Delta\mu = -RT \ln \alpha \tag{2.24}$$

(Q) In the chemical vapour deposition, chemical reactions are involved. How can you formulate the irreversibility of chemical reactions? What would be the driving force for chemical reactions?

2.5 Irreversibility for Chemical Reactions

Consider the chemical reaction where gases A and B react with each other and produce gas 2C (Gaskell 1995). Initially there are 1 mol of A and 1 mol of B with C being absent. If 2x mole of C is formed by chemical reaction after time t, the number of moles of both A and B would be 1 – x as shown by the following equation.

$$\begin{array}{llll} & A_{(g)} + B_{(g)} & \Leftrightarrow & 2C_{(g)} \\ \text{at t = 0,} & 1 \quad\quad 1 & & 0 \\ \text{at t = t,} & 1-x \quad 1-x & & 2x \end{array} \tag{2.25}$$

The Gibbs free energy of the system becomes

$$G' = (1-x)\mu_A + (1-x)\mu_B + 2x\mu_C = \mu_A + \mu_B + x(2\mu_C - \mu_A - \mu_B) \tag{2.26}$$

The derivative of G′ with respect to x can be positive, zero or negative. If the derivative is positive, the reaction to produce C would increase the Gibbs free energy of the system, which is not allowed. This means that the reaction would be backward. C is unstable and decomposes into B and C. The sum of chemical potential on the right hand side of the chemical reaction is larger than that on the left side.

$$\left(\frac{\partial G'}{\partial x}\right) = 2\mu_C - \mu_A - \mu_B > 0 \rightarrow \mu_A + \mu_B < 2\mu_C \tag{2.27}$$

If the derivative is zero, the chemical reaction is in equilibrium. The sum of chemical potential on the right hand side is equal to that on the left side.

$$\left(\frac{\partial G'}{\partial x}\right) = 2\mu_C - \mu_A - \mu_B = 0 \rightarrow \mu_A + \mu_B = 2\mu_C \tag{2.28}$$

If the derivative is negative, the reaction to produce C would decrease the Gibbs free energy, which is allowed. This means that the reaction would be forward. C is stable and would be formed at the expense of B and C. The sum of chemical potential on the right hand side of the chemical reaction is smaller than that on the left side.

$$\left(\frac{\partial G'}{\partial x}\right) = 2\mu_C - \mu_A - \mu_B < 0 \rightarrow \mu_A + \mu_B > 2\mu_C \tag{2.29}$$

This conclusion is generally valid for any chemical reaction. If the sum of chemical potentials on the left side is larger than that on the right side, the forward reaction will be irreversible. If the sum of chemical potentials on the left side is smaller than that on the right side, the backward reaction will be irreversible. If the sum of chemical potentials on the left side is equal to that on the right side, the chemical reaction is in equilibrium.

If the reactants are on the left side and the products are on the right side, the driving force for the forward reaction can be written as

$$\Delta\mu = \sum \mu_i^p n_i^p - \sum \mu_i^r n_i^r \tag{2.30}$$

This driving force changes with time or the chemical reaction. Consider the simple reaction of (2.25).

$$\Delta\mu = 2\mu_C - \mu_A - \mu_B = 2\mu_C^o + 2RT\ln p_C - \mu_A^o - RT\ln p_A - \mu_B^o - RT\ln p_B \tag{2.31}$$

$$\Delta\mu = \Delta\mu^o + RT\ln\frac{p_C^2}{p_A p_B} \tag{2.32}$$

Since C is initially absent, P_C is equal to zero. Then the driving force in (2.32) becomes infinite. This infinite driving force comes from the entropy of mixing, which goes to infinity when something new is formed infinitesimally. This means that the driving force for existence from non-existence is infinite. In other words, nature tries to experience all the possible things it can.

The driving force for the chemical reaction in (2.32) continues to decrease as the reaction goes on until $\Delta\mu$ becomes zero, where the equilibrium state is reached. If the chemical reaction proceeds further, it will increase $\Delta\mu$, which is not allowed by the second law of thermodynamics. Therefore, once the equilibrium is reached, the irreversible chemical reaction stops. However, the reversible chemical reaction continues, satisfying the constraint of the dynamic equilibrium that the rates of the forward and backward reactions are the same.

2.6 Driving Force for Deposition in the Chemical Vapour Deposition Process

Consider a chemical reaction similar to (2.25) but this time the component C is not gas but solid.

$$A_{(g)} + B_{(g)} \Leftrightarrow C_{(s)} \tag{2.33}$$

This reaction would be one example of chemical vapour deposition. The driving force for the forward reaction of (2.33) is expressed as

$$\Delta\mu = \Delta\mu^o + RT\ln\frac{1}{p_A p_B} \tag{2.34}$$

where $\Delta\mu^o$, p_A, and p_B are respectively standard chemical potential change (or standard molar Gibbs free energy change), partial pressure of A and partial pressure of B. It should be noted that since the chemical potential of the condensed phase is equal to that of its standard state, $p_C = 1$. Equation (2.34) represents the driving force for the forward chemical reaction, which should be distinguished from the driving force for deposition of C (Hwang and Yoon 1994a).

(Q) What would be the driving force for deposition of solid C as a result of the chemical reaction of (2.33)?

This question may not be easy to answer. It would be better to answer an easier question as follows. Consider the A–B binary alloy, where the α solid solution is stable at high temperature and the β solid solution is stable at low temperature. If the alloy of the composition $X_B^{\alpha_o}$ is supercooled from the α one phase region into the two phase region of α + β, the α phase will be decomposed as follows.

$$\alpha \rightarrow \alpha' + \beta \tag{2.35}$$

(Q) What would be the driving force for precipitation of β?

In Fig. 2.1, the molar Gibbs free energies of the α and β phases for the A–B alloy are shown with respect to X_B. The final equilibrium state of α′ + β is defined by the common tangent $\overline{QS}$. The molar Gibbs free energy at the composition $X_B^{\alpha_o}$ in the equilibrium state is denoted as R. And the driving force for the reaction of (2.35) is $\overline{OR}$ in Fig. 2.1. However, this driving force should be distinguished from the driving force for precipitation of β. The driving force $\overline{OR}$ is dissipated in the decomposition of α into α′ and β. However, we do not use this value as a driving force for nucleation of β. In other words, this driving force is not used to calculate the critical nucleus or the nucleation barrier of β. The driving force for precipitation of β should be evaluated at the composition of β, which is $X_B^{\beta_o}$ in Fig. 2.1. The molar Gibbs free energy of β at $X_B^{\beta_o}$, which is $G_m^\beta\left(X_B^{\beta_o}\right)$, is designated as Q in Fig. 2.1. At the composition of $X_B^{\beta_o}$, we need to determine the Gibbs free energy of the state which is supersaturated with respect to precipitation of β. This supersaturated state should be in equilibrium with the initial state, which is the α phase at the composition of $X_B^{\alpha_o}$. The tangent on the molar Gibbs free energy curve of the α phase at $X_B^{\alpha_o}$ represents the molar Gibbs free energy of varying composition in equilibrium with the α phase at $X_B^{\alpha_o}$. For example, the intercept of the tangent at the axis of $X_B = 1$ represents $\mu_B^\alpha(X_B^{\alpha_o})$, which is the chemical potential of B in equilibrium with the α phase at $X_B^{\alpha_o}$. Likewise, the intercept of the tangent at the axis of

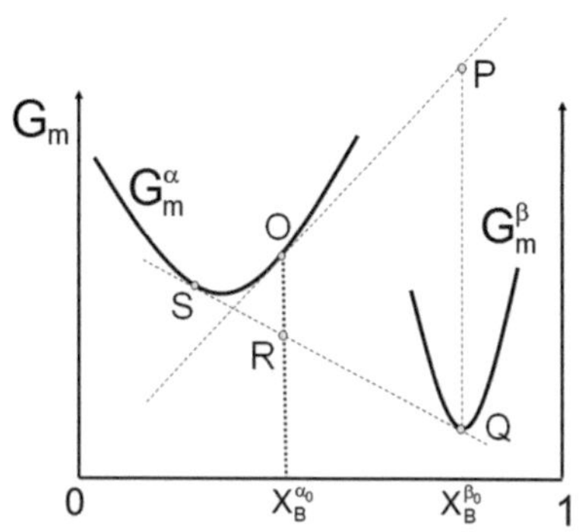

Fig. 2.1 Molar Gibbs free energy of α and β phases in the binary A–B alloy, illustrating the driving force for precipitation of β at $X_B^{\beta_o}$ from the α matrix at $X_B^{\alpha_o}$ (Hillert and Aaronson 1975)

$X_B = 0$ represents $\mu_A^\alpha(X_B^{\alpha_o})$, which is the chemical potential of A in equilibrium with the α phase at $X_B^{\alpha_o}$. Therefore, the molar Gibbs free energy at the composition of $X_B^{\beta_o}$ in equilibrium with the α phase at $X_B^{\alpha_o}$ is determined by the intercept of the tangent with the axis of $X_B = X_B^{\beta_o}$. This value is designated as P in Fig. 2.1. Therefore, the driving force for precipitation of β is $\overline{PQ}$ in Fig. 2.1. It should be noted that the driving force for precipitation of β changes as the composition of the α matrix changes with the precipitation process because the tangent should be made at the new composition of the α matrix.

With this thermodynamic scheme in mind, consider the driving force for deposition of the solid C in the chemical reaction of (2.33). It should be noted that the driving force should be evaluated at the composition of the solid C. The driving force should be the difference of the chemical potential of C between solid and gas. If the standard state for gas is used as a reference state, the chemical potential of C in solid, $\mu_{C,solid}$, is expressed as

$$\mu_{C,solid} = \mu_{C,gas}^o + RT \ln p_C^{eq} \tag{2.36}$$

where P_C^{eq} is the equilibrium vapour pressure of C at the substrate temperature.

Since the driving force for deposition is the difference in the chemical potential C between solid and gas in the vicinity of the growing film, the remaining problem is to identify the chemical potential of C in gas, $\mu_{C,gas}$, made by the chemical reaction of (2.33). If the steady state partial pressure of C in the vicinity of the growing film is denoted as $P_{C,gas}$, the driving force is expressed as

$$\Delta\mu_C^{gas\to solid} = \mu_{C,solid} - \mu_{C,gas} = \mu^o + RT\, ln\, p_C^{eq} - \mu^o - RT\, ln\, p_{C,gas} \tag{2.37}$$

Arranging (2.37) gives (Hwang and Yoon 1994a)

$$\Delta\mu_C^{gas\to solid} = -RT\, ln \frac{p_{C,gas}}{p_C^{eq}} \tag{2.38}$$

And $P_{C,gas}/P_C^{eq}$ is the supersaturation ratio α for deposition. To determine α or driving force for deposition, we need to determine $p_{C,gas}$. The steady state value of $p_{C,gas}$ in the CVD process depends on the kinetics of chemical reactions under the given processing conditions with temperature as a major factor. If plasma is used in the CVD process, it drastically increases the kinetics of chemical reactions.

Think about the extreme case where the chemical reaction of (2.33) hardly occurs in the CVD reactor. This would happen if the reactor temperature is too low to decompose the reactant gases. Then, there will be no deposition because $p_{C,gas}$ would be practically zero. $p_{C,gas}$ would increase with the extent of the chemical reaction. Depending on the extent of chemical reaction, $p_{C,gas}$ can have the three ranges of values as follows.

$$0 \leq p_{C,gas} < p_C^{eq} \tag{2.39}$$

$$p_{C,gas} = p_C^{eq} \tag{2.40}$$

$$p_C^{eq} < p_{C,gas} \leq p_{C,gas}^{max} \tag{2.41}$$

where $p_{C,gas}^{max}$ represents the maximum value that $p_{C,gas}$ can have. In the range of (2.39), the supersaturation ratio $\alpha = p_{C,gas} / p_C^{eq}$ is less than 1 and the driving force is for etching rather than for deposition. In (2.40), α is 1 and the driving force is zero, indicating neither deposition nor etching occurs. Only in the range of (2.41), α is larger than 1 and the driving force is for deposition.

(Q) How can $p_{C,gas}^{max}$ *be determined?*

$p_{C,gas}^{max}$ represents the partial pressure of C that is obtained by the maximum extent of reaction or the maximum metastable state without precipitating the solid. It can be determined by minimizing the Gibbs free energy excluding the condensed phase of C. In other words, $p_{C,gas}^{max}$ can be determined by minimizing the Gibbs free energy of the gas phase only (Hwang and Yoon 1994a). Therefore, the maximum supersaturation ratio and driving force can be expressed as

$$\alpha = p_{C,gas}^{max} / p_C^{eq} \tag{2.42}$$

$$\Delta\mu = -RT\, ln\left(p_{C,gas}^{max} / p_C^{eq}\right) \tag{2.43}$$

In the real practice of CVD, if $P_{C,gas} \ll P_{C,gas}^{max}$, most reactant gases flow out of the CVD reactor without undergoing an appreciable extent of chemical reactions, resulting in very low efficiency. Therefore, such a CVD condition would not be adopted. In most processes, process parameters would be adjusted to maximize the chemical reactions of reactants so that the condition of $P_{C,gas} \approx P_{C,gas}^{max}$ may be achieved. In this sense, $P_{C,gas} \approx P_{C,gas}^{max}$ may be a reasonable approximation. Since $P_{C,gas}^{max}$ can be determined by minimization of the Gibbs free energy of the gas phase only, $\alpha = p_{C,gas}^{max} / p_C^{eq}$ can be calculated with respect to thermodynamic parameters such as temperature, pressure and composition.

Equations (2.42) and (2.43) as a reasonable approximation can be used if the nucleation occurs in the gas phase. In most CVD processes, however, nucleation and growth are believed to occur on the substrate or surface. Besides, it is known that most flux contributing to the film growth does not directly come from the gas phase but come from the diffusion of adatoms on the surface. First, the gas atoms are adsorbed on the terrace. And then the adatoms diffuse to the ledge of monoatomic height. Since monoatomic ledges have abundant kinks, the adatoms will find kink sites, where they become incorporated into the crystal. This means that we need to evaluate the supersaturation ratio in terms of the adatoms on the surface. The composition of the adsorbed species may be different from that of the gas phase because the desorption energies are different for each species.

(Q) How can you formulate the supersaturation ratio and the driving force for deposition on the surface?

2.7 Consideration of the Substrate or Surface

Considering the surface, $p_{C,gas}^{max}$ and p_C^{eq} in (2.42) should be replaced by the corresponding adsorbed concentrations. If the equilibrium between the gas phase and the surface is maintained, the ratio in (2.42) does not change and thus (2.43) does not change because the chemical potentials of the species in the gas phase are the same as those on the adsorbed species. If the process is in the steady state, it is safely assumed that the equilibrium between the gas and the surface is maintained. If the equilibrium between the gas phase and the surface is not maintained, the adsorbed concentration may not be in the steady state but depend on time. Then, the supersaturation ratio and the driving force are different from (2.42) and (2.43) respectively. In such a non-steady state, it is difficult to determine the superstation ratio and the driving force.

When the species to be deposited is a compound or a solid solution by the CVD process, the partial pressure or the equilibrium vapour pressure is often difficult to define. For example, compounds such as SiC, SiO_2 and Si_3N_4, do not have vapour species corresponding to the same stoichiometry of the solid phase.

(Q) How can you formulate the supersaturation ratio and the driving force for deposition of compounds or solid solution?

2.8 Driving Force for Deposition of Compound or Solid Solution

The compound A_aB_b does not have the vapour species of the same stoichiometry, where A and B are the element. If (2.42) is applied, the supersaturation ratio for deposition of A_aB_b is expressed as

$$\alpha = \frac{p_{A_aB_b,gas}^{max}}{p_{A_aB_b}^{eq}} \tag{2.44}$$

The problem is that $p_{A_aB_b,gas}^{max}$ and $p_{A_aB_b}^{eq}$ cannot be determined because the compound A_aB_b does not exist in the gas phase. However, the ratio of $p_{A_aB_b,gas}^{max}$ to $p_{A_aB_b}^{eq}$ can be determined by considering the gas species that are in equilibrium with the hypothetical gas compound A_aB_b. Consider the two hypothetical equilibria: one for the metastable gas phase equilibrium and the other for the stable equilibrium.

$$aA^{gas,eq} + bB^{gas,eq} = A_aB_b^{gas,eq} \tag{2.45}$$

$$aA^{eq} + bB^{eq} = A_aB_b^{eq} \tag{2.46}$$

From these two equilibria of (2.45) and (2.46), the following relationship is derived.

$$\alpha = \frac{p_{A_aB_b,gas}^{max}}{p_{A_aB_b}^{eq}} = \frac{\left(p_{A,gas}^{max}\right)^a \left(p_{B,gas}^{max}\right)^b}{\left(p_A^{eq}\right)^a \left(p_B^{eq}\right)^b} \tag{2.47}$$

And $p_{A,gas}^{max}$, $p_{B,gas}^{max}$, p_A^{eq} and p_B^{eq} can be determined by the metastable gas phase equilibrium and the stable equilibrium of the chemical reaction. From (2.47), the driving force for deposition can also be determined. (Hwang and Yoon 1994b)

This result can be extended to the solid solution by assuming that the stoichiometry coefficients a and b can continuously change within the solid solution range with the constraint of a + b = 1, which is needed when the driving force is evaluated per mole of the solid solution.

Let's take an example of SiC deposition using the reactant gases of SiH_4 and CH_4. The overall chemical reaction would be as follows.

$$SiH_4 + CH_4 = SiC + 4H_2 \tag{2.48}$$

If the thermodynamic scheme of (2.47) is applied to the reaction in (2.48), the supersaturation ratio is expressed as

$$\alpha = \frac{p_{SiC,gas}^{max}}{p_{SiC}^{eq}} = \frac{p_{SiH_4,gas}^{max}\, p_{CH_4,gas}^{max} \left(p_{H_2}^{eq}\right)^4}{p_{SiH_4}^{eq}\, p_{CH_4}^{eq} \left(p_{H_2,gas}^{max}\right)^4} \tag{2.49}$$

The supersaturation ratio can be expressed in a more simplified form by considering a following reaction satisfying the law of mass action.

$$Si + C = SiC \tag{2.50}$$

If the thermodynamic scheme in (2.47) is applied to (2.50), the supersaturation ratio is expressed as

$$\alpha = \frac{p_{SiC,gas}^{max}}{p_{SiC}^{eq}} = \frac{p_{Si,gas}^{max}\, p_{C,gas}^{max}}{p_{Si}^{eq}\, p_{C}^{eq}} \tag{2.51}$$

which is much simpler than (2.49)

2.9 Calculation Examples Using Thermo-Calc

2.9.1 Si–Cl–H System

At the temperature of 1323 K and the pressure of 10,666 Pa with the gas mixture of $SiH_4 : HCl : H_2 = 0.14 : 0.92 : 99$, the following conditions are set for calculation by Thermo-Calc (Sundman et al. 1985).

Set-Conditions: T = 1323, P = 10,666, N(Si) = 1.4E–1, N(Cl) = 9.2E–1, N(H) = 199.36

Here, N represents the number of moles. The Gibbs free energy minimization under these conditions produces the following amount of solid silicon and gases. Here, SI1, which has a mole fraction of 7.97411E–10, corresponds to the equilibrium vapour pressure of silicon at the given temperature.

H2	9.91047E–01	CL2H2SI1	3.78938E–06	SI1	7.97511E–10
CL1H1	8.72066E–03	CL4SI1	7.88797E–07	CL2	5.88841E–13
CL2SI1	2.12707E–04	H4SI1	8.60203E–08	SI2	2.60034E–13
CL3SI1	5.55192E–06	CL1	3.56902E–08	SI3	3.14674E–14
H1	4.83980E–06	CL1SI1	9.72339E–09	H6SI2	6.50693E–15
CL3H1SI1	4.57379E–06	H1SI1	2.01285E–09		

SI1_S#1			Status ENTERED	Driving force		0.0000E+00
Number of moles		1.1722E–01	Mass 3.2921E+00	Mole fraction		
SI	1.00000E+00	CL	0.00000E+00	H		0.00000E+00

The Gibbs free energy minimization by suspending condensed phases of silicon in the calculations produces the following amount of gases. Here, SI1, which has a mole fraction of 9.40131E−10, corresponds to the maximum partial pressure of silicon in the supersaturated gas phase, where condensed phases of silicon did not precipitate.

H2	9.92254E−01	H1	4.84275E−06	SI1	9.40131E−09
CL1H1	6.34253E−03	CL4SI1	2.59578E−06	SI3	5.15678E−11
CL2SI1	1.32491E−03	H4SI1	1.01663E−06	SI2	3.61445E−11
CL3SI1	2.51360E−05	CL1SI1	8.33246E−08	H6SI2	9.07768E−13
CL2H2SI1	2.36321E−05	CL1	2.59417E−08	CL2	3.11097E−13
CL3H1SI1	2.07202E−05	H1S1	2.37455E−08		

The supersaturation ratio for deposition of Si can be determined as

$$\alpha = \frac{p_{Si}^{\max}}{p_{Si}^{eq}} = \frac{9.40131 \times 10^{-9}}{7.97411 \times 10^{-10}} = 11.79$$

2.9.2 Si–C–H System

At the temperature of 1400 K and the pressure 1000 Pa with the gas mixture of $SiH_4 : CH_4 = 1 : 1$, the following conditions are set for calculation by Thermo-Calc (Sundman et al. 1985).

Set-Conditions: T = 1400, P = 1000, N(Si) = 1, N(C) = 1, N(H) = 8

The Gibbs free energy minimization under these conditions produces the following.

H2	9.99951E−01	C2H4	3.00948E−13	C3H4_1	1.09232E−19
H	4.87045E−05	C1H2	6.00656E−15	C	1.05200E−19
C1H4	1.64031E−07	C2H3	1.17540E−15	C3H6_2	1.59231E−21
SI	9.98049E−08	C2SI1	1.02915E−15	C3H6_1	2.53376E−24
H1SI1	5.25792E−08	H6SI2	1.31302E−16	C2	1.05475E−24
H4SI1	1.16175E−08	C2H6	5.95508E−17	C3	2.51092E−25
C1H3	1.67072E−10	C2H1	1.49582E−17	C3H8	5.92461E−26
SI2	8.01027E−11	C1SI1	6.96977E−18	C4	1.00000E−30
C1SI2	6.87487E−11	C2H5	2.43303E−18	C4H12SI1	1.00000E−30
C2H2	3.82934E−11	C1H1	1.39138E−18	C5	1.00000E−30
SI3	1.48834E−11	C3H4_2	3.16405E−19		

The Gibbs free energy minimization by suspending all condensed phases related to silicon, carbon and silicon carbide in the calculations produces the following amount of gases.

H2	8.30213E−01	C1H3	5.15943E−06	C1SI1	5.66556E−11
C1SI2	1.11284E−01	C3H4_1	5.12173E−06	C3	1.70796E−11
C1H2	5.29783E−02	SI2	3.17637E−06	H6SI2	2.97979E−12
C1H4	4.61561E−03	H4SI1	1.59469E−06	C3H8	1.91491E−12
C2H4	3.45682E−04	C2H3	1.48171E−06	C5	5.42009E−13
C2SI1	3.41494E−04	C3H6_2	6.19877E−08	C1H1	5.17529E−14
SI3	1.17524E−04	C2H6	5.67915E−08	C	4.29435E−15
H	4.43787E−05	C2H1	2.27117E−08	C4	2.46159E−15
SI	1.98744E−05	C2H5	2.54647E−09	C2	1.75758E−15
C3H4_2	1.48358E−05	C1H2	2.03572E−10	C4H12SI1	5.68974E−16
H1SI1	9.54028E−06	C3H6_1	9.86377E−11		

If the supersaturation ratio for deposition of silicon carbide is calculated based on (2.49)

$$\alpha_1 = \frac{p^{max}_{SiC,gas}}{p^{eq}_{SiC}} = \frac{p^{max}_{SiH_4,gas}\, p^{max}_{CH_4,gas} \left(p^{eq}_{H_2}\right)^4}{p^{eq}_{SiH_4}\, p^{eq}_{CH_4} \left(p^{max}_{H_2,gas}\right)^4}$$

$$= \frac{1.59469 \times 10^{-6} \times 4.61561 \times 10^{-3} \times (0.999951)^4}{1.16175 \times 10^{-8} \times 1.64031 \times 10^{-7} \times (0.830213)^4} = 8{,}128{,}785$$

If the supersaturation ratio for deposition of silicon carbide is calculated based on (2.51),

$$\alpha_2 = \frac{p^{max}_{SiC,gas}}{p^{eq}_{SiC}} = \frac{p^{max}_{Si,gas}\, p^{max}_{C,gas}}{p^{eq}_{Si}\, p^{eq}_{C}}$$

$$= \frac{1.98744 \times 10^{-5} \times 4.29435 \times 10^{-15}}{9.98049 \times 10^{-8} \times 1.052 \times 10^{-19}} = 8{,}128{,}752$$

Therefore, the supersaturation ratio (α_1) calculated based on (2.49) is equal to that (α_2) calculated based on (2.51) within a round-off error.

2.10 Validity of the Thermodynamic Scheme

Figure 2.2 shows the CVD phase diagram of the Si–Cl–H system at 10,666 Pa and composition of N(H) = 200 and N(Cl) = 1. The line of the phase diagram divides into regions of gas and gas + solid. In the gas region, only the gas phase is stable and the solid cannot precipitate. Therefore, the supersaturation ratio for deposition should be less than unity ($\alpha < 1$). Along the line, the supersaturation ratio should be unity ($\alpha = 1$). In the gas + solid region, both gas and solid are stable and the solid can precipitate. Therefore, the supersaturation ratio should be larger than unity ($\alpha > 1$).

The supersaturation ratio, $\alpha = 1$, based on (2.42) corresponds to the phase boundary of gas/(gas + solid) in Fig. 2.2. And, the regions of $\alpha < 1$ and $\alpha > 1$ represent respectively the one-phase region of gas and the two-phase region of gas + solid. This means that the thermodynamic scheme for the supersaturation ratio for CVD given by (2.42) is valid, agreeing with the CVD phase diagram.

Equation (2.42) provides additional information in the CVD phase diagram as shown in Fig. 2.3, where the lines of iso-supersaturation ratio, $\alpha = 0.01$, $\alpha = 0.1$, $\alpha = 10$, and $\alpha = 100$ are drawn. Note that a melting point of the solid silicon is 1687 K above which the solid silicon is metastable with respect to its liquid.

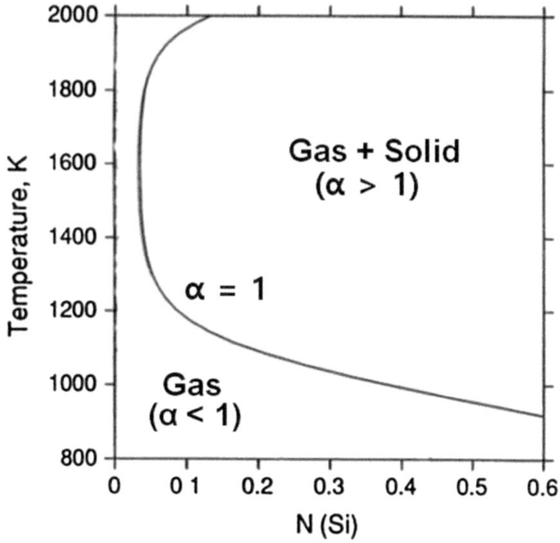

Fig. 2.2 The CVD phase diagram of the Si–Cl–H system at 10,666 Pa and composition of N(H) = 200 and N(Cl) = 1

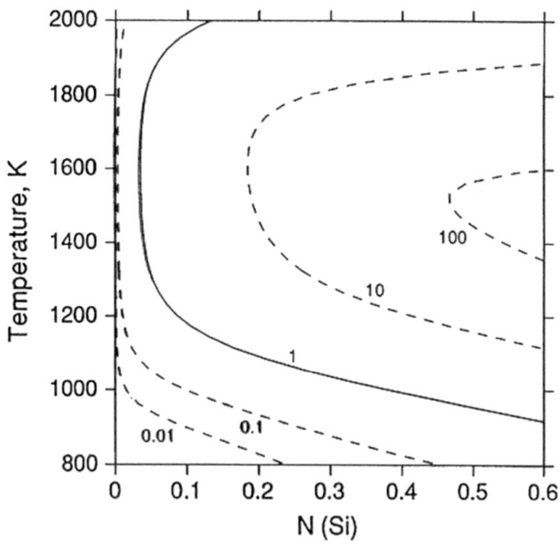

Fig. 2.3 The CVD phase diagram of the Si–Cl–H system at 10,666 Pa and composition of N(H) = 200 and N(Cl) = 1. The *dashed lines* of the iso-supersaturation ratios, 0.01, 0.1, 10 and 100 are evaluated from (2.42) (Hwang and Yoon 1994b)

Figure 2.3 shows a relatively sensitive change of the iso-supersaturation ratio with N(Si) at ~1500 K, indicating the gradient of the undersaturation with N(Si) is high. By this thermodynamic scheme, lines of iso-supersaturation or driving force for etching or deposition can be evaluated with respect to any independent CVD thermodynamic variables.

2.11 Maximum Film Growth Rate in the CVD Process

Since $p^{max}_{C,gas}$ in (2.41) represents the maximum partial pressure of 'C' for the given CVD process, the flux for the maximum deposition rate can be evaluated using the Hertz-Knusen equation as

$$J_C^{max} = \frac{p_{C,gas}^{max}}{\sqrt{2\pi mkT}} \quad \text{mols/s cm}^2 \tag{2.52}$$

This equation is under the assumption that the metastable equilibrium among gas species without condensed phases is maintained at the given temperature. In order to consider the supercooling effect of the gas in the presence of the temperature gradient, $p^{max}_{C,gas}$ at the temperature, from which the gas is expected be supercooled, should be used.

Using (2.52), not only the maximum film growth rate can be estimated, but also the minimum partial pressure of the species to be deposited can be estimated if the film growth rate is given. However, in estimating the maximum film growth rate using (2.52), it is implicitly assumed that the gas phase nucleation does not occur and the film grows by individual atoms or molecules. If the gas phase nucleation occurs and thereby gas phase generated nanoparticles begin to contribute as a building block to film growth, (2.52) is not valid.

2.12 Driving Force for Deposition in the Diamond Chemical Vapour Deposition Process

One of typical processing conditions for diamond deposition by hot filament CVD (HFCVD) is the gas mixture of 1 % CH_4–99 % H_2, the reactor pressure of 2700 Pa, the substrate temperature of 1200 K and the hot filament temperature of 2300 K. By the hot filament, the reactant gases will be decomposed into many species including the solid carbon.

$$CH_4 + H_2 \rightarrow C_{gas} + CH + CH_2 + \cdots + C_{solid} \tag{2.53}$$

where C_{gas} and C_{solid} represent respectively the gas and solid carbon, which can be graphite or diamond depending on the kinetics.

The minimization of the Gibbs free energy of the mixture gas at 2700 Pa and 1200 K excluding all the condensed phases produces the metastable gas phase equilibrium which specifies each mole fraction of species as follows.

$$\Rightarrow C^* + C_2H_2^* + C_2H^* + CH_4^* + CH_3^* + C_2H_4^* + \cdots \tag{2.54}$$

where * represents the metastable gas phase equilibrium. If p_i^* is defined to represent the metastable partial pressure of the species i, p_C^* represents the maximum partial pressure of carbon that can be achieved metastably at the given thermodynamic condition.

The minimization of the Gibbs free energy excluding the stable graphite would produce the metastable equilibrium which specifies each mole fraction of species including diamond as follows.

$$\Rightarrow C^{eq,d} + C_2H_2^{eq,d} + C_2H^{eq,d} + CH_4^{eq,d} + CH_3^{eq,d} + C_2H_4^{eq,d} + \cdots + \text{diamond} \quad (2.55)$$

where all the gas species are in metastable equilibrium with diamond. Likewise, $p_C^{eq,d}$ represents the equilibrium vapour pressure of diamond at 1200 K.

The minimization of the Gibbs free energy without excluding any condensed phase would produces the stable equilibrium which specifies each mole fraction of species including graphite as follows.

$$\Rightarrow C^{eq} + C_2H_2^{eq} + C_2H^{eq} + CH_4^{eq} + CH_3^{eq} + C_2H_4^{eq} + \cdots + \text{graphite} \quad (2.56)$$

where all the gas species are in stable equilibrium with graphite. Likewise, p_C^{eq} represents the equilibrium vapour pressure of graphite at 1200 K.

Then, the maximum supersaturation ratio and driving force for graphite deposition are expressed respectively as

$$\alpha = \frac{p_C^*}{p_C^{eq}} \quad (2.57)$$

and

$$\Delta\mu^{gas \rightarrow gra} = -RT\ln\left(\frac{p_C^*}{p_C^{eq}}\right). \quad (2.58)$$

The calculation under the condition of T = 1200 K and P = 2700 Pa by Thermo-Calc (Sundman et al. 1985) based on the scientific group thermodata Europe (SGTE) substance database shows that $\alpha = 23.3$ and $\Delta\mu^{gas \rightarrow gra} = -31413$ J/mole (Hwang et al. 1994).

Similarly, the maximum supersaturation ratio and the driving force for diamond deposition are expressed as

$$\alpha = \frac{p_C^*}{p_C^{eq,d}} \quad (2.59)$$

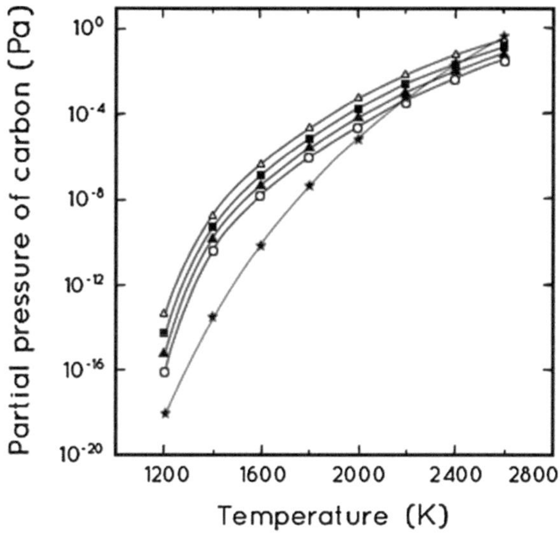

Fig. 2.4 Variation of the partial pressure of carbon in the gas phase with temperature for four different ratios of carbon to hydrogen (total pressure, 2700 Pa); Δ, C:H = 1:50; ■, C:H = 1:100; ▲, C:H = 1:200; O,1:400; ★, equilibrium vapour pressure of graphite or diamond (Hwang et al. 1994)

$$\Delta\mu^{\text{gas}\to\text{dia}} = -\text{RT}\ln\left(\frac{p_C^*}{p_C^{\text{eq,d}}}\right) \tag{2.60}$$

The calculation shows that $\alpha = 11.7$, $\Delta\mu^{\text{gas}\to\text{dia}} = -24540$ J/mole and $\Delta\mu^{\text{dia}\to\text{gra}} = -6873$ J/mole. As seen from the supersaturation ratio, the equilibrium vapour pressure of diamond is twice larger than that of graphite. Based on this thermodynamic scheme, the metastable partial pressure of carbon can be determined with varying temperature and various mixtures of carbon and hydrogen, which is shown in Fig. 2.4 at a reactor pressure 2700 Pa. In Fig. 2.4, the equilibrium vapour pressure of graphite or diamond is also shown for comparison. It should be noted that since the vertical axis of pressure is a logarithmic scale, the slight difference in equilibrium vapour pressure between graphite and diamond is not distinguishable.

The gas mixtures of 0.5 % CH_4–99.5 % H_2, 1 % CH_4–99 % H_2, 2 % CH_4–98 % H_2 and 3 % CH_4–97 % H_2, which cover the composition range mostly used in the diamond HFCVD process, have the ratios of C to H of 1/402, 1/202, 1/102 and 1/69, respectively. For the ratio of C : H = 1 : 200, the maximum partial pressure of carbon is higher than its equilibrium vapour pressure over the temperature range between roughly 890 and 2290 K. This means that in the temperature range of $\sim$890 K < T < $\sim$2290 K, the driving force is for deposition of solid carbon and in the temperature range of T < $\sim$890 K, and T > $\sim$2290 K, the driving force is for etching of solid carbon.

This deposition condition appears to agree with the actual deposition at least to an extent. First, in the diamond HFCVD process, the deposition hardly occurs below the substrate temperature of 900 K if the gas mixture of 1 % CH_4–99 % H_2 is used. The general tendency is that the quality of diamonds is degraded when the filament temperature decreases or the concentration of CH_4 increases. It is experimentally observed that under the condition where the solid carbon deposits on the hot filament, poor quality diamonds are formed on the substrate (Sommer et al. 1989). In order for carbon to deposit on the hot filament, the driving force should be for deposition at the hot filament temperature. Considering the above calculation, this condition is achieved when the filament temperature is decreased or the concentration of CH_4 is increased. These results indicate that in the steady state of the diamond HFCVD process, most of the reactant gases decompose.

The experimental observation that under the condition where the solid carbon deposits on the hot filament, poor quality diamonds are deposited is not a problem of thermodynamics but related with the growth mechanism of diamond. The carbon deposited filament surface can have quite a different property in its catalytic effect, which is not clearly understood. It would be better to treat this problem after the growth mechanism of diamonds is studied in Chap. 6.

From (2.60), the chemical potential of carbon in the gas can be written as

$$\mu^{gas} = \mu^{dia} + RT\ln\left(\frac{p_C^*}{p_C^{eq,d}}\right) = \mu^{dia} + RT\ln a_C^* \tag{2.61}$$

This means that if diamond is set as a reference state, the supersaturation ratio for diamond deposition becomes an activity of the gas phase. Using (2.61), the carbon activities of gas, diamond and graphite can be compared with temperature at a gas mixture of 1 % CH_4–99 % H_2 at a reactor pressure of 2700 Pa, which is shown in Fig. 2.5.

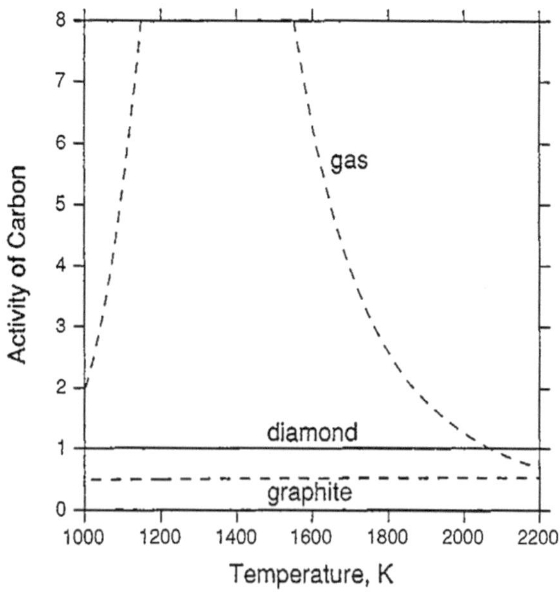

Fig. 2.5 Temperature dependence of carbon activities of gas, diamond and graphite at a gas mixture of 1 % CH_4–99 % H_2 at 2700 Pa (Hwang et al. 1996)

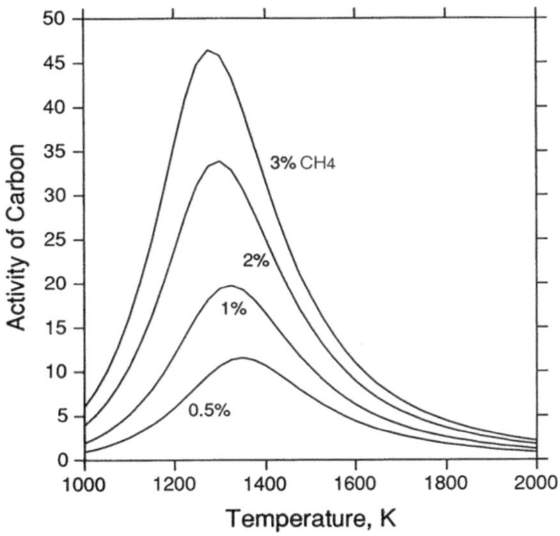

Fig. 2.6 Temperature dependence of carbon activity of gas for four different gas mixtures of 0.5 % CH_4–99.5 % H_2, 1 % CH_4–99 % H_2, 2 % CH_4–98 % H_2 and 3 % CH_4–97 % H_2 at 2700 Pa. The supersaturation ratio is with respect to graphite (Yoon and Hwang 1995)

The activity of graphite is about a half of that of diamond. The carbon activity of the gas has maximum at ~1350 K. At temperature higher than ~1350 K, the activity of the gas decreases with increasing temperature and at temperature lower than ~1350 K, it decreases with decreasing temperature. Figure 2.5 shows that the activities of gas and diamond intersect at temperature at ~2050 K, which means at temperature lower than this intersection, the driving force is for deposition of diamond and at temperature higher than this intersection, the driving force is for etching of diamond. The temperature at which the activities of gas and graphite intersect, is a little higher than 2200 K.

The carbon activity of the gas can be evaluated as a function of temperature for different concentration of CH_4 as shown in Fig. 2.6. As expected, the activity, which is the same as the supersaturation ratio for graphite deposition, increases with increasing concentration of CH_4.

Based on the thermodynamic scheme of (2.59) and (2.60), the lines of iso-supersaturation ratio or iso-activity of carbon in the gas can be calculated as a function of the atomic fraction of carbon and temperature as shown in Fig. 2.7. Since the supersaturation ratio is evaluated for diamond deposition, diamond can be deposited only in the region of $\alpha > 1$. Figure 2.7 can be divided into three regions. The first region is outside the graphite line, where only gas is stable. In this region, neither graphite nor diamond can be deposited. The second region is between the graphite and diamond lines, where graphite and gas are stable. In this region, graphite can be deposited but diamond cannot be deposited. The third region is inside the diamond line. In this region, graphite and gas are stable but diamond is metastable. In order that metastable diamond may be deposited, it is necessary that the thermodynamic condition should be in the third region. In addition to the diamond line of $\alpha = 1$, Fig. 2.7 shows other lines of $\alpha = 0.1$ and 10. Comparison of

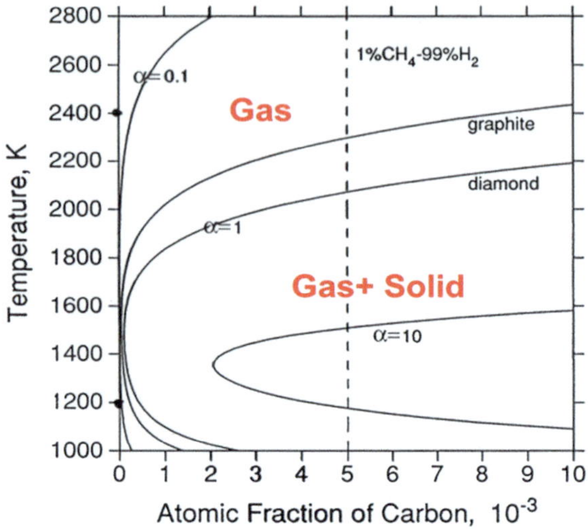

Fig. 2.7 Lines of iso-supersaturation ratio for diamond deposition as a function of temperature and atomic fraction of carbon in the C–H system (Hwang 1994)

the three lines of $\alpha = 0.1$, 1 and 10 gives the information as to how steeply the supersaturation or the undersaturation varies with temperature and composition. The vertical dashed line represents the composition of 1 % CH_4–99 % H_2. At this composition, diamond deposition is possible below ~2050 K in agreement with Figs. 2.4 and 2.5.

The line of $\alpha = 1$, which corresponds to the phase boundary of the CVD phase diagram, represents a solubility limit of carbon in the gas phase. In the region of $\alpha < 1$, carbon is soluble in the gas phase in the form of all kinds of hydrocarbon such as CH, CH_2, and CH_3. In the region of $\alpha > 1$, carbon in the gas phase exceeds the solubility limit of the gas phase and diamond can be precipitated. Figure 2.7 shows that the solubility limit is minimum at ~1350 K and increases with decreasing temperature. This increase of the solubility limit with decreasing temperature is rather an unusual case and called a retrograde solubility; normally the solubility limit decreases with decreasing temperature.

The list of commands to calculate the iso-activity lines with diamond as a reference state shown in Fig. 2.7 by Thermo-Calc is as follows. A brief explanation for each command is given.

$TC
go ges → Go to Gibbs Energy System Module
read chd → Read the data file saved with a file name of chd in the GES module
go poly-3 → Go to Poly-3 modules
"Set the thermodynamic condition so that the degree of freedom may be zero."
set-condition t = 1200 p = 2700 x(c) = 0.0049261 n = 1
→ temperature in K, pressure in Pa, the atomic fraction of carbon, x (c) = 0.0049261, corresponds to the composition of 1 % CH_4–99 % H_2, n = 1 means that the total number of mole is equal to 1.

set-automatic-starting value
Yes → accept an automatic starting value
compute-equilibrium → compute Gibbs free energy minimization for the given condition
set-reference-state c diamond → set reference state carbon as diamond
change-status phase c_s = suspend → change status of the phase solid carbon to be suspended
change-status phase graphite = suspend → change status of the phase graphite to be suspended
change-status phase diamond = suspend → change status of the phase diamond to be suspended
set-axis-value 2 t 1000 2800 → Designate y axis as temperature with its range from 1000 to 2800 K.
list-equilibrium → list the values of equilibrium calculations
TERMINAL → Show the result on the computer terminal
VWCs
compute-equilibrium → Compute Gibbs free energy minimization for the given condition
add
set-condition X(c) = none, acr(c) = 1 → Set the constraint of the relative activity with a defined reference state being equal to 1.
compute-equilibrium → Compute Gibbs free energy minimization satisfying the constraint of the set-condition
add

The region where diamond and gas are stable in Fig. 2.7 is the two phase region, where the lever rule can be applied to determine the equilibrium amount of diamond precipitation. When the gas mixture of 1 % CH_4–99 % H_2 is used, the atomic fraction of 1 % CH_4–99 % H_2 is 5×10^{-3} as shown by the vertical dashed line in Fig. 2.7. If the minimum carbon solubility is 10^{-4} at ~1350 K, the mole fraction of diamond can be determined by the lever rule as

$$\text{Mole fraction of diamond} = \frac{5 \times 10^{-3} - 10^{-4}}{1 - 10^{-4}} = 0.0049 \qquad (2.62)$$

Then, the mole fraction of gas is 1 − 0.0049 = 0.9951. Since the mole fraction of diamond is evaluated at the minimum solubility of carbon at ~1350 K, this mole fraction is maximum and would decrease with decreasing temperature.

This decrease in the amount of precipitation with decreasing temperature is a unique feature of the retrograde solubility. This fact has a very special meaning in the deposition of diamond and silicon by HFCVD, where the temperature decreases monotonously from the hot filament to the substrate. In such cases, if the gas phase nucleation takes place, the equilibrium amount of gas phase nuclei would decrease toward the substrate temperature, considering the temperature gradient. This means that in the presence of gas phase nuclei, the driving force is for etching at the substrate temperature.

If any film grows on the substrate in this situation, it can be inferred that the film grow by the gas phase nuclei because the driving force is for etching. In other words, the gas phase nuclei deposit on the substrate as a film and simultaneously the deposited film would undergo etching atomically. Simultaneous deposition and etching are often macroscopically observed in the CVD process (Badzian et al. 1988; Salvadori et al. 1992; Kumomi and Yonehara 1990). This phenomenon provides the indirect evidence that the film should grow 100 % by gas phase nuclei because atomic contribution is negative by etching. This aspect will be explained in more detail in Chaps. 6 and 7.

If the gas phase nucleation takes place, which turns out to be very general in the thin film process by CVD or even by some PVD, the information about the equilibrium amount of precipitation is important because this amount is related with the deposition rate. Figure 2.8 shows the temperature dependence of the equilibrium amounts of diamond and graphite that can be precipitated from the gas phase when the gas mixture of 1 % CH_4–99 % H_2 is used at the reactor pressure of 2700 Pa (Hwang and Yoon 1996). In the entire range of temperature, the equilibrium fraction of graphite is larger than that of diamond in agreement with the fact that graphite is more stable than diamond.

Figure 2.8 provides an important fact that the equilibrium amount of diamond to be precipitated in the gas phase decreases with decreasing temperature, which can explain the well-known but extremely puzzling experimental observation of simultaneous diamond deposition and graphite etching in the diamond CVD process.

The reactor pressure is an important process parameter. What would be the effect of pressure on the C–H phase diagram? As mentioned earlier, the line of $\alpha = 1$

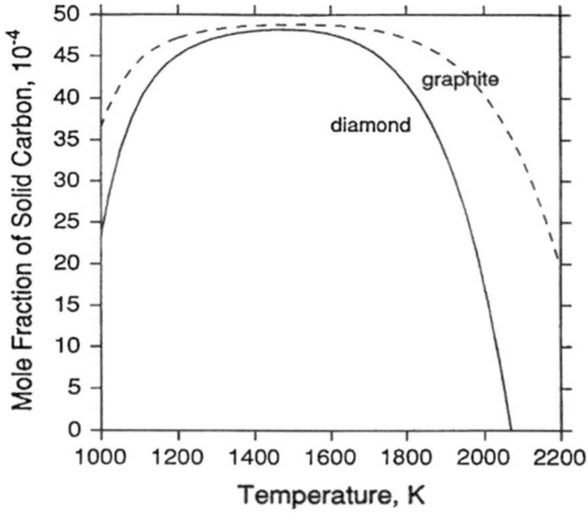

Fig. 2.8 The temperature dependence of the equilibrium mole fraction of diamond and graphite when the gas mixture of 1 % CH_4–99 % H_2 is used at 2700 Pa. The *dashed* and *solid lines* are for graphite and diamond, respectively (Hwang and Yoon 1996)

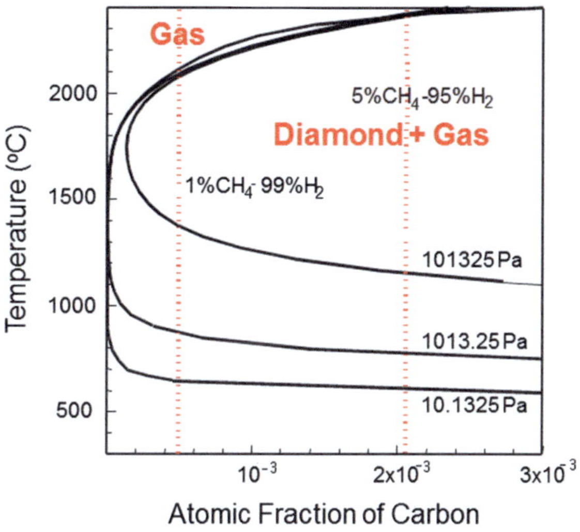

Fig. 2.9 Metastable CVD phase diagram of the C–H system: in the *left* of the curve, only a gas phase is stable and in the *right* of the curve, two phases of gas + diamond are stable (Gueroudji and Hwang 2000)

corresponds to the phase diagram, which represents the thermodynamic limit of deposition. In the C–H system using a gas mixture of CH_4 and H_2, it is known that there is a lower limit of the substrate temperature that diamond can be deposited and this temperature increases with increasing reactor pressure. The effect of the pressure on the line of $\alpha = 1$ or the metastable diamond CVD phase diagram is shown in Fig. 2.9, which divides the region into the gas phase and the gas + diamond phase (Gueroudji and Hwang 2000). The curves drawn in Fig. 2.9 correspond to the temperature dependence of the solubility limit of carbon separating a diamond deposition domain from its etching or non-deposition domain at different values of the total pressure in the system.

At 101,325 Pa for a gas mixture of 1 % CH_4–99 % H_2 indicated by the left vertical dashed line in Fig. 2.9, the lower limit of temperature for diamond deposition is ~1400 K. It should be noted that the substrate temperature cannot be increased too much because diamond starts to transform to graphite. This temperature is ~1500 K, which would be the practical upper limit of the deposition temperature. With increasing the methane concentration to 5 % CH_4–95 % H_2 indicated by the right vertical dashed line in Fig. 2.9, the deposition domain can be extended and the lower limit is ~1200 K. In experimental reality, however, increasing the methane concentration tends to degrade the quality of diamond or to produce the non-diamond phase. Normally, the methane concentration in typical conditions is less than 3 % in the C–H system. Figure 2.9 shows that the lower limit of the substrate temperature for diamond deposition can go down to ~900 K at 1013.25 Pa and ~600 K at 10.1325 Pa for the gas mixture of 1 % CH_4–99 % H_2. However, the reactor pressure cannot be reduced much in the HFCVD process because the evaporation rate of tungsten hot filaments increases, which will degrade the diamond film quality.

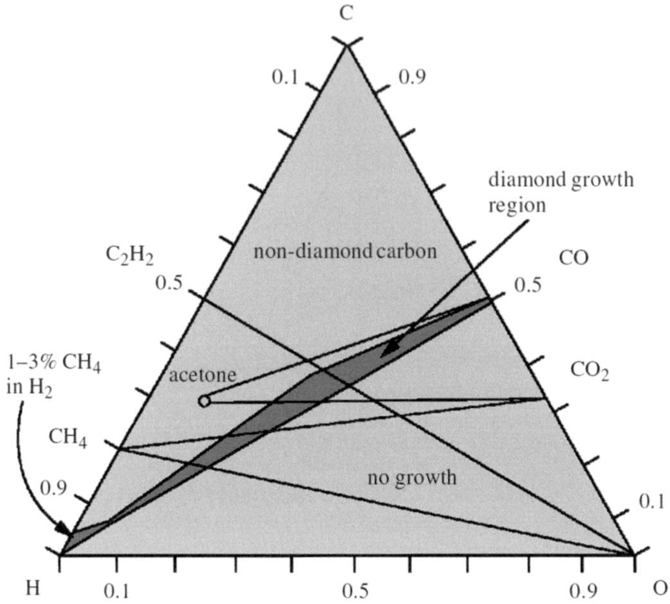

Fig. 2.10 A simplified form of the Bachmann C–H–O diagram for diamond deposition. Below the H–CO tie line, no growth was achieved. Above the H–CO tie line, non-diamond carbon is generally deposited except in a narrow window close to the tie line, which produces polycrystalline diamond (May 2000)

The gas mixture of CH_4 and H_2 is just one example to deposit diamond. It turned out that various sources can be used to deposit diamond. For examples, not only pure methanol and pure acetone, but various gas mixtures containing carbon, hydrogen, and oxygen such as CH_4–H_2–O_2, CH_4–H_2–H_2O, CH_4–CO–H_2, CH_4–CO_2–H_2, CO–H_2, C_2H_2–O_2, C_2H_4–O_2, C_2H_6–O_2, C_2H_2–CO_2, C_2H_2–O_2–H_2, C_2H_2–O_2–H_2, CH_3COCH_3–H_2, CH_3COCH_3–O_2, CH_4–CH_3COCH_3–O_2, CO–CO_2–H_2, and CO–H_2–O_2. Besides, in addition to HFCVD, various CVD processes such as microwave plasma, DC plasma jet, and flame could be used. By examining vast compiled reports on diamond CVD, Bachmann et al. (1991) found out that the diamond deposition is highly correlated with some composition range in the C–H–O system regardless of the sources of reactants nor the method of processes. Diamond deposition was possible only in the narrow composition range centred at the ratio of C/O = 1 with varying hydrogen in the C–H–O system. In the carbon-rich side of the line of C/O = 1, non-diamond is generally deposited and in the oxygen-rich side, there was no growth. This aspect is shown in Fig. 2.10.

There must be some reason behind this high correlation of the diamond deposition behaviour with the specific composition range. Before going further, it is worth thinking why this is so.

(Q) What do you think is the reason for this deposition behaviour?

Among the possibility we can think of, the thermodynamics should come first. In other words, this deposition behaviour might be related with the thermodynamic driving force for diamond deposition. If the thermodynamic scheme for the supersaturation ratio for diamond deposition given in (2.59) is applied, the lines of iso-supersaturation ratio can be evaluated at 1200 K in the C–H–O system as shown in Fig. 2.11.

The supersaturation ratio for diamond deposition is unity almost along the line of C/O = 1. In the carbon-rich side of this line, the supersaturation ratio increases steeply with increasing carbon. This tendency is strong especially when the hydrogen content is small. In the oxygen-rich side, the supersaturation ratio is less than unity and the driving force is for etching of diamond. If Fig. 2.11 is compared with Fig. 2.10, the diamond deposition behaviour is related with the supersaturation ratio. No growth in the oxygen-rich side of the line of C/O = 1 might be related with the supersaturation ratio being less than unity. In other words, this region is undersaturated and the driving force is for etching of diamond. And non-diamond deposition is related with the high supersaturation ratio. It may be asked why non-diamond deposits under the condition of the high supersaturation ratio. This problem is related with the diamond deposition mechanism, which will be explained in Chap. 6.

The thermodynamic analysis made so far shows that the diamond deposition is thermodynamically sound in that diamond deposition occurs under the thermodynamic condition where the driving force is for deposition of diamond. The deposition of less stable diamond instead of stable graphite is not a problem of thermodynamics but a problem of kinetics, which will be explained in Chap. 5.

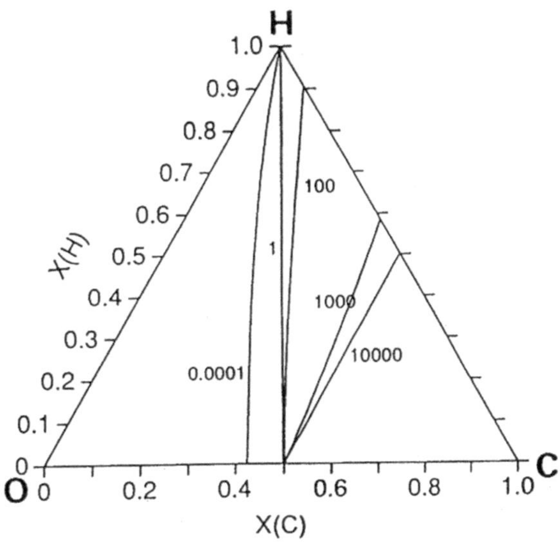

Fig. 2.11 The ternary composition diagram of the C–H–O system at 1200 K under 2700 Pa showing five different *lines* of the iso-supersaturation ratio for diamond deposition (Hwang 1994)

2.13 Kinetics

The important thermodynamic functions such as E (internal energy), H (enthalpy), G (Gibbs free energy), and S (entropy) are state functions, which does not depend on the path. The information about two out of the three state variables P, T and V is enough to determine these state functions. For example, if P and T are known for states A and B, the Gibbs free energy, G, for each state can be determined. Then, we can determine the direction of irreversibility or whether A $\rightarrow$ B or B $\rightarrow$ A. If $G_A > G_B$, the state A irreversibly changes to the state B and vice versa. This is illustrated in Fig. 2.12.

Although thermodynamics can powerfully predict the direction of irreversibility of A $\rightarrow$ B from the knowledge of $G_A > G_B$, it does not say anything about how long it will take for state A to change to state B. Time is not a parameter in thermodynamics. How long the process will take is not a problem of thermodynamics but a problem of kinetics.

(Q) If thermodynamics allows the irreversibility of A $\rightarrow$ B as shown in Fig. 2.12a and the process is thermally activated, what would be the factors that determine the rate?

The important factor is the kinetic barrier for the process A $\rightarrow$ B. If the kinetic barrier is ΔG^*, the probability to overcome the kinetic barrier would be identical to the Boltzman probability for the thermally-activated process. Therefore, the probability would be proportional to $\exp(-\Delta G^*/kT)$.

There is a case where there exist multiple kinetic paths. For example, consider the situation where $G_A > G_B > G_C$. In this case, nature has two choices, either A $\rightarrow$ B or A $\rightarrow$ C.

(Q) In this case, which path will nature choose, A $\rightarrow$ B or A $\rightarrow$ C?

Since $G_A > G_B > G_C$, the magnitude of $\Delta G_{A \rightarrow C}$ is larger than that of $\Delta G_{A \rightarrow B}$. Does this mean that nature chooses the path A $\rightarrow$ C instead of the path A $\rightarrow$ B? Thermodynamics does not say anything about this possibility. Larger driving force does not guarantee faster kinetics. The determining factor is the kinetic barrier. If the kinetic barrier of the A $\rightarrow$ B path is smaller than that of the A $\rightarrow$ C path, nature will choose the path of A $\rightarrow$ B. In this sense, it can be said that nature

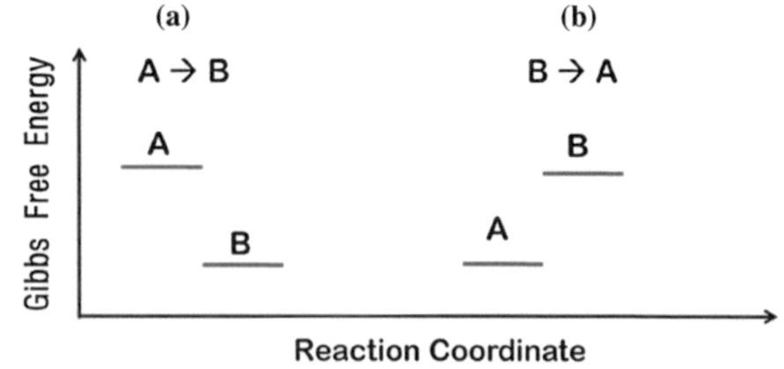

Fig. 2.12 The direction of irreversibility of **a** A $\rightarrow$ B or **b** B $\rightarrow$ A is determined by the relative magnitude of Gibbs free energy at constant temperature and pressure

proceeds in the most probable way. This kinetic law can be compared with the second law of thermodynamics, which says that nature does not proceed in a less probable way.

The kinetic statement that nature proceeds in the most probable way exactly describes what nature does whereas the thermodynamic statement that nature does not proceed in a less probable way rather vaguely describes what nature does. For the nature's way of doing, the kinetic statement provides a necessary and sufficient condition but the thermodynamic statement provides a necessary condition. Considering this, the kinetic statement sounds much more attractive and powerful than the thermodynamic one.

(Q) Then, why do we bother to learn thermodynamics? What is the advantage of thermodynamics over kinetics?

The advantage of thermodynamics lies in the fact that important thermodynamic functions are a state function, which makes thermodynamics very simple, powerful and generally formulated. This is why we bother to learn thermodynamics, which is worth our laborious efforts. This is highly in contrast with kinetics which depends on the path. This means that there is no general way to formulate the kinetics. This is why the general formulation for kinetics is not available and each kinetics should be approached case by case.

References

Bachmann PK, Leers D, Lydtin H (1991) Towards a general concept of diamond chemical vapour deposition. Diam Relat Mater 1(1):1–12. doi:10.1016/0925-9635(91)90005-U

Badzian AR, Badzian T, Roy R, Messier R, Spear K (1988) Crystallization of diamond crystals and films by microwave assisted CVD (Part II). Mater Res Bull 23(4):531–548

Gaskell DR (1995) Introduction to the thermodynamics of materials. Taylor & Francis, Washington, DC

Gueroudji L, Hwang NM (2000) Thermodynamic limits for the substrate temperature in the CVD diamond process. Diam Relat Mater 9(2):205–211. doi:http://dx.doi.org/10.1016/S0925-9635(00)00232-6

Hillert M, Aaronson H (1975) Lectures on the theory of phase transformations. Amer Inst Mining, Metallurgical and Petroleum Engrs, New York, p 1–44

Hwang NM (1994) Thermodynamic analysis of the chemical vapor deposition of diamond in the C-H, C-O and C-H-O systems. J Cryst Growth 135(1):165–171. doi:http://dx.doi.org/10.1016/0022-0248(94)90738-2

Hwang N, Yoon D (1994a) Driving force for deposition in the chemical vapour deposition process. J Mater Sci Lett 13(19):1437–1439

Hwang NM, Yoon DY (1994b) Thermodynamic approach to the chemical vapor deposition process. J Cryst Growth 143(1–2):103–sss109. doi:10.1016/0022-0248(94)90372-7

Hwang NM, Yoon DY (1996) Thermodynamic approach to the paradox of diamond formation with simultaneous graphite etching in the low pressure synthesis of diamond. J Cryst Growth 160(1–2):98–103. doi:http://dx.doi.org/10.1016/0022-0248(95)00549-8

Hwang NM, Hahn JH, Bahng GW (1994) Thermodynamic approach to the C–H–O deposition diagram in the diamond chemical vapour deposition process. Diam Relat Mater 3(1–2):163–167. doi:http://dx.doi.org/10.1016/0925-9635(94)90051-5

Hwang NM, Hahn JH, Yoon DY (1996) Chemical potential of carbon in the low pressure synthesis of diamond. J Cryst Growth 160(1–2):87–97. doi:http://dx.doi.org/10.1016/0022-0248(95)00548-X

Kumomi H, Yonehara T (1990) Coarsening phenomenon of Si clusters. In: MRS Proceedings, Cambridge Univ Press, p 83

May PW (2000) Diamond thin films: a 21st-century material. Phil Trans R Soc A 358(1766):473–495

Salvadori M, Brewer M, Ager J, Krishnan K, Brown I (1992) The effect of a graphite holder on diamond synthesis by microwave plasma chemical vapor deposition. J Electrochem Soc 139 (2):558–560. doi:10.1149/1.2069256

Sommer M, Mui K, Smith FW (1989) Thermodynamic analysis of the chemical vapor deposition of diamond films. Solid State Commun 69(7):775–778. doi:http://dx.doi.org/10.1016/0038-1098(89)90829-6

Sundman B, Jansson B, Andersson J-O (1985) The thermo-calc databank system. Calphad 9(2):153–190. doi:http://dx.doi.org/10.1016/0364-5916(85)90021-5

Yoon DY, Hwang NM (1995) Thin film fabrications: thermodynamics of chemical vapor deposition and its application to diamond deposition. Adv Mater Proc 1:905–911

Chapter 3
Nucleation

Nucleation is an irreversible process, where a new phase is initiated. It is an incremental clustering process; starting from a monomer, the cluster becomes a dimer, a trimer and so on. In order for a new phase to nucleate, supersaturation (supercooling or superheating) is necessary. The thermodynamic driving force does not guarantee triggering the nucleation. For nucleation to take place, the kinetic barrier of nucleation should be overcome. If the kinetic barrier cannot be overcome, the unstable or metastable state can be maintained without limitation. The good example would be the diamond in an engagement ring, which is unstable with respect to graphite at ambient pressure and temperature but does not transform to graphite.

3.1 Classical Nucleation Theory

(Q) What would be the origin of the kinetic barrier of nucleation?

The origin of the nucleation barrier is the interface, which is accompanied by the newly-formed nucleus. In this sense, nucleation is a result of the competition between the interface energy to resist the creation of new interfaces and a driving force for the formation of a more stable phase. In the solid-state transformation, the additional barrier is the strain energy. When the nucleation occurs on any defect such as the interface, the total interface energy accompanied by nucleation can be diminished. This results in decreasing the nucleation barrier and called the heterogeneous nucleation. In an extreme case, the nucleation barrier disappears: when a solid is superheated, usually there is no barrier for melting because the sum of the solid-liquid and the liquid-vapor interface energies is smaller than the solid-vapor interface energy.

Since nucleation is a thermally-activated process, its rate would be proportional to the Boltzman factor and can be expressed as the form,

© Springer Science+Business Media Dordrecht 2016

N.M. Hwang, *Non-Classical Crystallization of Thin Films and Nanostructures in CVD and PVD Processes*, Springer Series in Surface Sciences 60,
DOI 10.1007/978-94-017-7616-5_3

$$R = A\ \exp(-\Delta G^*/kT) \tag{3.1}$$

where A is the pre-exponential factor and ΔG^* is the kinetic barrier of nucleation. The expressions for A and ΔG^* can be derived with a framework of the classical nucleation theory, where the free energy of the formation of small clusters is determined by using macroscopic quantities such as the interface energy.

The free energy of the system consists of two contributions. One is from the driving force or the volume free energy change (Δf) for the phase transition and the other is from the interface. The free energy of formation of the spherical nucleus with the radius r is expressed as

$$\Delta G = \frac{4\pi r^3}{3}\Delta f + 4\pi r^2 \sigma \tag{3.2}$$

where σ is the interface energy. If ΔG is plotted against r, it looks like Fig. 3.1.

The radius at which the barrier of the nucleation curve is maximum is called the critical radius (r*). r* can be determined by differentiating (3.2) with respect to r and setting the derivatives equal to zero.

$$r^* = -\frac{2\sigma}{\Delta f} \tag{3.3}$$

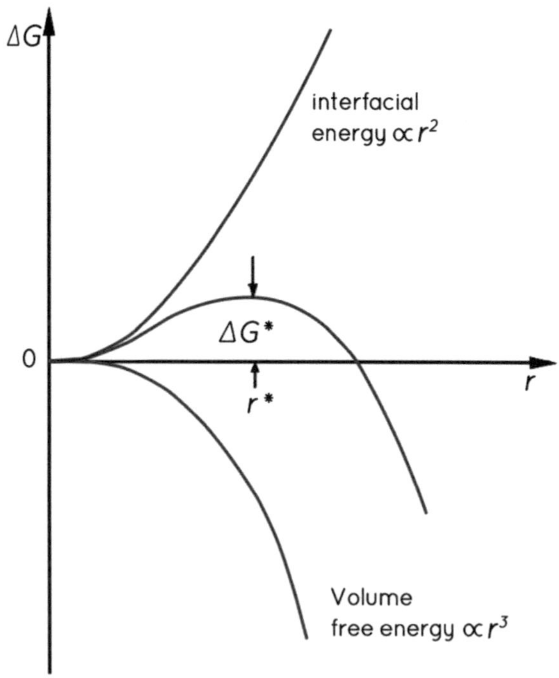

Fig. 3.1 The nucleation curve which consists of contributions from the volume free energy and the interface energy (Porter et al. 2009)

The nucleation barrier can be determined by putting (3.3) into (3.2) as

$$\Delta G^* = \frac{16\pi\sigma^3}{3(\Delta f)^2} \tag{3.4}$$

(Q) Why is the critical nucleus determined under the condition of ΔG being maximum instead of ΔG = 0?

The nucleus is a thermodynamically open system because a material is transferred between the system and the surrounding. The criterion for the irreversible material transfer is the chemical potential difference, which is determined by differentiating (3.2) with respect to r. The derivative, which is the chemical potential difference $\Delta\mu$, is positive for $r < r^*$, zero for $r = r^*$, and negative for $r > r^*$. Therefore, the cluster of $r < r^*$, which is called an embryo, is unstable and shrinks. However, the cluster of $r > r^*$, which is called a nucleus, is stable and grows. The cluster of $r = r^*$ is called a critical nucleus. In order for the cluster to grow, its size should be larger than the critical nucleus or the nucleation barrier should be overcome.

Equation (3.4) tells that the nucleation barrier increases with increasing σ and decreasing Δf. The probability of nucleation decreases as the nucleation barrier increases. Great efforts have been made to derive a quantitative relationship between the nucleation rate and the nucleation barrier.

3.2 Rate of Nucleation

Consider the following consecutive reactions, where i-mers grow by capturing monomers.

$$A_1 \rightarrow A_2 \rightarrow A_3 \rightarrow \ldots \rightarrow A_{m-1} \rightarrow A_m \tag{3.5}$$

The equilibrium concentrations of i-mers can be determined as follows.

$$\begin{aligned}
n_2 &= n_1 \exp\left(-\frac{\Delta G^{1\rightarrow 2}}{kT}\right) \\
n_3 &= n_2 \exp\left(-\frac{\Delta G^{2\rightarrow 3}}{kT}\right) \\
n_4 &= n_3 \exp\left(-\frac{\Delta G^{3\rightarrow 4}}{kT}\right) \\
&\vdots \\
n_m &= n_{m-1} \exp\left(-\frac{\Delta G^{m-1\rightarrow m}}{kT}\right) \\
n_m &= n_1 \exp\left(-\frac{\Delta G^{1\rightarrow 2} + \Delta G^{2\rightarrow 3} + \cdots + \Delta G^{m-1\rightarrow m}}{kT}\right)
\end{aligned} \tag{3.6}$$

where n_m is the concentration of m-mers per unit volume. If these equations are multiplied all together, terms on the left side and pre-exponential terms on the right side cancel out, producing the concentration of m-mers as

$$n_m = n_1 \exp\left(-\frac{\Delta G^{1\rightarrow m}}{kT}\right) \tag{3.7}$$

Therefore, the number concentration of critical nuclei N* can be expressed as

$$N^* = n_1 \exp\left(-\frac{\Delta G^*}{kT}\right) \tag{3.8}$$

Since the addition of one more monomer to each of these clusters will convert them into stable nuclei, the nucleation rate, defined as number of nucleus formed per unit volume in unit time, is given by the products of N*, the critical surface area $4\pi(r^*)^2$ and the flux of monomer impingement (monomers/cm^2 s).

There are many versions for the expression of the nucleation rate. According to Becker and Döring (1935), the nucleation rate for the solid from the vapour is expressed as

$$I = \omega A^* \left(\frac{\Delta G^*}{kT}\right) \exp\left(-\frac{\Delta G^*}{kT}\right) \tag{3.9}$$

where I is the number of stable nuclei created per second, ω is the probability per unit time per unit area of capturing a vapour atom, A* is the area of the critical embryo. A more advanced treatment by Mutaftschiev (1993, Chap. 4) produces

$$I = q_{rep} \frac{\omega A^*}{i^*} \left(\frac{\Delta G^*}{3\pi kT}\right)^{1/2} \exp\left(-\frac{\Delta G^*}{kT}\right) \tag{3.10}$$

where q_{rep} is the replacement factor and i* is the number of atoms in the critical nucleus. Although there are many versions of the nucleation rate, the common feature is that the exponential term is the same and only the pre-exponential form varies. The pre-exponential factor in (3.10) is typically of the order of 10^{25} for a gas at atmospheric pressure (Mutaftschiev 1993, Chap. 4).

The important aspect of nucleation is that the nucleation rate is relatively insensitive to the change of the pre-exponential factor but depends exponentially on the nucleation barrier and the rate varies sensitively to its change.

The volume free energy change Δf in (3.4) can be expressed in terms of the supersaturation ratio P/P_{eq} as

$$\Delta f = -\frac{kT}{\Omega} \ln\left(\frac{P}{P_{eq}}\right) \tag{3.11}$$

where Ω is the atomic volume.

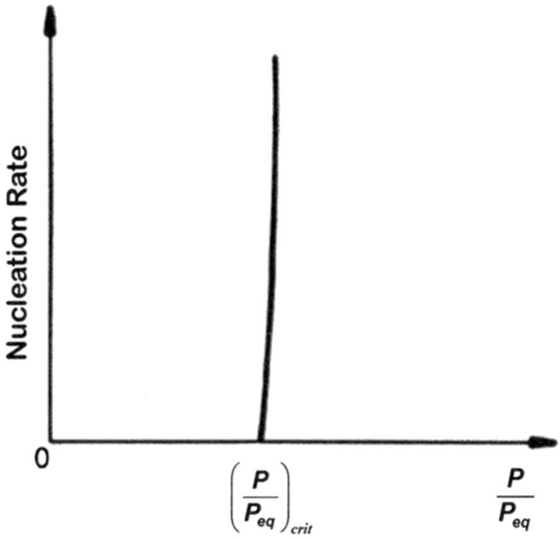

Fig. 3.2 Dependence of the nucleation rate on the supersaturation ratio, showing the step function like behaviour (Hirth and Pound 1963)

Considering (3.4) and (3.11), the nucleation rate also varies sensitively with the supersaturation ratio. As shown in Fig. 3.2, below a critical value of supersaturation raio, the nucleation rate is practically zero and above it, the rate increases abruptly to very high values. The nucleation rate varies with the supersaturation ratio like a step function.

This value of the supersaturation ratio represents the onset of nucleation and has a special meaning. Normally, the onset of nucleation is represented by the nucleation rate of one per second per cubic centimetre, which corresponds to 10^6 per second per cubic meter. Practically, the critical supersaturation ratio is defined as the value to produce this onset of nucleation. The nucleation barrier corresponding to this critical supersaturation ratio is called the critical nucleation barrier.

In the solidification, the critical supersaturation ratio should be changed to critical supercooling. If the heterogeneous nucleation can be suppressed, the critical supercooling can be determined experimentally by detecting the onset of nucleation. Then, the critical nucleation barrier can be determined, which again determines the interface energy according to (3.4). By decreasing the droplet size of metallic melt, Turnbull (1952) could inhibit the heterogeneous nucleation and determine the critical supercooling for the onset of homogeneous nucleation of solid from the supercooled liquid, thereby determining the solid-liquid interface energy of metals as shown in Table 3.1.

Table 3.1 Experimentally determined solid/liquid interfacial free energies (Values selected from Turnbull 1952) (Porter et al. 2009)

Material	T_m/K	$\gamma_{SL}/\mathrm{mJ\,m^{-2}}$
Sn	505.7	54.5
Pb	600.7	33.3
Al	931.7	93
Ag	1233.7	126
Au	1336	132
Cu	1356	177
Mn	1493	206
Ni	1725	255
Co	1763	234
Fe	1803	204
Pd	1828	209
Pt	2043	240

3.3 Heterogeneous Nucleation

Since the nature tends to choose the path of the lowest kinetic barrier, the homogeneous nucleation, which has a high kinetic barrier, is rather exceptional and seldom occurs in reality. Instead, the heterogeneous nucleation, which has a much lower kinetic barrier, normally occurs.

When a spherical nucleus is formed on the substrate with a contact angle θ as shown in Fig. 3.3, the critical nucleus would be the same as that of homogeneous nucleation (3.3) but the nucleation barrier decreases by the shape factor $S(\theta) = (2 - 3\cos\theta + \cos^3\theta)/4$.

Therefore,

$$\Delta G^*_{het} = S(\theta)\Delta G^*_{\mathrm{hom}} \tag{3.12}$$

This aspect is illustrated in Fig. 3.4. To get a rough idea how much the nucleation barrier is reduced by heterogeneous nucleation, let's substitute some numerical values for θ. For θ = 10, 30, and 90, S(θ) becomes respectively $\sim 10^{-4}$, ~0.02, and 0.5. The nucleation barrier is reduced remarkably by heterogeneous nucleation.

Also in heterogeneous nucleation, the nucleation rate varies like a step function with the supersaturation ratio even though the critical supersaturation ratio is much smaller than that of the homogeneous nucleation. This aspect is shown in Fig. 3.5. If the supersaturation ratio is less than the critical value for the onset of nucleation, the nucleation rate is practically zero.

If the scratch is made on the substrate, the nucleation can take place at the groove. If the groove angle and the contact angle are known, the shape factor can be determined. Sometimes, it is observed that nucleation takes place on the edge or the top of the pyramidal shape. In this case also, the shape factor can be determined.

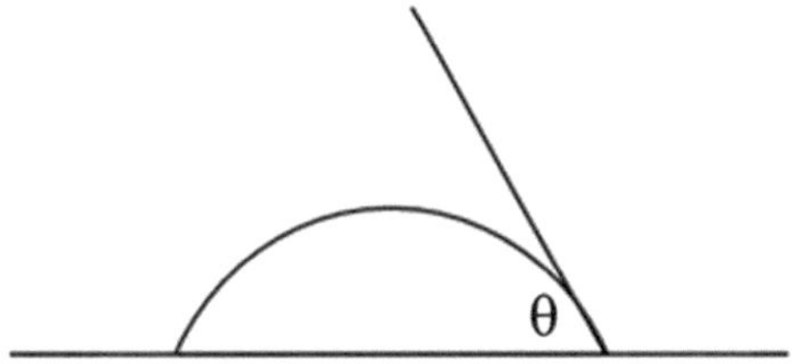

Fig. 3.3 Spherical nucleus on the substrate with a contact angle θ

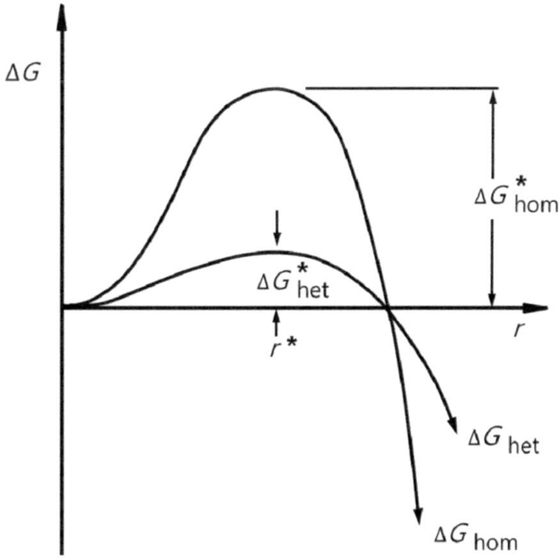

Fig. 3.4 The free energy of solid clusters for homogeneous and heterogeneous nucleation. Note r* is the same for both cases (Porter et al. 2009)

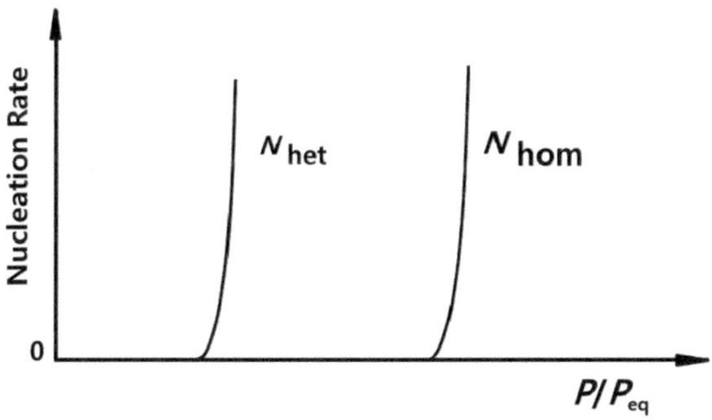

Fig. 3.5 Dependence of the nucleation rate on the supersaturation ratio for homogeneous and heterogeneous nucleation (Porter et al. 2009)

(Q) What would be the shape factor $S(\theta)$ of heterogeneous nucleation when the nucleus is on the substrate in Fig. 3.6? The contact angle is 90° in all cases and the angles of the substrates are (a) 270°, (b) 180° and (c) 90°.

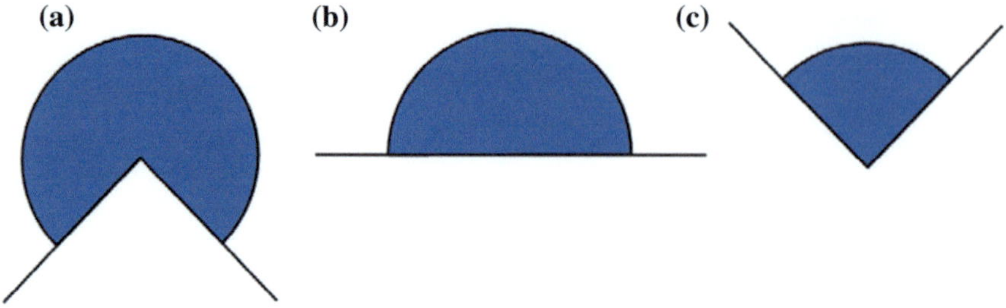

Fig. 3.6 Heterogeneous nucleation on the substrates of three different geometries (Hwang et al. 1996)

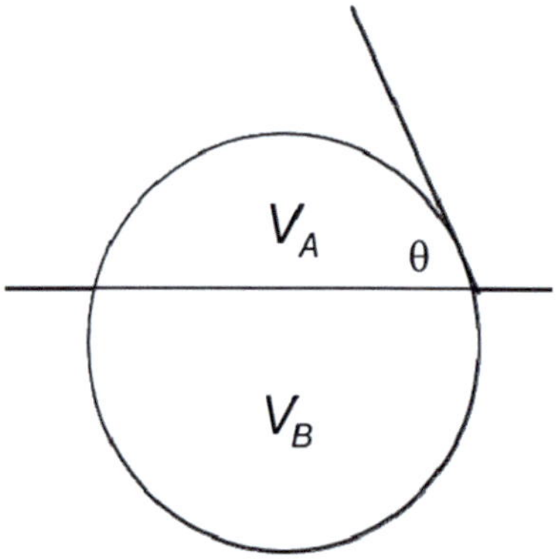

Fig. 3.7 The heterogeneous nucleus on the substrate has a volume V_A and the homogenous nucleus of the same radius has a volume $V_A + V_B$

The shape factor S(θ) is given by the ratio of the volume of the heterogeneous nucleus to that of the homogeneous nucleus of the same radius. In other words, the following relation holds.

$$\frac{\Delta G^*_{het}}{\Delta G^*_{\text{hom}}} = \frac{V^*_{het}}{V^*_{\text{hom}}} = S(\theta) \tag{3.13}$$

If the heterogeneous nucleus on the substrate with a contact angle θ has a volume V_A as shown in Fig. 3.7, the homogeneous nucleus with the same radius has a volume $V_A + V_B$. Then $S(\theta) = V_A/(V_A + V_B)$. The relation in (3.13) generally holds and can be used to determine the shape factor of the heterogeneous nucleation on the surface of more complex geometries. This relation can also be used to determine the shape factor of the heterogeneous nucleation for the non-spherical equilibrium shape.

The ratios of the volume of the heterogeneous nucleus to that of homogeneous nucleus of the same radius are 3/4, 1/2 and 1/4 respectively for Fig. 3.6a–c, whose S(θ) is respectively 0.75, 0.5 and 0.25. Therefore, the nucleation barrier decreases in the order of edge, flat and groove.

The effect of the surface geometry on the nucleation rate is worth somewhat detailed quantitative analysis because it is related with some puzzling nucleation behaviour in the diamond CVD, which will be described shortly.

There are many suggested equations of the nucleation rate including (3.9) and (3.10) explained above. However, they commonly have the same exponential dependence and differ mainly in the pre-exponential term. If the relative nucleation rate is evaluated, the pre-exponential terms cancel out, leaving only the ratio of the exponential terms.

$$\frac{I_{\text{hetero}}}{I_{\text{homo}}} = \exp\left\{-\frac{1}{kT}\left(\Delta G^*_{\text{hetero}} - \Delta G^*_{\text{homo}}\right)\right\} \tag{3.14}$$

The relative nucleation rate given by (3.14) would be much more reliable than the absolute nucleation rate because it does not have the uncertain pre-exponential term. Let's assume the following values: $\alpha = 10$, $T = 1200$ K, $v = 3.41 \times 10^{-6}$ M^3/mole, $\sigma = 3.7$ J/M^2, where α, T, v and σ are respectively the supersaturation ratio, substrate temperature, molar volume and surface energy. These numerical values are chosen to examine the nucleation behavior in the diamond CVD process. It is assumed that the nucleus has a spherical shape. Considering (3.4), the nucleation barrier for homogeneous nucleation is expressed as

$$\Delta G^*_{\text{homo}} = \frac{16\pi v^2 \gamma^3}{3(RT \ln \alpha)^2} \tag{3.15}$$

Putting the numerical values into (3.15) gives

$$\Delta G^*_{\text{homo}} = \frac{2.96 \times 10^{-8}}{1.58 \times 10^9} = 1.87 \times 10^{-17}\,\text{J} \tag{3.16}$$

The relative nucleation rates for the three geometries in Fig. 3.6 are respectively given as

$$\frac{I_{\text{hetero}}}{I_{\text{homo}}} = \exp\left\{\frac{\Delta G^*_{\text{homo}}}{4kT}\right\}, \quad \exp\left\{\frac{\Delta G^*_{\text{homo}}}{2kT}\right\} \quad \textit{and} \quad \textit{exp}\left\{\frac{3\Delta G^*_{\text{homo}}}{4kT}\right\} \tag{3.17}$$

Putting the value in (3.16) into (3.17) gives 10^{123}, 10^{245} and 10^{368}, respectively. This analysis reveals the extremely strong tendency of heterogeneous nucleation. If these three relative nucleation rates are normalized by 10^{123}, they become respectively 1, 10^{123} and 10^{245}. This indicates that the heterogeneous nucleation rate strongly depends on the surface geometry of the substrate.

Since the nucleation barrier decreases with increasing supersaturation ratio, it is expected that the relative nucleation rate would decrease. The thermodynamic analysis on the CVD diamond process in the previous chapter indicates that the

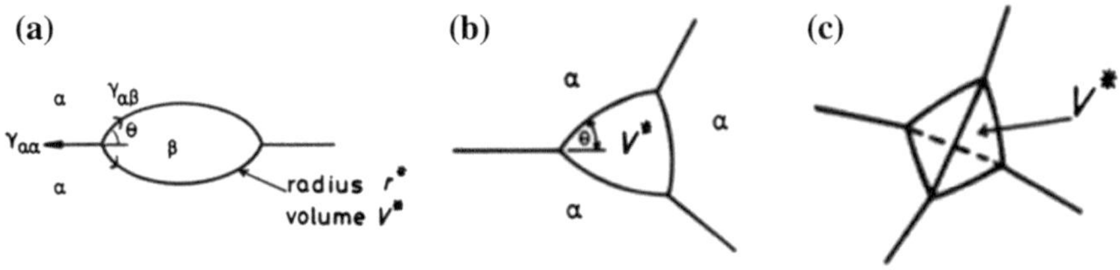

Fig. 3.8 Shapes of critical nuclei on the grain boundary, grain edge and grain corner (Porter et al. 2009)

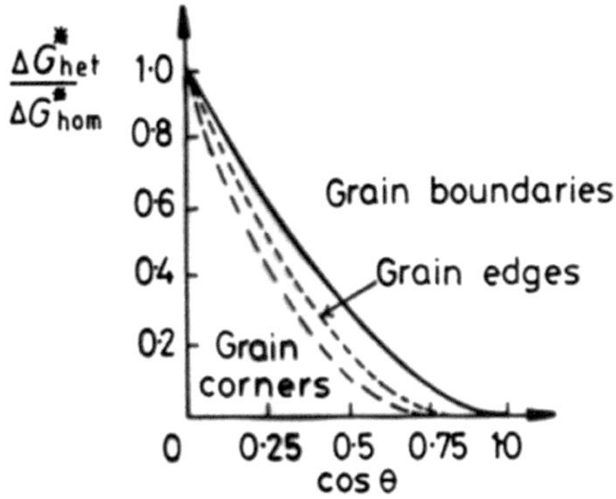

Fig. 3.9 The effect of θ on the nucleation barrier on the grain boundary, edge and corner relative to homogeneous nucleation (Cahn 1956; Porter and Easterling 1981)

supersaturation ratio cannot be higher than 100 under typical processing conditions. If $\alpha = 100$ is used with other parameters being the same as above, the relative nucleation rates become 10^{31}, 10^{62} and 10^{93}. If normalized by 10^{31}, they become 1, 10^{31} and 10^{62}. Still, the values are extremely large.

According to this analysis, if there is a crevice on the wall of the mould, where liquid metal is poured for solidification, the crevice would provide a strong heterogeneous site with a very low nucleation barrier. Nucleation would start from such a crevice, which is known as the crevice nucleation.

In a polycrystalline structure, grain boundaries, grain edges and grain corners are respectively 2-dimensional, 1-dimensional and 0-dimensional defects, which provide sites for heterogeneous nucleation. The nucleus shape on them is shown in Fig. 3.8. The nucleation barrier on them can be evaluated using the scheme of (3.13).

S(θ), which varies with contact angles, are compared among grain boundaries, edges and corners in Fig. 3.9.

(Q) Consider a cone-shaped surface with the apex angle of 60° as shown in Fig. 3.10. If the liquid nucleus is formed at this site, what would be the contact

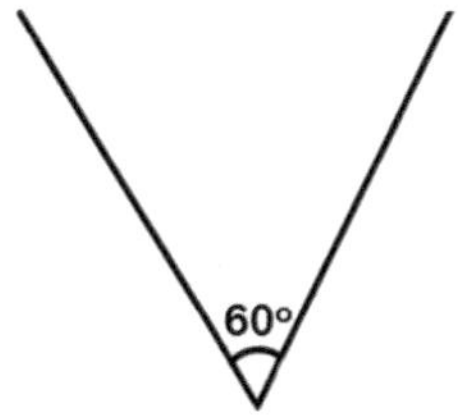

Fig. 3.10 Cone-shaped surface with the apex angle of 60°

angle to make the radius of the liquid nucleus infinite? If the liquid nucleus has the contact angle 30° with the surface, what would be the shape of the curvature? What does the shape of the curvature mean?

The liquid nucleus would have the uniform curvature. Since the sum of the triangle interior angles is 180°, the nucleus would have an infinite radius if the contact angle is 60°. If the contact angle is less than 60°, the curvature will be inward or negative. Therefore, if the nucleus has a contact angle of 30°, the curvature looks like Fig. 3.11. The negative curvature means that the interface energy does not act as a nucleation barrier but rather help the nucleation. This means that the nucleus would be stable even in the absence of supersaturation.

The contact angle represents the wettability of the nucleus on the substrate. The small and large contact angles represent respectively good and bad wetting. However, if the equilibrium shape is non-spherical like diamond, the contact angle cannot be defined. Therefore, the wettability cannot be defined by the contact angle.

(Q) Assume that the equilibrium shape of the nucleus is a cube. If nucleation takes place on the substrate, how can we determine the wettability? What would be the shape of the nucleus for good wetting and bad wetting?

One way to define the wettability is the shape factor. The small shape factor represents good wetting while the large one represents bad wetting. Figure 3.12 shows the case of heterogeneous nucleation of the cube equilibrium shape. Figure 3.12a is the case where the interface energy between the nucleus and the substrate is low, resulting in good wetting and a small shape factor whereas Fig. 3.12b is the case where the interface energy is high, resulting in bad wetting and a large shape factor. The shape factor in both Fig. 3.12a, b is $V_A/(V_A + V_B)$.

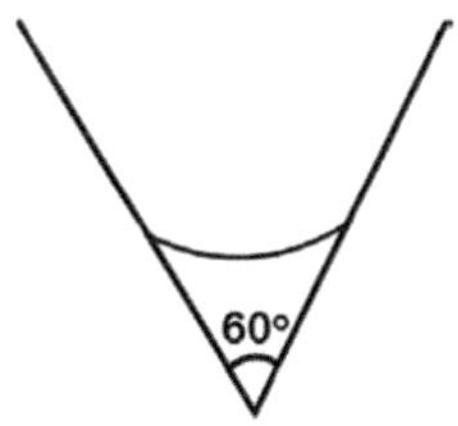

Fig. 3.11 The negative curvature of the nucleus with the contact angle of 30°

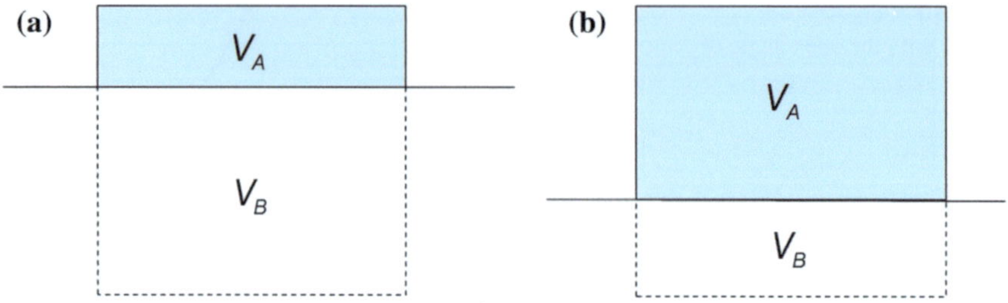

Fig. 3.12 Heterogeneous nucleation of the nucleus with a cube equilibrium shape with **a** good wetting and **b** bad wetting

3.4 Preferential Nucleation Along Scratches on the Substrate in the Diamond CVD Process

In the diamond CVD process, the pre-treatment of the substrate surface has a strong effect on the diamond initiation. Usually, the pre-treatment is done by diamond paste, which makes scratches along the substrate surface. Then, diamond nucleation occurs preferentially along these scratches on the substrate as shown in Fig. 3.13.

(Q) What do you think is the reason for the preferential diamond formation along the scratches?

You might think that the preferential diamond formation is attributed to the heterogeneous nucleation along the groove made by the scratches. However, this problem is more complicated.

The pre-treatment leaves behind diamond debris, which may act as diamond nuclei or enhance diamond formation. It is known that pre-existing diamond debris enhances diamond formation in the nearby area. Therefore, it is not clear whether preferential diamond formation along scratches is due to pre-existing diamond debris or the strong heterogeneous site of the grooves made by scratches. In order to rule out the possibility of leaving behind diamond debris, SiC paste instead of diamond paste can be used. It was observed that diamonds were preferentially formed along the scratches also when SiC paste was used.

Therefore, it appeared that scratches provide the geometrical effect to the diamond nucleation, which can be understood by the well-established theory of heterogeneous nucleation. However, there were arguments as to the location of diamond nucleation: whether diamond nucleation occurs on the edge or the valley of the scratches. Since the theory of heterogeneous nucleation supports the preferential nucleation on the valley, some scientists argued that the location should be on the valley. However, other scientists argued that the location should be on the edge according to their observation by SEM

In order to solve this dispute, Dennig et al. (1992), Dennig and Stevenson (1991) prepared the silicon substrate with a well-defined geometry by the anisotropic etching technique used in microelectronics. Contrary to the expectation from the

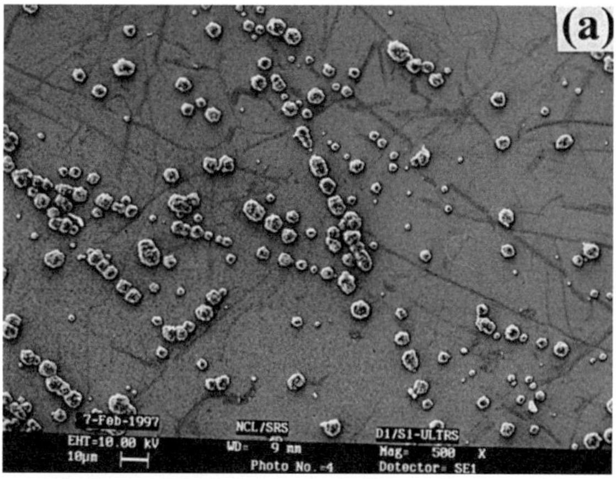

Fig. 3.13 Role of surface features in CVD diamond nucleation on surface-pretreated substrates (Karve et al. 1998)

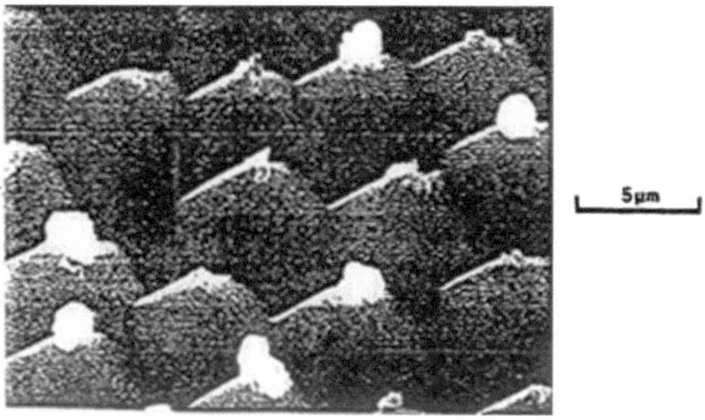

Fig. 3.14 SEM image showing the preferential diamond nucleation on the tips of pyramids etched into ⟨100⟩ high resistivity silicon, 45° tilt (Dennig and Stevenson 1991)

theory of heterogeneous nucleation, diamonds nucleated preferentially on the tips of the pyramids as shown in Fig. 3.14.

This somewhat puzzling nucleation behaviour was reproduced by Ramesham and Ellis (1992) and Givargizov et al. (1996). Therefore, it is evident that the effect of the geometry on nucleation experimentally observed in the diamond CVD process contradicts with the theory of heterogeneous nucleation. To my knowledge, no satisfactory explanation has been given yet based on the classical nucleation and growth. This contradiction is another example supporting that diamond crystals do not grow by classical crystallization but grow by non-classical crystallization.

Figure 3.14 shows clearly that the edge provides the preferential site for diamond formation. Then, a question arises as to how it is possible that the sharp edge should have an advantage in diamond formation than the valley. How can the edge have an advantage over the valley? One possibility is that the edge would be more advantageous than the valley in attracting the depositing source if the source is electrically charged because the sharp edge provides the higher field gradient than the valley. As we all know, the lightning rod has a sharp point to produce the high electric field gradient.

Before going into the next section, consider the following question.

(Q) If nucleation takes place in the gas phase, would it be homogeneous nucleation or not?

3.5 Ion-Induced Nucleation: Heterogeneous Nucleation in the Gas Phase

It might be thought that if nucleation occurs in the gas phase, it would be homogeneous nucleation. It turned out that the homogeneous nucleation in the gas phase is very difficult to achieve. In relation to this, it is worth to mention early experiments on nucleation by Wilson (1897, 1899, 1904, 1900) who invented the Wilson cloud chamber, which is a simple adiabatic expansion system. By controlling the adiabatic expansion, Wilson could induce the supersaturation of water or alcohol vapour. With this chamber, he could artificially produce mist or cloud. He tried to test the Volmer theory of the homogeneous nucleation rate. In the first expansion, homogeneous nucleation did not occur but nucleation occurred heterogeneously on dust in air. After several expansions, the dust-free clean vapour could be prepared but still nucleation was not homogeneous. It turned out that nucleation took place on ions. Even after repeated expansions, ion-induced nucleation could not be avoided. Later it was found out that ions are continuously generated by natural radioactivity and cosmic rays and a steady state concentration of $\sim 10^3$ ion pairs cm^{-3} is maintained in air (Hirth and Pound 1963).

Nucleation on ions was so sensitive that it could be used to identify the location of ions because nuclei on ions instantly grow into visible sizes. Wilson cloud chamber is better known for this application. The principle is as follows. High energy particles are passed into a supersaturated medium, where nucleation is suppressed because of the nucleation barrier. When the high energy particles collide with the supersaturated medium, ionization of the medium occurs along the path of high energy particles. The ions act as a strong heterogeneous site for nucleation and trigger the nucleation. The nuclei on ions instantly grow into the visible size, providing the visible track of the high energy particles. If particles are electrically charged like an alpha or beta particle, their track is affected by the electric field or magnetic field. From the radius of the deflection, their mass can be determined. When any uniform magnetic field is applied across the cloud chamber, charged particles will be deflected. From the analysis of the deflection, the mass and the sign of charge can be determined.

Several Nobel prizes were awarded for studies related to Wilson cloud chamber experiments. Wilson was awarded a Nobel prize in 1927 for his work on the cloud chamber. Anderson discovered the positron in 1932 and muon in 1936 in cosmic rays using cloud chambers as detectors. For his discovery of the positron, Anderson

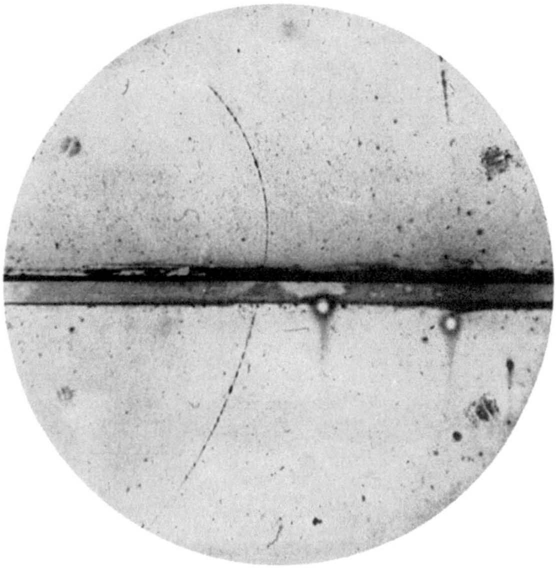

Fig. 3.15 Cloud chamber photograph of the first positron ever observed (Wikipedia common) (Anderson 1933)

was awarded a Nobel prize in 1936. The kaon was discovered in 1947 using the cloud chamber. In 1948 Blackett received a Nobel prize for his work on cosmic rays using his invention of the counter-controlled cloud chamber. Cloud chambers played a prominent role in the experimental particle physics from 1920s to the 1950s until the invention of the bubble chamber. Glaser received the Nobel prize in 1960 for his invention of the bubble chamber. Figure 3.15 shows the cloud chamber photograph of the first positron observed by Anderson (1933).

Ion-induced nucleation is still actively studied because it is important in atmospheric nucleation (Curtius et al. 2006). The Gibbs free energy change for the formation of clusters of the radius r from the gas phase to condense around an ion is represented in the classical nucleation theory by the Thompson equation.

$$\Delta G = \frac{4\pi r^3}{3}\Delta f + 4\pi r^2\sigma + \frac{q^2}{2}\left(1 - \frac{1}{\varepsilon}\right)\left(\frac{1}{r} - \frac{1}{r_o}\right) \quad (3.18)$$

Here q is the unit charge, ε is the dielectric constant and r_o is the radius of the core ion. The first two terms in (3.18) are for the neutral nucleation, which is identical to (3.2). The third term is the electrostatic energy arising from nucleation on an ion. The term of $1/r_o$ represents the electrostatic energy of an ion and it has a minus sign because it is replaced by a charged cluster.

The schematic of the Gibbs free energy change as a function of the cluster radius is shown in Fig. 3.16. The electrostatic energy in (3.18) is inversely proportional to the radius and has hyperbola functional dependence on the radius. This means that clustering occurs spontaneously even in the absence of supersaturation. Because of this spontaneous clustering, the Gibbs free energy curve has an energy well, which

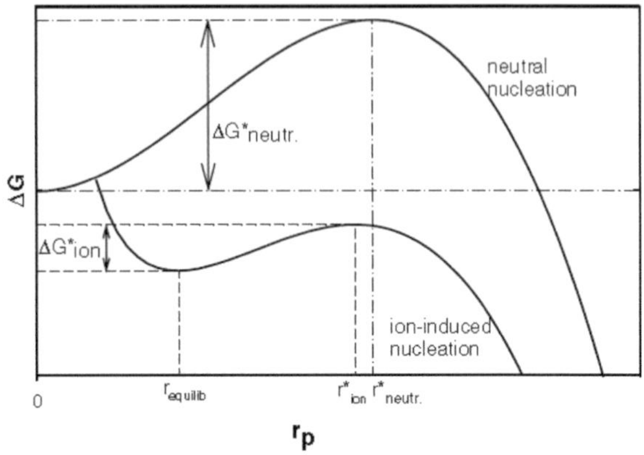

Fig. 3.16 Gibbs free energy as a function of cluster radius. Nucleation barriers (ΔG*) and radius of the critical nucleus (r*) for the neutral and ion-induced nucleation pathway are indicated (Curtius et al. 2006)

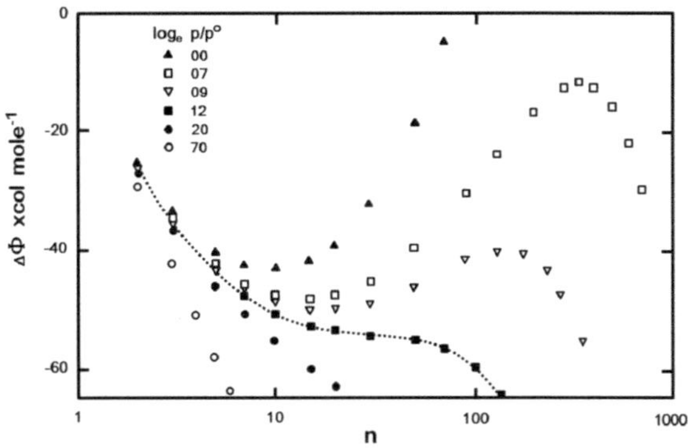

Fig. 3.17 Energy barriers to ion-induced nucleation as a function of cluster size. The parameters are the natural logarithm of the supersaturation ratio ln(p/p°) as shown in the key (Castleman Jr et al. 1978)

corresponds to $r_{equilib}$ in Fig. 3.16. Because of this spontaneous clustering, the nucleation barrier for nucleation on an ion ΔG^*_{ion} becomes much smaller than that of neutral nucleation ΔG^*_{neutr}. as shown in Fig. 3.16.

Castleman et al. (1978) estimated the barrier of ion-induced nucleation of the water vapour as a function of the number of molecules as shown in Fig. 3.17. The values of $\log_e$ p/p°, 0.0, 0.7, 0.9, 1.2, 2.0 and 7.0 in the figure correspond to the supersaturation ratios, 1, 2, 2.46, 3.32, 7.34 and 10.97, respectively. It should be noted that in the absence of supersaturation corresponding to the data of filled triangles in Fig. 3.17, the clusters containing about 10 molecules are stable. The nucleation barrier decreases with increasing supersaturation ratio from 2 to 2.46. When the supersaturation ratio is larger than 3.32, which corresponds to the data of filled squares, the nucleation barrier disappears. In such cases, nucleation would be spontaneous.

Considering that the ion-induced nucleation could not be avoided even in dust-free clean air at room temperature in the Wilson cloud chamber experiment, it is probable that ion-induced nucleation might occur in thin film reactors, where the appreciable supersaturation exists for precipitation of the material to deposit.

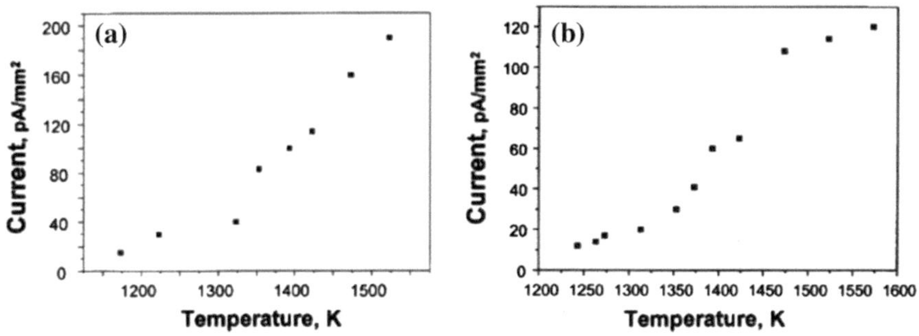

Fig. 3.18 Temperature dependence of the electric current detected on a Faraday cup during thermal evaporation of **a** gold and **b** copper in the tungsten basket (Barnes et al. 2002; Jeon et al. 2003)

Especially, in the plasma CVD process, by which all kinds of films and nanostructures are made, lots of electrons and ions are generated. Also in the hot wire CVD (HWCVD) process, an appreciable amount of electrons and ions are generated. It should be noted that hot tungsten filament is used to generate electrons for electron gun.

The best way to confirm the generation of electric charge in the thin film reactor is to measure the current between the conducting substrate and the ground. Such measurements were done by Hwang and his colleagues (Jeon et al. 2001, 2003; Barnes et al. 2002; Lee and Hwang 2008; Song et al. 2008). The thin film reactor generating electric charge is least expected in a thermal evaporation reactor. Surprisingly, however, an appreciable amount of current could be detected without exception in all the thin film reactors that were measured.

Barnes et al. (2002) measured the current generated during thermal evaporation of gold on the Faraday cup placed 40 mm below the tungsten basket. The detected current was found to be quite sensitive to the processing condition. Usually, a poorer vacuum increased the positive current. A negative current began to be detected above 1573 K. The result is shown in Fig. 3.18a.

Jeon et al. (2003) measured the current generated during thermal evaporation of copper on the Faraday cup placed 40 cm below the tungsten basket as a function of the basket temperature as shown in Fig. 3.18b. They observed that the current was negligible when the basket temperature was lower than 1223 K but an appreciable current was measured only when the evaporation rate was appreciable. This result indicates that the generation of charge is accompanied by the evaporated atoms or clusters, which implies that evaporated atoms or clusters might carry an electric charge.

In order to check the generation of electric charge during ZrO_2 CVD, Jeon et al. (2001) measured current using a Faraday cup with varying evaporation temperature of the $ZrCl_4$ precursor as shown in Fig. 3.19. Negative bias of −20 V was applied for repelling negative charges. The current was measured at two locations in the reactor: one is near the precursor and the other on the substrate position. In the first

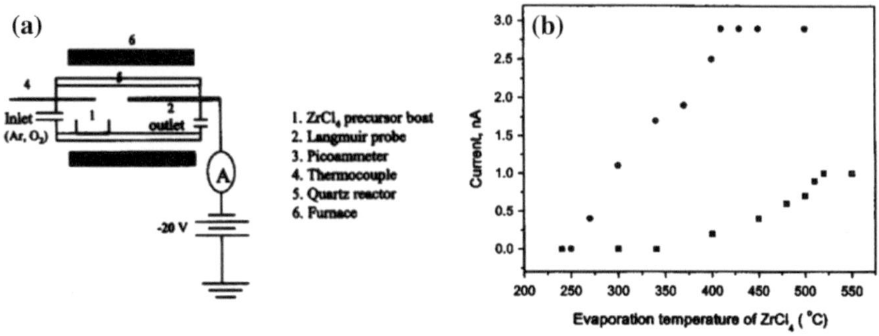

Fig. 3.19 **a** Schematic for measurements of the current inside the CVD reactor by a Faraday cup and **b** the measured current as a function of evaporation temperature of $ZrCl_4$. The *filled squares* and *circles* represent respectively the current produced near the evaporation temperature of $ZrCl_4$ and at the location of film deposition (Jeon et al. 2001)

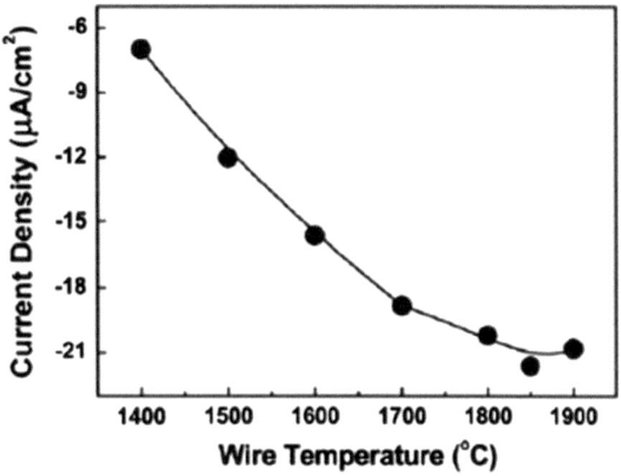

Fig. 3.20 Temperature dependence of the measured current during silicon HWCVD under 66.7 Pa in a gas mixture of 20 % SiH_4 + 80 % H_2 (Song et al. 2008)

case, the temperature was ~100 °C higher than that of the $ZrCl_4$ precursor and in the second case, the temperature was fixed at 1000 °C.

Song et al. (2008) measured the generation of current in the silicon HWCVD process as a function of the wire temperature as shown in Fig. 3.20. A square-shaped platinum plate with an area of 0.5 cm^2 was placed 3.0 cm below the hot wire. The current was measured between the platinum plate and the ground. The current is negative. Its magnitude increased with increasing wire temperature.

The amount of the measured current differs by a few orders of magnitude depending on the type of CVD. In the plasma enhanced CVD (PECVD), HWCVD and thermal CVD processes, the measured current was in the ranges of milli-amperes, micro-amperes and nano-amperes per cm^2. In the thermal evaporation process, the measured current was respectively in the range of pico-amperes per cm^2. Considering that one ampere corresponds to the flux of ~10^{19} charges per second, even one pico-ampere per cm^2 corresponds to the flux of $\sim 10^7$ charges per second per cm^2. This means that an enormous amount of charges are generated in the thin film reactors.

Therefore we naturally arrived at the following question.

(Q) The ion-induced nucleation in the cloud chamber is much established but ion-induced nucleation in the plasma and HWCVD processes is hardly known. Why is this so?

When the ion-induced nucleation takes place in the cloud chamber, the amount of supersaturated medium for nuclei to grow is enormously large compared to the concentration of ions. As a result, the nuclei instantly grow into visible sizes, which should be larger than the minimum wave length of visible light ($\sim$300 nm). In the plasma or HWCVD process, however, the amount of supersaturated medium is small compared with the huge amount of electrons and ions. In this situation, the number of charged nuclei is so large and the supersaturated medium is divided into so many nuclei that the charged nuclei do not grow much but maintain their nanometer size during their residence time in the gas phase. In other words, even if ion-induced nucleation takes place extensively, the nuclei are invisible. Therefore, it is important to make the experimental confirmation that the charged gas phase nuclei should exist in the gas phase of the plasma and HWCVD processes. This aspect will be dealt in Chaps. 9 and 10.

3.6 Photo-Induced Nucleation: Heterogeneous Nucleation in the Gas Phase

It should be noted that nucleation in the gas phase is induced not only by ions but also by other sources such as photo-excited species. Photo-induced nucleation is known to be much more powerful than ion-induced nucleation in triggering nucleation in the gas phase (Katz et al. 1977; Wen et al. 1978). In the case of nucleation of pure water from its supercooled vapor, photo-induced nucleation was detected at supersaturations as low as 1.00042 (Wen et al. 1978).

According to the study on the formation of dust particles in capacitively-coupled reactive plasma by Watanabe et al. (2001), clusters of $\sim 10^{11}$ cm^{-3}, whose number density is much higher than a plasma ion density of $\sim 10^9$ cm^{-3}, are found at t $\sim$50 ms after the discharge initiation. Since the number of clusters is two orders of magnitude larger that the number of ions, initial clusters must be neutral, indicating that they are not formed by ion-induced nucleation. They further suggested that species contributing to the initial growth of clusters are principally produced in the radical production region around the plasma/sheath boundary near the RF electrode. Considering the result of Watanabe et al. (2001) and that radicals are photo-excited species, the initiation of clusters in the plasma CVD process appears to be formed by photo-induced nucleation.

In the diamond CVD process, it is experimentally observed that diamond is formed even for the gas mixture of 0.2 % CH_4–99.8 % H_2 (Badzian et al. 1988; Kobashi et al. 1988). It is common that the diamond film is prepared by a gas mixture of 0.5 % CH_4–99.5 % H_2. Since the diamond surface energy (3.7 JM^{-2}) is very large and the supersaturation ratio from the gas mixture 0.2 % CH_4–99.8 % H_2 are so low, even the ion-induced nucleation is not expected (Choi et al. 1997). It is possible that photo-induced nucleation takes place and the neutral clusters are charged by the attachment of electrons in the diamond CVD process.

On the other hand, it should be noted that the bubble chamber experiment (Peyrou 1967) could not be explained by ion-induced nucleation because the electrostatic energy involved in bubble nucleation inhibits nucleation rather than inducing it. Therefore, the enhanced nucleation in the bubble chamber experiment has remained unsolved. At the time of bubble chamber experiments, photo-induced nucleation was not discovered. Bubble nucleation might be more properly explained by photo-induced nucleation because photo-excited atoms or molecules are accompanied in the ionizing environment.

References

Anderson CD (1933) The positive electron. Phys Rev 43(6):491–494

Badzian AR, Badzian T, Roy R, Messier R, Spear K (1988) Crystallization of diamond crystals and films by microwave assisted CVD (Part II). Mater Res Bull 23(4):531–548

Barnes MC, Jeon I-D, Kim D-Y, Hwang NM (2002) Generation of charged clusters during thermal evaporation of gold. J Cryst Growth 242(3–4):455–462. doi:http://dx.doi.org/10.1016/S0022-0248(02)01417-3

Becker R, Döring W (1935) Kinetische behandlung der keimbildung in übersättigten dämpfen. Ann Phys 24:719–752

Cahn JW (1956) The kinetics of grain boundary nucleated reactions. Acta Metall 4(5):449–459. doi:http://dx.doi.org/10.1016/0001-6160(56)90041-4

Castleman Jr AW, Holland PM, Keesee RG (1978) The properties of ion clusters and their relationship to heteromolecular nucleation. J Chem Phys 68:1760–1767. doi:http://dx.doi.org/10.1063/1.435946

Choi K, Kang S-JL, Jang HM, Hwang NM (1997) Nucleation behavior in the presence of charge in the CVD diamond process. J Cryst Growth 172(3–4):416–425. doi:http://dx.doi.org/10.1016/S0022-0248(96)00759-2

Curtius J, Lovejoy ER, Froyd KD (2006) Atmospheric ion-induced aerosol nucleation. Space Sci Rev 125(1–4):159–167. doi:10.1007/s11214-006-9054-5

Dennig PA, Stevenson DA (1991) Influence of substrate topography on the nucleation of diamond thin films. Appl Phys Lett 59(13):1562–1564. doi:http://dx.doi.org/10.1063/1.106283

Dennig PA, Shiomi H, Stevenson DA, Johnson NM (1992) Influence of substrate treatments on diamond thin film nucleation. Thin Solid Films 212(1–2):63–67. doi:http://dx.doi.org/10.1016/0040-6090(92)90501-2

Givargizov EI, Aksenova LL, Galstyan VG, Kiselev AN, Kuznetsov AV, Muratova VI, Rakova EV, Stepanova AN (1996) Growth of diamond particles on sharpened Si tips. J Cryst Growth 162(1–2):73–78. doi:http://dx.doi.org/10.1016/0022-0248(95)00945-0

Hirth JP, Pound GM (1963) Condensation and evaporation. Pergamon Press, Oxford

Hwang NM, Hahn JH, Yoon DY (1996) Charged cluster model in the low pressure synthesis of diamond. J Cryst Growth 162(1):55–68
Jeon I-D, Gueroudji L, Kim D-Y, Hwang N-M (2001) Temperature dependence of the deposition behavior of yttria-stabilized zirconia CVD films: Approach by charged cluster model. J Korean Ceram Soc 38(3):218–224
Jeon I-D, Barnes MC, Kim D-Y, Hwang NM (2003) Origin of positive charging of nanometer-sized clusters generated during thermal evaporation of copper. J Cryst Growth 247(3–4):623–630. doi:10.1016/S0022-0248(02)02058-4
Karve P, Sainkar SR, Kshirsagar ST (1998) Role of surface features in CVD diamond nucleation on surface-pretreated substrates. Mater Lett 34(3–6):387–391. doi:http://dx.doi.org/10.1016/S0167-577X(97)00208-5
Katz JL, Wen F, McLaughlin T, Reusch RJ, Partch R (1977) Nucleation on photoexcited molecules. Science 196(4295):1203–1205
Kobashi K, Nishimura K, Kawate Y, Horiuchi T (1988) Synthesis of diamonds by use of microwave plasma chemical-vapor deposition: Morphology and growth of diamond films. Phys Rev B 38(6):4067
Lee J-I, Hwang N-M (2008) Generation of negative-charge carriers in the gas phase and their contribution to the growth of carbon nanotubes during hot-filament chemical vapor deposition. Carbon 46(12):1588–1592. doi:http://dx.doi.org/10.1016/j.carbon.2008.07.006
Mutaftschiev B (1993) Handbook of crystal growth, vol 1a. North-Holland, Amsterdam
Peyrou C Bubble chamber principles. In: Bubble and spark chambers: Volume II; Principles and Use, 1967. p 20
Porter DA, Easterling KE (1981) Phase transformations in metals and alloys. Van Nostrand Reinhold, New York
Porter DA, Easterling KE, Sherif M (2009) Phase transformations in metals and alloys. 3rd edn. CRC Press,
Ramesham R, Ellis C (1992) Selective growth of diamond crystals on the apex of silicon pyramids. J Mater Res 7(05):1189–1194. doi:10.1557/JMR.1992.1189
Song J-H, Park SH, Hwang N-M, Kim HJ (2008) Effect of negatively charged species on the growth behavior of silicon films in hot wire chemical vapor deposition. Thin Solid Films 516 (15):5122–5126. doi:http://dx.doi.org/10.1016/j.tsf.2008.02.019
Turnbull D (1952) Kinetics of solidification of supercooled liquid mercury droplets. J Chem Phys 20(3):411–424. doi:http://dx.doi.org/10.1063/1.1700435
Watanabe Y, Shiratani M, Koga K (2001) Formation kinetics and control of dust particles in capacitively-coupled reactive plasmas. Phys Scripta 2001 (T89):29
Wen F, McLaughlin T, Katz JL (1978) Photoinduced nucleation of water vapor. Science 200 (4343):769–771
Wilson CTR (1897) Condensation of water vapour in the presence of dust-free air and other gases. Phil Trans R Soc A 189:265–307. doi:10.2307/90698
Wilson CTR (1899) On the condensation nuclei produced in gases by the action of rontgen rays, uranium rays, ultra-violet light, and other agents. Phil Trans R Soc A 192:403–453. doi:10.2307/90785
Wilson CTR (1900) On the comparative efficiency as condensation nuclei of positively and negatively charged ions. Phil Trans R Soc A 193:289–308. doi:10.2307/90798
Wilson CTR (1904) LXXV. The condensation method of demonstrating the ionisation of air under normal conditions. Philos Mag Series 6 7 (42):681–690. doi:10.1080/14786440409463162

Chapter 4
Growth

Once nucleated, the nucleus starts to grow. The growth consists of two serial processes: the atomic attachment on the interface and the transport of atoms to the interface. If the kinetics of atomic attachment is slower than that of atomic transport, the growth kinetics is controlled by the atomic attachment process, which is called the interface (or surface) controlled growth. If the kinetics of the atomic transport is slower than that of atomic attachment, the growth kinetics is controlled by the atomic transport, which is called the transport (or diffusion) controlled growth. Depending on whether the process is interface-controlled or transport-controlled, a quite different growth morphology is evolved.

4.1 Rough and Smooth Interfaces

The kinetics of atomic attachment depends on the interface structure on an atomic scale. In relation to the interface structure, there are two kinds of equilibrium shape of condensed phases. One is spherical and the other is non-spherical, being faceted. The equilibrium shape of liquid is spherical except a very special case. The spherical equilibrium shape implies that the interface energy is isotropic, independent of orientation. On the other hand, the equilibrium shape of a crystal can be spherical or faceted as schematically shown in Fig. 4.1.

The interface structure of a spherical equilibrium shape is atomically disordered or rough as shown in Fig. 4.1a. The atomically-disordered structure implies that entropy is dominant over enthalpy. Such an interface is called a rough interface and has an isotropic interface energy. There is no barrier for attachment or detachment of atoms or molecules because the interface has plenty of kink sites. The growth proceeds normally by diffusion control. Therefore, if an interface is atomically rough, adhesive type growth or continuous growth occurs.

On the other hand, the interface structure of a cubic equilibrium shape is atomically ordered as shown in Fig. 4.1b. The atomically-ordered structure implies

© Springer Science+Business Media Dordrecht 2016
N.M. Hwang, *Non-Classical Crystallization of Thin Films and Nanostructures in CVD and PVD Processes*, Springer Series in Surface Sciences 60,
DOI 10.1007/978-94-017-7616-5_4

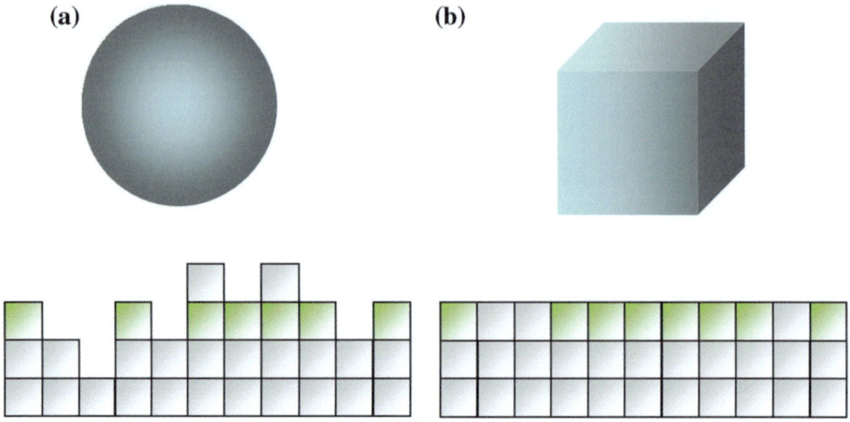

Fig. 4.1 **a** *Spherical* and **b** *cubic* equilibrium shapes with their schematic interface structure on an atomic scale

that enthalpy is dominant over entropy. Such an interface is called a smooth interface. It is also called a singular interface. The term 'singular' comes from the fact that the orientation of such an interface corresponds to singularity or cusp in the Wulff-plot. It has an anisotropic interface energy. There is an appreciable barrier for atomic attachment because the interface is atomically flat and has no kink sites. The growth proceeds normally by interface control.

Since quite a different growth morphologies are evolved between diffusion-controlled and interface-controlled growth, a more specific growth process needs to be examined.

4.2 Diffusion-Controlled Growth

When the growth is controlled by the diffusion of atoms to the growing interface and the diffusion is in the steady state, the growth rate can be described by the Fick's first law, which is expressed as

$$J = -D\frac{\partial C}{\partial x} \tag{4.1}$$

where J is the diffusion flux, D is diffusivity, C is concentration and x is the position. The growth rate increases in proportion to the concentration gradient $\partial C/\partial x$. During growth, a small protrusion can be formed by fluctuation at the growth front as shown in Fig. 4.2. Then, the protruded area has a larger concentration gradient, which results in a higher growth rate than the other region. Then, the concentration gradient becomes even larger, which results in an even higher growth

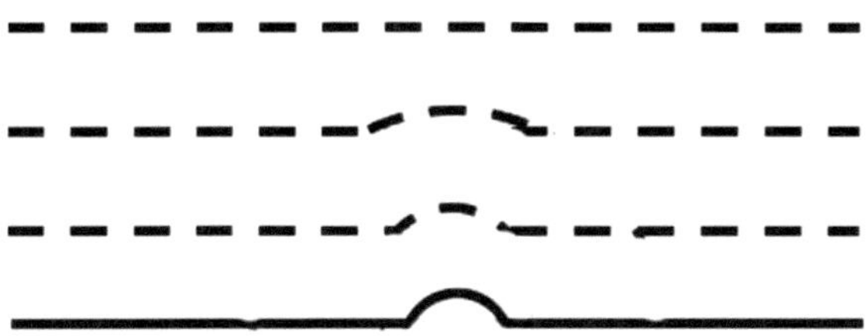

Fig. 4.2 The growing interface with a protrusion (*solid line*) and the iso-concentration profiles (*dashed lines*) in the diffusion-controlled growth

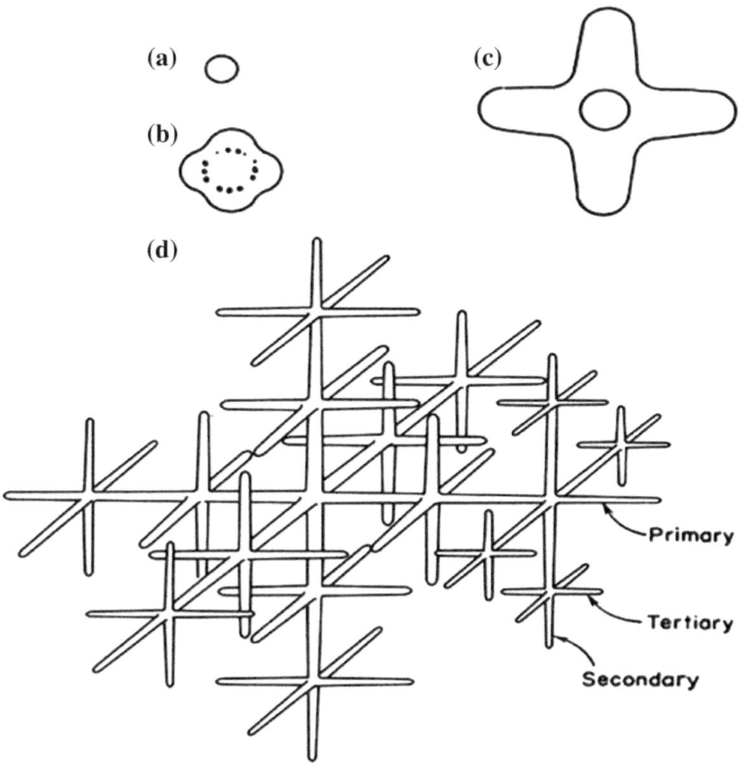

Fig. 4.3 The development of dendrites: **a** a *spherical* nucleus, **b** the interface becomes unstable, **c** primary arms develop in crystallographic directions (⟨100⟩ in cubic crystals), **d** secondary and tertiary arms develop (Reed-Hill 1973)

rate. As time goes on, the growth rate on the protruded area becomes higher and higher. In other words, the growth is accelerated locally.

Such accelerated growth results in so called 'dendritic growth'. Although Fig. 4.2 illustrates the one dimensional growth, the actual growth would be three dimensional. The protruded spine is called the dendrite arm. Another dendritic growth starts on the primary dendritic arm, producing the secondary dendrite arm, on which the third dendrite arm develops. This aspect is illustrated in Fig. 4.3. Therefore, the growth front of the rough interface, which undergoes the diffusion-controlled growth, is unstable and the dendritic growth would occur. In order to inhibit the dendritic growth, the driving force should be made very small.

4.3 Interface-Controlled Growth

Unlike the rough interface, there is an appreciable barrier for atomic attachment on the smooth interface. Because of this barrier, the growth tends to be interface-controlled. Since the smooth interface is quite common in crystal growth, it is important to understand how the growth occurs when the interface is smooth.

4.3.1 Terrace-Ledge-Kink (TLK) Model

The microscopic elementary process of the attachment and detachment of building units (atoms or molecules) was introduced by Kossel (1927) and by Stranski (1928) based on the terrace, ledge and kink (TLK) model of the smooth interface. Consider the (100) surface of a simple cubic crystal (Kossel crystal) in a supersaturated medium (Fig. 4.4a). Suppose the building unit (atom) has 6 bonds to be saturated as shown in Fig. 4.4b. It is assumed that the bond interaction is only with the nearest neighbor.

The site 1 has 5 broken bonds. If the atom lands on the site 1, 5 broken bonds are replaced with 1 broken bond, resulting in the decrease of 4 broken bonds. Similarly, if the atom lands on the site 2, which has 4 broken bonds, 4 broken bonds are replaced with 2 broken bonds, resulting in the decrease of 2 broken bonds. If the

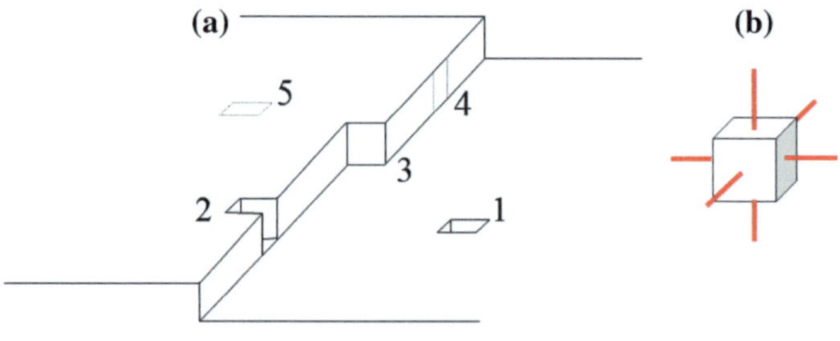

Fig. 4.4 **a** (100) surface of a simple cubic Kossel crystal with terrace, ledge and kink and **b** the building unit (atom) with 6 bonds to be saturated

atom lands on the site 3 (kink), 3 broken bonds are replaced with 3 broken bonds without any energy decrease or increase. Therefore, atoms transfer reversibly at the kink site. If the atom lands on the site 4 (ledge), 2 broken bonds are replaced with 4 broken bonds, resulting in the increase of 2 broken bonds. If the atom lands on the site 5 (terrace), 1 broken bond is replaced with 5 broken bonds, resulting in the increase of 4 broken bonds.

Consider a situation where the driving force for growth is small enough. Since the derivative of energy with respect to displacement is a force, the sites 1 and 2 exert an attractive force for the incoming atoms whereas the ledge 4 and the terrace 5 exert repulsive forces. It should be noted that the kink 3 does not have a repulsive force. The sites 1 and 2 disappear in an early stage of growth because they are most favourable for atomic attachment. Then, only the kink site 3 continues to remain as a favourable site for incoming atoms.

It is assumed that the ledge of monoatomic height is disordered or roughened at ambient temperature. This means that the step of monoatomic height has plenty of kinks. This assumption is valid under most conditions of crystal growth. Therefore, the step of a monoatomic ledge is regarded as kinks. Therefore, the ledge 4 in Fig. 4.4a plays the same role as the kink 3.

Because a specific site is attractive or repulsive, the crystal growth occurs by atomic self-assembly. When a crystal is growing, millions of atomic building blocks are thrown into the growing surface. If there is no atomic self-assembly, the atoms would be packed randomly, resulting in an amorphous structure. However, when atoms are thrown into the terrace such as the site 5 in Fig. 4.4a, they are not accepted but rejected by the repulsive force. Only the sites 3 and 4 accept atoms. Because of such atomic self-assembly, a crystalline phase of highly regular arrays of atoms can be grown.

When an atom lands on the terrace such as the site 5 in Fig. 4.4a, it temporally resides and makes atomic jumps there for a mean residence time, during which it travels a mean residence distance. If the atom reaches the site 3 or 4 during the mean residence time, it becomes a part of crystal; otherwise, the atom goes back to the liquid or gas phase. Therefore, the atom on the terrace does not belong to the crystal and is called an 'adatom', which is abbreviation of an adsorbed atom. The chemical potential of an adatom is not the same as that of the crystal but the same as that in the supersaturated medium in the vicinity of the growing surface. If the spacing between ledges is less than the mean residence distance, most atoms which land on the terrace will contribute to growth. Then the growth kinetics of the smooth interface would be similar to that of the rough interface. This is known as 'kinetic roughening', which will be treated in Sect. 4.3.4.

(Q) How can crystal grow when there exists no step of monoatomic height on the terrace?

4.3.2 Growth by 2D Nucleation

Assume that the driving force for crystal growth per atom is given by

$$\Delta f = -kT \ln\left(\frac{P}{P_{eq}}\right) \tag{4.2}$$

If one atom lies on the terrace as shown in Fig. 4.5a, the Gibbs free energy is

$$\Delta G = \Delta f + 4\phi > 0 \tag{4.3}$$

where φ is the broken bond energy.

The probability to overcome this barrier is exp(−ΔG/kT), which is far less than one. This is why an adatom is unstable. The concentration of adatoms on the terrace is proportional to exp(−ΔG/kT). If 4 atoms lie together on the terrace as shown in Fig. 4.5b, the Gibbs energy is

$$\Delta G = 4\Delta f + 8\phi = 4(\Delta f + 2\phi) \tag{4.4}$$

Although the Gibbs energy in (4.4) may be larger than that in (4.3), the excess broken bonds per atom in (4.4) is smaller than that in (4.3). This tendency increases even further for 9 atoms lying together on the terrace in Fig. 4.5c. In this case, the Gibbs free energy is

$$\Delta G = 9\Delta f + 12\phi = 9\left(\Delta f + \frac{4}{3}\phi\right) \tag{4.5}$$

In Fig. 4.6, ΔG is plotted against the number of atoms. This curve looks similar to the nucleation curve of Fig. 3.1 in Chap 3. However, clusters in Fig. 4.6 is two dimensional, meaning that the clusters are epitaxial and have no interface energy with the surface. This is why it is called 2-dimensional (2D) nucleation.

As in the case of 3-dimensional (3D) nucleation, the expressions for the critical nucleus and the 2D nucleation barrier can be derived. Consider a disc-shaped nucleus of monoatomic height h with radius r as shown in Fig. 4.7.

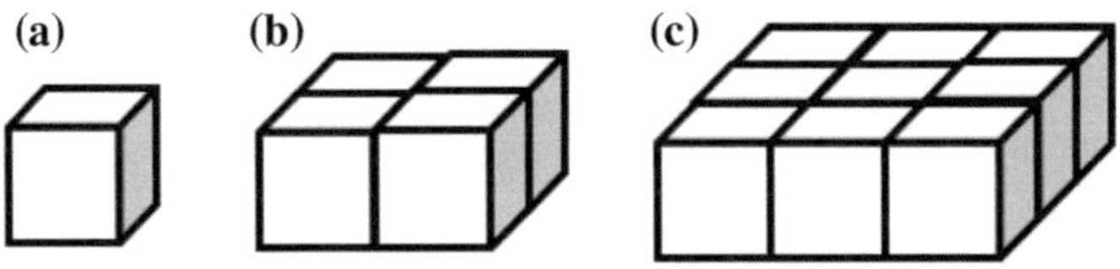

Fig. 4.5 Atoms on the terrace; **a** a single atom, **b** 4 atoms and **c** 9 atoms

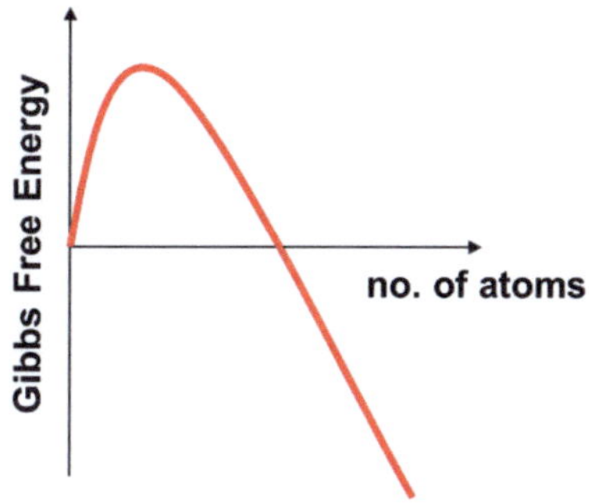

Fig. 4.6 Curve of Gibbs free energy versus number of atoms for 2D nucleation

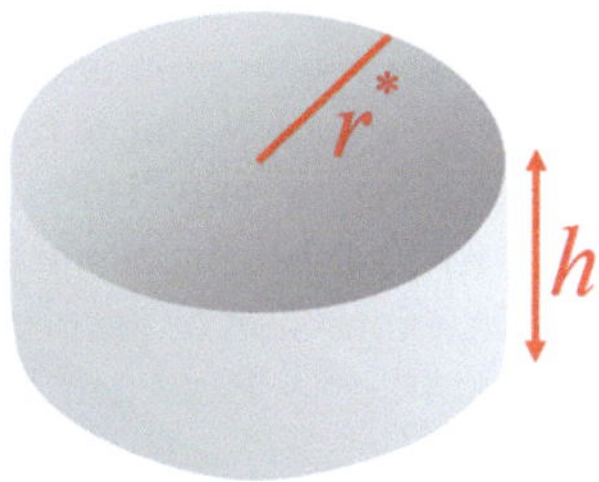

Fig. 4.7 Disc-shaped 2D nucleus of monoatomic height *h* with radius *r*

The surface of the disc-shaped 2D nucleus lying epitaxially on the terrace does not provide the extra energy because it replaces the underneath surface. The extra interface energy comes from the peripheral edge of the disc, which corresponds to the monoatomic step. It should be noted that edge, ledge and step have the same meaning. If the edge energy is ε, which is also called the step free energy, the Gibbs free energy becomes

$$\Delta G = \pi r^2 h \cdot \Delta f_v + 2\pi r \varepsilon \tag{4.6}$$

where Δf_v represents the driving force for crystal growth per volume.

Once a 2D nucleus is formed, a crystal can grow by 2D spreading of the nucleus parallel to the interface. Such growth by 2D nucleation was first suggested by Kossel (1934) and Stranski (1928). The 2D nuclei provide the step of monoatomic height, which is the kink-generating source. Therefore, the growth can continue if 2D nucleation continues to occur. The disc-shape means that the step is round with a plenty of kinks in agreement with the assumption that the step of monoatomic height is roughened and isotropic. If the shape of the 2D nucleus is rectangular, the step is not roughened. In this case, the step does not have kinks. For this step to grow parallel to the interface, one-dimensional nucleation is needed. However, the barrier for one-dimensional nucleation is much smaller than that for 2D nucleation and can be neglected when the growth occurs by 2D nucleation.

Differentiating (4.6) with respect to r and setting the derivatives equal to zero, the critical radius is obtained as

$$r^* = \frac{\varepsilon}{h\Delta f_v} \tag{4.7}$$

Here, Δf_v should be regarded as a magnitude of driving force, which has a positive value. Putting (4.7) into (4.6) gives the barrier for 2D nucleation as

$$\Delta G^* = \frac{\pi\varepsilon^2}{h\Delta f_v} \tag{4.8}$$

As in the 3D nucleation, the rate of 2D nucleation is practically zero below a certain supersaturation and increases abruptly to high values above it as shown in Fig. 4.8, where N and α represent respectively the rate of 2D nucleation and the supersaturation ratio. This supersaturation ratio is called the critical supersaturation ratio for 2-dimensional nucleation $(P/P_e)_{crit}$.

The expression of the critical supersaturation ratio for the onset of 2D nucleation was derived by Hirth and Pound (1963) as

$$\left(\frac{P}{P_e}\right)_{crit} = \exp\left(\frac{\pi\varepsilon^2\Omega}{65hk^2T^2}\right) \tag{4.9}$$

From this expression, the critical supersaturation ratio for 2D nucleation can be estimated. If the 2D nucleation continues, the layer-by-layer growth would continue. In order for 2D nucleation to occur, however, the supersaturation ratio higher than $(P/P_e)_{crit}$ should be provided. Below the critical supersaturation ratio, 2D nuclei would not be formed and no source for growth steps is available on the surface. However, it was experimentally observed repeatedly that the crystal with a smooth interface could grow at supersaturation ratio much smaller than $(P/P_{eq})_{crit}$.

(Q) What is your explanation for this experimental observation?

This means that there must be some other kink-generating source, which turned out to be a screw dislocation existing in the crystal.

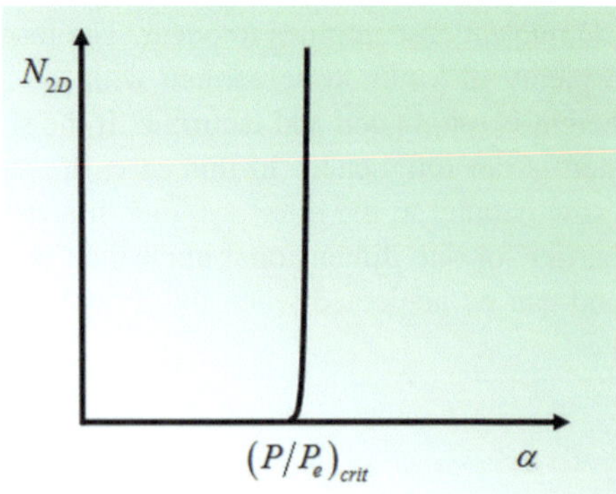

Fig. 4.8 Dependence of the 2D nucleation rate on the supersaturation ratio

4.3.3 *Growth by Spiral Steps*

The spiral growth theory was put forward by Frank (1949). Screw type dislocations provide a spiral step of monoatomic height, which is self-perpetuating. Therefore, in the presence of screw dislocations, a crystal can continue to grow at the supersaturation ratio much smaller than $(P/P_{eq})_{crit}$. Frank's spiral growth theory was verified by the observation of spiral step patterns with monoatomic step height in various crystals. Crystals grown with a low supersaturation were always found to have 'growth spirals' on the growing surfaces.

The term 'supersaturation' is used when the crystal grows in a supersaturated medium, which is the precipitation reaction. The term 'supercooling or undercooling' of the interface (ΔT_i) is used when the crystal grows by solidification. When ΔT_i is small, the driving force is proportional to ΔT_i.

$$\Delta G = \frac{L}{T_m} \cdot \Delta T_i \tag{4.10}$$

where L and T_m are respectively the latent heat and melting temperature.

The growth rate for the ideal growth of the rough interface is proportional to ΔT_i and expressed as

$$v = k_1 \cdot \Delta T_i \tag{4.11}$$

The kinetics of screw dislocation-assisted growth was formulated by Burton et al. (BCF) (1951). According to the BCF theory, the rate of spiral growth is proportional to the square of ΔT_i and expressed as

$$v = k_2 \cdot (\Delta T_i)^2 \tag{4.12}$$

The rate of 2D nucleation is proportional to $\exp(-\Delta G^*/kT)$ and (4.8) shows that ΔG^* is inversely proportional to Δf_v, which is proportional to ΔT_i. Therefore, the growth rate by 2D nucleation increases exponentially with ΔT_i and expressed as

$$v = k_3 \cdot \exp\left(-\frac{k_4}{\Delta T_i}\right) \tag{4.13}$$

For the crystal with a smooth interface, when ΔT_i is less than the critical undercooling for the onset of 2D nucleation, the spiral growth would be dominant with its rate increasing with the square of ΔT_i. When ΔT_i is larger than the critical undercooling, the 2D nucleation begins to occur and the growth occurs dominantly by 2D nucleation.

(Q) At small undercooling, ice crystals grow as a hexagonal plate or column, indicating that they have a smooth interface. At large undercooling, however, ice

crystals undergo dendritic growth, as commonly shown in a snow flake. The dendritic growth indicates the diffusion-controlled process, which is expected for a rough interface. How can you explain the dendritic growth of ice crystals at large undercooling?

4.3.4 Kinetic Roughening

Figure 4.9 schematically shows the dependence of the growth rate on ΔT_i for the ideal growth of a rough interface as well as for the growth by screw dislocation and by 2D nucleation of a smooth interface. In Fig. 4.9, with increasing ΔT_i, the growth rate by the screw dislocation approaches the ideal growth rate of the rough interface. In this case, the spacing between spiral steps is so small that the smooth interface behaves like a rough interface in accommodating incoming atoms or molecules. Similarly Fig. 4.9 shows that with increasing ΔT_i, the growth rate by 2D nucleation also approaches the ideal growth rate of the rough interface. In this case, the density of two-dimensional nuclei is so high that the smooth interface behaves like a rough interface. This phenomenon is called ‘kinetic roughening’.

Therefore, even if a crystal has a smooth interface, it can undergo dendritic growth under the condition that undercooling ΔT_i is so large that the growth rate approaches the ideal growth rate and ‘kinetic roughening’ of the interface occurs. The dendritic growth of ice crystals can be understood by ‘kinetic roughening’.

Since the kinetic roughening occurs soon after the onset of 2D nucleation, the supersaturation ratio for kinetic roughening can be approximated to (4.9) as (Hirth and Pound 1963)

$$\left(\frac{P}{P_e}\right)_{crit} = \exp\left(\frac{\pi\varepsilon^2\Omega}{65hk_B^2T^2}\right) \approx \left(\frac{P}{P_e}\right)_{kinetic\ roughening} \tag{4.14}$$

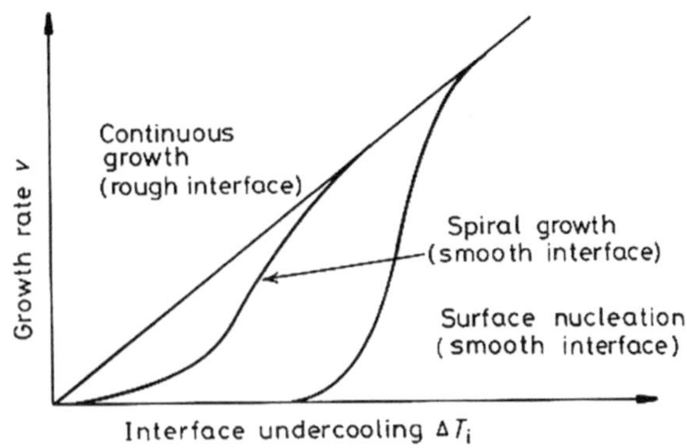

Fig. 4.9 Dependence of the growth rate on undercooling for continuous, spiral and 2D nucleation (Flemings 1974)

4.3.5 *Dissolution or Evaporation*

When the driving force is for dissolution or evaporation, the similar concept in Fig. 4.4 can be applied. In this case also, the kink site continues to remain as a favourable site for dissolution of outgoing atoms. However, growth and dissolution are not symmetric. In the growth process, if a layer completely sweeps the surface, no kink or step is available. For further growth, a new growth layer by 2D nucleation should be formed. In the dissolution process, however, even if one layer of the entire surface is dissolved, the corner of the crystal acts as a kink site. Once the corner atom dissolves, the next atom at the edge becomes the kink in succession, providing the kink perpetually. This means that there exists no barrier for dissolution, which was experimentally confirmed (Hirth and Pound 1963). Because there is no barrier for dissolving, the dissolved shape of a crystal has a rounded corner or edge. Natural diamonds underwent dissolution when they are transported from depths in the earth. Such dissolution is evidenced by the rounded dodecahedroidal faces (Sunagawa 1995).

4.4 Roughening Transition of the Interface

Equation (4.8) shows that the nucleation barrier sensitively varies not only with the driving force but also with the edge energy ε. This edge energy is different from surface energy. Figure 4.10 shows how surface energy and edge energy vary with temperature.

At zero K, both surface energy and edge energy are the same but the edge energy decreases more steeply with temperature than the surface energy. At some temperature, the edge energy becomes zero. This means that the barrier for 2D nucleation becomes zero, which means again that the interface has no barrier for atomic attachment, indicating that the interface changes to a rough interface. This is called roughening transition. Since this transition is induced by temperature, it is more specifically called thermal roughening. Roughening transition is a structural transition of the interface from ordered to disordered interface. In contrast with the

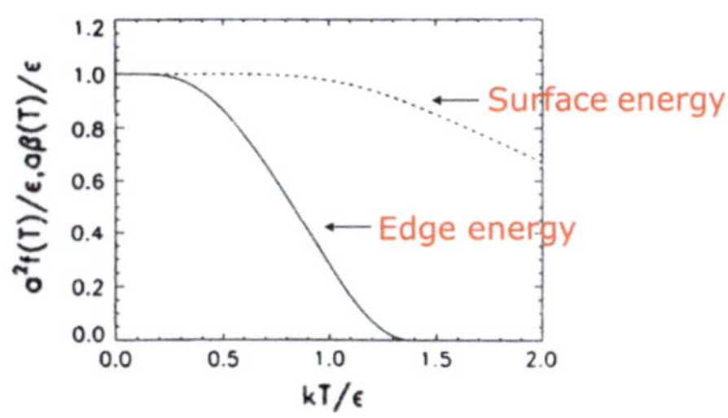

Fig. 4.10 Temperature dependence of *surface energy* and *edge energy* (van Beijeren 1977)

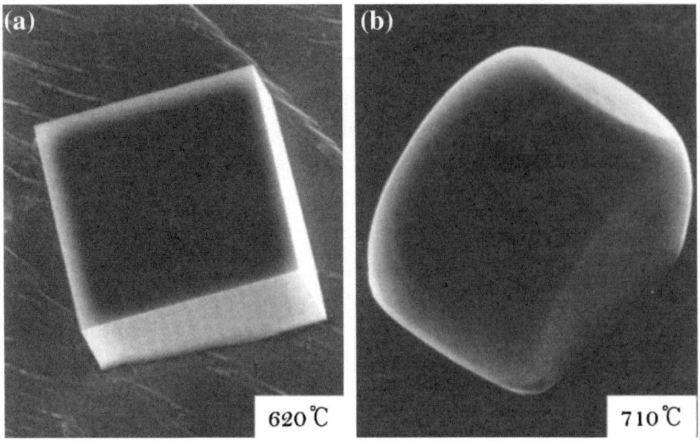

Fig. 4.11 Equilibrium shapes of a salt crystal (NaCl) at, **a** 620 °C and **b** 710 °C (Heyraud and Métois 1987)

liquid–solid transition, which is abrupt and a first order, roughening transition is gradual and an infinite order because the edge energy does not become abruptly zero but decreases gradually with temperature. It should be noted also that the edge energy is different for each facet plane of a crystal.

Figure 4.11 shows two different equilibrium shapes of a salt crystal (NaCl) at 620 and 710 °C. At the lower temperature of 620 °C, not only the faces are flat or faceted but also the corners and edges are sharp. At the higher temperature of 710 °C, however, although the six faces are flat, the corners and edges are round, indicating corners and edges are roughened.

The temperature dependence of the edge energy depends on the bond strength, which is related with enthalpy. Therefore, materials of ionic or covalent bonding such as oxide, nitride or carbide have higher temperature of roughening transition than metals. Since the enthalpy change for solid–vapor transition, which is the heat of sublimation, is much larger than that for solid–liquid transition, which is the heat of melting, the solid–vapor interface has higher temperature of roughening transition than the solid–liquid interface.

4.5 Equilibrium Shape

The crystal with a rough interface has isotropic interface energy. In this case, the equilibrium shape is spherical because a sphere has a minimum interface area. However, the crystal with a smooth interface has anisotropic interface energy with different interface energy for each plane. If γ_i is the interface energy per unit area of the ith crystal face and A_i is the area of the face, the total Gibbs free energy arising from the interface would be expressed as

$$G = \sum_{i=1}^{n} A_i \gamma_i \tag{4.15}$$

The equilibrium shape would be the shape which minimizes (4.15). The length of a vector drawn normal to a crystal face A_i is proportional to its interface energy γ_i: $h_i = \lambda/\gamma_i$. The vector h_i is the "height" of the ith face, drawn from the centre of the crystal to the face; for a spherical crystal this is simply the radius. This is known as the Gibbs-Wulff theorem. In 1953 Herring proved this theorem and suggested a method for determining the equilibrium shape of a crystal using a polar diagram called γ plot, where the length of the radius vector is proportional to the value of interface energy.

4.6 Growth Shape

Although the equilibrium shape of a crystal is determined thermodynamically, the growth shape is determined kinetically. As mentioned earlier, the crystal with a rough interface tends to grow dendritically, which is an extreme case of the growth shape. Under a given thermodynamic condition, there is only one equilibrium shape but there can be many growth shapes.

Since the edge energy is different for each face of a crystal, the crystal with a smooth interface has a different growth rate on each face. The face with low or zero edge energy would grow faster while that with high edge energy would grow slower. The face with a higher growth rate tends to disappear and the crystal tends to be enclosed by the slowest growing interface. In order to predict the growth shape, we need to know the relative growth rate for different faces. However, the information about the edge energy for each face is not available. Therefore, a more qualitative way based on a periodic bond chain (PBC) analysis, which was suggested by Hartman (1973), can be used. The PBC is an uninterrupted periodic chain of strong bonds between the crystallising units such as atoms, molecules, or ions. Strong bonds refer to short bonds between nearest neighbors. Interaction energies between a particle and its neighbors in the second nearest neighbors are usually much weaker.

F (flat) faces contain two or more PBCs in a slice of thickness d_{hkl}. S (stepped) faces contain only one PBC and K (kinked) faces contain no PBC. In Fig. 4.12, F, S, and K faces are illustrated in a hypothetical crystal.

K faces are like a rough interface and grow quickly. S faces have a step of a monoatomic height, which provides sufficient kinks. If the step of a monoatomic height is not roughened, S faces grow by one-dimensional nucleation, whose barrier is negligibly small. Therefore, S faces also grow fast enough and disappear during

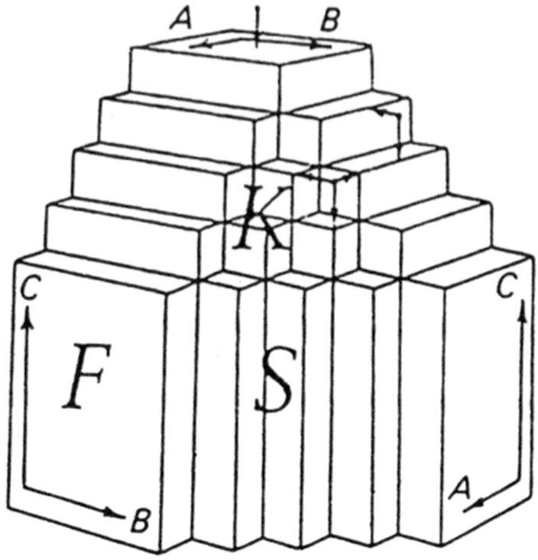

Fig. 4.12 Hypothetical crystal with three PBC's: A//[100], B//[010] and C//[001]. Flat faces are (100), (010) and (001). Stepped faces are (110), (101) and (011). Kinked face is (111) (Hartman 1973)

growth. F faces, which grow either by 2D nucleation or by screw dislocation, undergo layer-by-layer growth. They grow at the lowest rate and determine the growth shape of crystals.

For example, in Ag with a face-centered cubic lattice, strong bonds occur between nearest neighbors only. The atomic chains of nearest neighbors are in the directions $\langle 1/2, 1/2, 0 \rangle$. Therefore, there are 3 PBCs on the (111) plane, which are in the directions of [110], $[\bar{1}01]$ and $[01\bar{1}]$. There are 2 PBCs on the (100) plane, which are [110] and $[\bar{1}10]$. There is 1 PBC on the (110) plane, which is $[\bar{1}10]$. Therefore, {111} and {100} are F faces and {110} is an S face. Therefore, the growth shape of Ag crystals are expected to be bounded by {111} and {001} faces (Hartman 1973).

4.7 Growth Shape of Diamond Crystals

The growth shape of synthetic diamonds implies the possibility that they did not grow by the classical growth mechanism but grew by non-classical crystallization. In view of this, the growth shape of diamond crystal needs to be treated in detail. The growth shape of both natural and synthetic diamonds was analyzed in detail by Sunagawa (1987a, b, 1990, 1995).

According to the PBC analysis of diamond by Hartman (1973) (111) has 3 PBC's and is the only F face, (110) has only one PBC and is an S face whereas (100) has two broken PBC's and is a K face. Since (111) is the only F face, the growth shape of diamond is expected be an octahedron enclosed by (111) faces. Growth layers of triangular form develop freely, and the triangle should be in the same orientation as the triangle of the (111) face. The (110) face may appear but exhibits only striations parallel to the edge with the neighboring (111) face. The flat

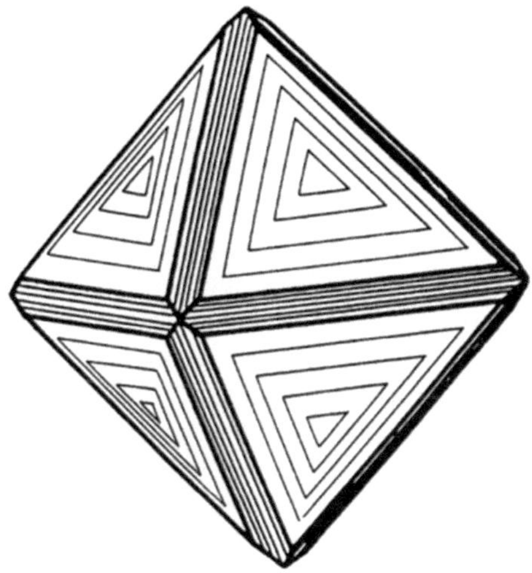

Fig. 4.13 Morphology and the surface microtopography of diamond crystal expected from the PBC analysis (Sunagawa 1987a)

(100) face is not expected to appear. Based on such a PBC analysis, Sunagawa (1987a, 1990) suggested the morphology and the surface microtopography of diamond crystal as shown in Fig. 4.13.

Sunagawa (1987a) compared the morphology of Fig. 4.13 with that of as-grown natural diamond crystals. Most natural diamond crystals have undergone dissolution to a degree while they were brought up in the magma to the surface and tend to have a rounded shape. Except this, as-grown natural diamond crystals of single crystalline type have exactly the same morphological characteristics as the morphology predicted by the PBC analysis shown in Fig. 4.13. They are mainly octahedral in habit, with occasional association of a smaller {110} face. But {100} faces never appear as flat faces on natural diamond crystals.

Sunagawa (1987a) summarized the characteristics of (111), (110) and (100) faces of natural diamond as follows.

{111} faces grew by a layer-by-layer growth mechanism throughout their growth history, which is clearly shown by growth banding revealed by etching gem quality natural diamond crystals. Straight crystallographic growth banding in {111} growth sectors is a universally observed feature on X-ray topographs of natural diamonds.

{110} faces usually appear on larger crystals due to dissolution of original octahedral crystals. On these crystals, the {110} faces are rounded and split into four curved {hhl} faces. They are not truly crystallographic dodecahedral faces, but dodecahedroidal faces. In contrast, on much smaller diamond crystals only a few micrometers across {110} faces often appear as flat and as true dodecahedral faces, their only obvious feature being striations parallel to an edge with the neighboring {111} face. No growth layers have been observed on these faces, although they are considered to have appeared by a growth process through piling up of steps of growth layers developing on {111} faces and not by a dissolution process.

{100} faces are always rugged, and never appear as flat crystallographic faces. They are not cubic faces, but cuboid faces. Cuboid crystals with depressed cuboid faces appear due to polycrystalline growth. Analysis of the so-called center cross pattern sometimes detected by X-ray topography, cathodoluminescence or by etching method at the centers of gem quality diamond crystals has indicated hummocky, non-straight growth banding in {100} growth sectors, in contrast to the straight banding seen in the {111} growth sectors of the same crystal.

Therefore, in the natural crystallization of single crystalline type diamonds, a layer-by-layer spiral growth took place only on {111} faces, not on {110} or {100} faces. The spiral growth mechanism indicates that diamond crystals grew atomically in a low supersaturated solution under diamond stable conditions.

However, the morphologies developed in the synthetic diamonds either by solution at high pressure and temperature or by CVD are quite in contrast. In the synthetic diamonds grown from metal solution at high temperature and pressure, {100} faces develop larger under pressure and temperature conditions far from the diamond-graphite line, i.e. at larger $\Delta\mu/kT$, whereas {111} faces predominate under conditions closer to the equilibrium line, i.e. at smaller $\Delta\mu/kT$. The striking difference between natural and synthetic diamonds grown in the diamond stable region is that {100} faces appear as flat faces upon which spiral growth layers develop and growth layers on both {111} and {100} are in general much thicker than those of monoatomic height observed on natural crystals.

Diamond crystals grown by CVD under diamond metastable condition (typically 700 °C < T < 1000 °C, P < 1 atm) show markedly different morphological characteristics from both natural and synthetic diamonds grown under the diamond stable conditions. They are similar to synthetic diamonds grown in the diamond stable condition in that {111} and {100} faces develop up to the size of habit controlling faces and have growth layers but are quite different from them in a few aspects. First, {100} faces develop larger as $\Delta\mu/kT$ decreases, i.e. as the supersaturation decreases. Secondly triangular growth layers on {111} faces have the opposite orientation to the triangle of each (111) face, which is again contrary to the case of crystals formed in the diamond stable region. Thirdly, multiply twinned particles are frequently encountered among metastably grown crystals.

Figure 4.14 schematically compares the morphological characteristics of natural and synthetic crystals grown from metallic solution in the diamond stable region and those grown by CVD in the metastable regions both for external forms and surface microtopographs. The morphology of diamonds grown by CVD deviates more than that of crystals grown from metallic solutions from the morphology deduced from the PBC analysis. Natural diamond has the closest morphological characteristics to those theoretically deduced.

Synthetic diamonds deviate from the morphology deduced from the PBC analysis in the following four aspects. First, {100} faces transform from S or K-faces to an F-face and the layer growth takes place on the face. Second, the growth layers are much thicker on synthetic crystals than on natural crystals. Third, crystals grown under metastable conditions behave oppositely in respect to the morphology versus $\Delta\mu/kT$ relation to those grown under stable conditions. Fourth,

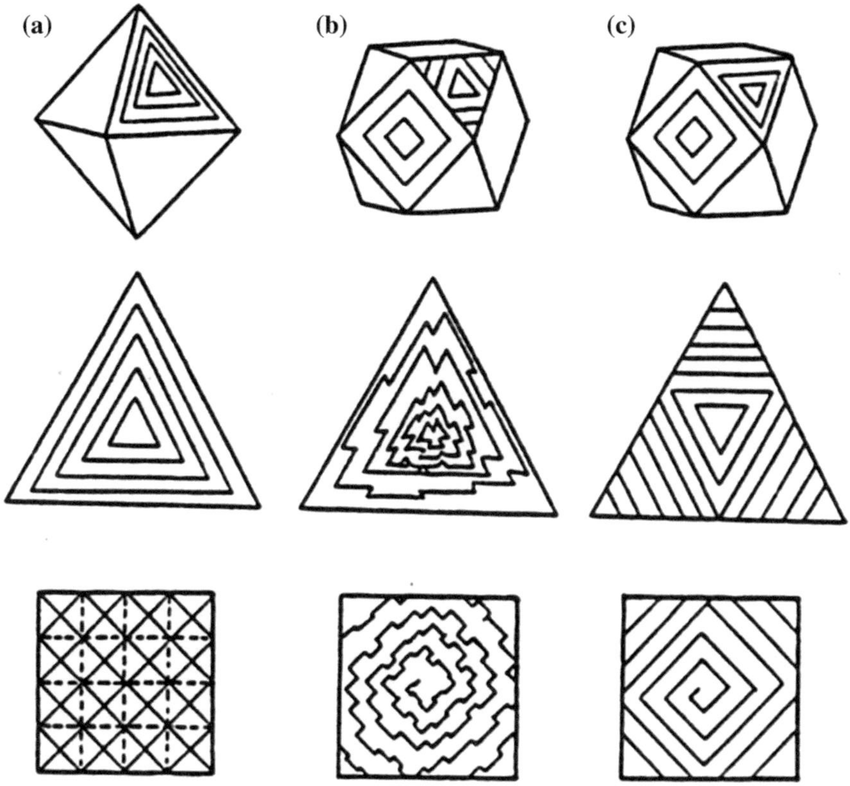

Fig. 4.14 Schematic comparison of morphological characteristics of natural diamonds (**a**), synthetic diamonds grown under diamond stable region (**b**) and those grown under diamond metastable region (**c**) [polyhedral habit (*upper*) and surface microtopographs of {111} (*middle*) and {100} (*lower*) faces] (Sunagawa 1987a)

the orientation of triangular growth layers on {111} faces are reversed on metastably grown crystals.

According to the PBC analysis of diamond structure, which is based on an assumption of atomic growth unit, the (100) face is an S or a K face, since it contains two broken PBC's in ⟨110⟩ directions. The development of the (100) face with growth layers is quite unexpected from the PBC analysis. This deviation implies that the growth mechanism of synthetic diamonds differs from that of natural diamonds. One possibility for this deviation is that some sort of surface reconstruction takes place to transform these broken PBCs to ordinary unbroken PBC's. This possibility of surface reconstruction was suggested by Van Enckevort et al. (1993). However, Sunagawa (1995) pointed out that surface reconstruction provides only two PBCs newly introduced to a (100) face, whereas a (111) contains three PBCs. Therefore, a (111) face should always be stronger than a (100) face under any growth conditions. However, in synthetic diamond growth under stable conditions, {100} can develop larger than {111} under a condition of a large driving force, which cannot be explained on the basis of a reconstruction model alone. Besides, the wavy macrosteps cannot be explained by the surface reconstruction.

In order to explain this deviation, Sunagawa (1987b, 1990) suggested that growth units of synthetic diamonds might be larger than atoms. Growth layers observed on synthetic crystals grown under diamond stable and metastable regions

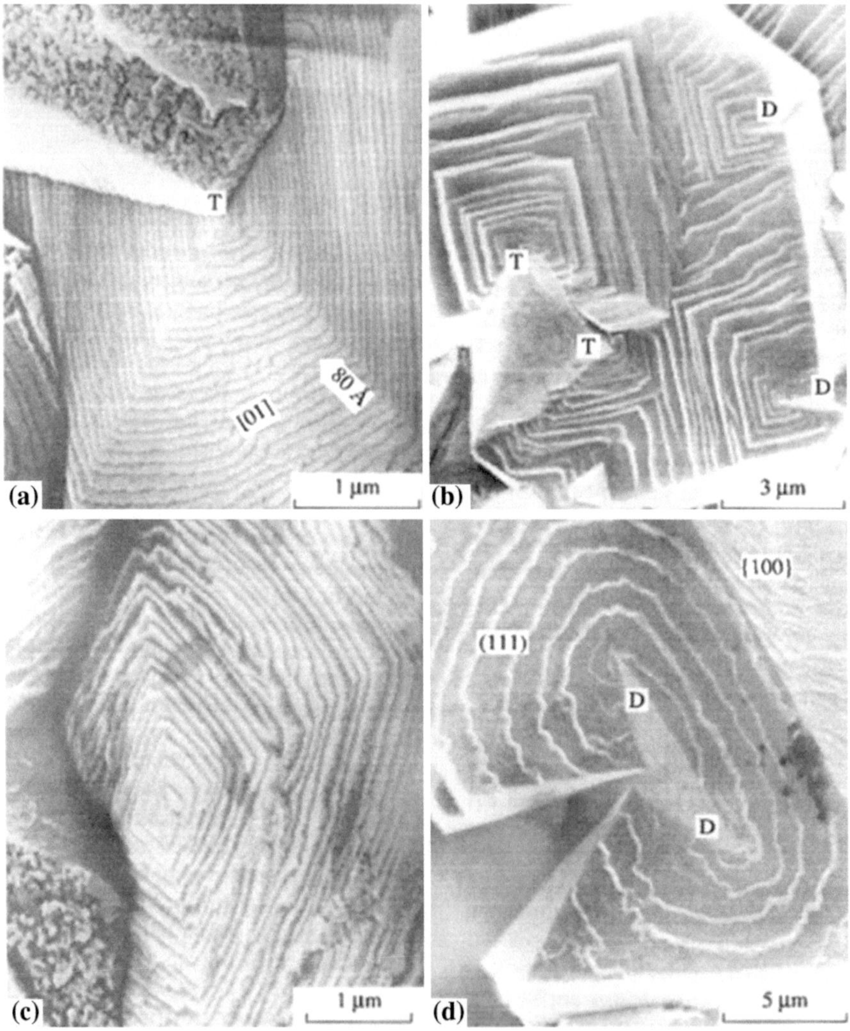

Fig. 4.15 SEM images of layered stepwise growth pyramids at diamond crystal faces {100} and {111} that characterize its growth according to the layered spiral mechanism generated by **a** tops (*T*) of overgrowing and interpenetrating crystals, **b, d** dislocation (*D*) steps on crystal faces resulting from nonparallel intergrowth, and **c** Franck-Read growth source (Samotoin 2001)

have in most cases step heights much higher than the monoatomic height, and their step morphology is wavy on their surfaces. It should be reminded that the step of monoatomic height is roughened and can be wavy but the step higher than monoatomic height is not roughened and should be straight. No molecularly thin growth layers are seen to coexist with macro-steps, which proves that macro-steps are not due to the bunching of thin steps. The observed morphological differences among three types of diamond crystals are probably due to the differences in size of the growth unit. It is probable that the growth unit becomes larger on deviating from the equilibrium condition, namely as $\Delta\mu/kT$ increases.

Figure 4.15 shows the SEM images of diamond crystal faces {100} and {111} that characterize its growth (Samotoin 2001). Based on Fig. 4.15 and other results on the microstructure evolution of diamond crystal faces, Samotoin (2001) suggested that the diamond crystal growth from hydrocarbon gas phase occurs at a level of clusters rather than atoms or molecules. He further suggested that only by assuming the diamond growth by clusters, pyramids with the step height equal to a few lattice periods can be formed on the crystal surface, and these steps are not split into lower steps. Based on the experimental observation that the smallest height of the growth steps of diamonds is about ~80 Å at all sources of layer growth, Samotoin (2004) concluded that crystals should grow by clusters comparable in size to this height.

4.8 Growth Mode and Morphology of Crystals

Depending on the mechanism by which the crystal grows, the different growth morphology is evolved. Based on his extensive observation of the growth morphology of crystals from the solution with the driving force, Sunagawa (1987b, 1990) illustrated how the growth mode and the growth morphology change with the driving force in Fig. 4.20, where the slopes for spiral growth and 2D nucleation growth (2DNG) are drawn much less steeper than the realistic slopes for convenience.

The curve ‘a’ corresponds to the growth rate versus driving force relation for the spiral growth mechanism. The curves ‘b’ and ‘c’ are respectively for the 2DNG mechanism and for the adhesive type continuous growth mechanism, which represents the ideal growth on the rough interface. The intersection of the curves a and b is denoted by σ^*. Another intersection of the curves b and c is denoted by σ^{**}. When the driving force is less than σ^*, the crystal grows mainly by spiral growth. In this region, the expected morphology is polyhedral bounded by flat low index crystal faces. When the driving force is between σ^* and σ^{**}, the crystal grows mainly by 2DNG, hopper crystals are expected to form. Hopper crystals are evolved by the so-called Berg effect which arises because supersaturation at the center of a face on a growing crystal is much lower than that at the edges and corners (Berg 1938). Berg effect can be understood by considering the cube crystal surrounded by the supersaturated medium. Since iso-concentration lines would be roughly spherical around the cube crystal, the concentration gradient would be higher near

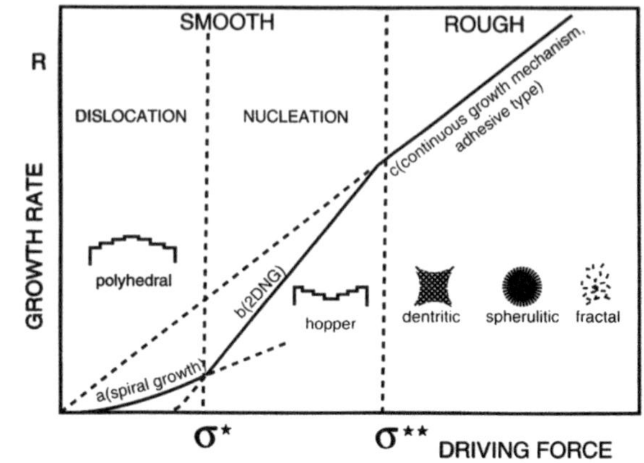

Fig. 4.16 Schematic diagram explaining the relationship of interface structure (*rough* and *smooth*), growth mechanism ((*a*) spiral growth, (*b*) 2DNG (*c*) adhesive type continuous growth) and morphologies. The morphological variations of crystals are related by growth rate *R* versus driving force (Sunagawa 1987a)

the corner but lower at the center of each face of the cube. When the driving force is larger than σ^{**} in Fig. 4.16, dendritic crystals and polycrystalline spherulitic morphologies are expected to occur.

Under much higher driving force conditions, nucleation repeatedly occur in front of the tips due to concentration fluctuation owing to nucleation, and a connected pattern appears. This corresponds to a fractal pattern. Patterns shown by diffusion limited aggregate (DLA) are such examples. It is to be noted that dendrite is a form made by a single crystal, whereas spherulite or DLA is a form made by aggregate of polycrystals.

4.9 Secondary Nucleation on the Growing Interface

In order to understand the microstructure evolution of films and nanostructures, it is important to have a clear concept of secondary nucleation. Nucleation is an irreversible process consisting of a series of atomic (molecular or ionic) additions. For nucleation to occur, the build-up of some degree of supersaturation is needed. On the rough interface, however, supersaturation does not build up because there exists no barrier for atomic attachment. Therefore, secondary nucleation cannot not occur on the rough interface.

For this reason, it is relatively easy for a material with the rough interface to be grown as a single crystal. The growth is of an adhesive or continuous type. For a rough interface, growth twins are hardly formed. For example, the solid–liquid interface of most metals are rough and growing a single crystal in a molten state is relatively easy: turbine blades of the single crystal nickel-based superalloy are commercially produced. In contrast, growth twins tend to be formed so easily on the smooth interface that special efforts must be made to prevent their formation. In addition to twins, the secondary nucleation would produce grain boundaries. The solid–liquid interface of materials with covalent or ionic bonding such as silicon, gallium arsenide and sapphire is smooth. Therefore, the single crystal growth of these materials would be rather difficult in a molten state. Although the secondary nucleation and the twin formation are allowed in the growth of a polycrystalline silicon ingot for solar cell applications, such formation is not allowed in a single crystal ingot for wafer fabrications.

Because the rough interface has no secondary nucleation, it is rather difficult to refine the grain size during solidification of cast metals, which normally consist of huge columnar structures. The grain size is determined by the impingement through the growth of independent nuclei. In order to refine the grain size, nucleation sites should be increased to increase the number of independent nuclei, for example, by adding the inoculating agent, which acts as a strong heterogeneous nucleation site.

Although most of the solid–liquid interfaces of metals are rough, most of the solid–vapor interfaces are smooth because the heat of vaporization is much larger than the heat of melting. Even the interface of metals is smooth if the deposition temperature is not sufficiently high as in the case of most thin film processes. In the case of the growth of thin films with a smooth interface, there are two factors that determine the grain size. One is the impingement of independent nuclei. The other is through secondary nucleation. As the frequency of secondary nucleation increases, the grain size would decrease. Let's examine how the grain size is determined by secondary nucleation.

Consider Fig. 4.9 but the abscissa is now the supersaturation ratio instead of supercooling. The smooth interface, which has an appreciable barrier for atomic attachment, has different growth modes depending on the supersaturation ratio as shown in Fig. 4.9. At low supersaturation, the growth does not occur in the absence of defects such as screw dislocations. In this regime of low supersaturation, only defect-assisted growth can occur. 2D nucleation would hardly take place until the supersaturation increases to the critical supersaturation ratio $(P/P_e)_{crit}$ for the onset of 2D nucleation. Above this value, the rate of 2D nucleation increases abruptly to a high value like a step function (Hirth and Pound 1963) and then 'kinetic roughening' would soon occur as shown in Fig. 4.9.

Once kinetic roughening occurs, the growth behaviour of the smooth interface changes to that of the rough interface. Therefore, above the supersaturation ratio of kinetic roughening, the secondary nucleation would not take place. It should be noted that the secondary nucleation has a higher barrier than the 2D nucleation and would not take place below the critical supersaturation ratio for the onset of 2D nucleation. Therefore, secondary nucleation can take place in a very narrow supersaturation range higher than the onset of 2D nucleation and lower than the onset of kinetic roughening.

In order to estimate the minimum grain size, which can be achieved by secondary nucleation in the thin film process, the ratio of 2D to secondary nucleation should be evaluated at the supersaturation ratio of kinetic roughening, which is given by (4.14). By putting typical values, Hirth and Pound (1963) estimated the ratio of the 2D to secondary nucleation rate for the stacking fault, which has the interface energy of 0.02 J/M^2 and for the incoherent grain boundary, which has the interface energy of 0.2 J/M^2. The estimations are as follow.

$$\left(\frac{I_{2D}}{I_{3D}}\right)_{kinetic\ roughening} \cong 4, \quad \text{for}\ \sigma = 0.02\,\text{J/M}^2(\text{Stacking Fault}) \tag{4.16}$$

$$\left(\frac{I_{2D}}{I_{3D}}\right)_{kinetic\ roughening} \cong 10^8, \quad \text{for}\ \sigma = 0.2\,\text{J/M}^2(\text{Grain Boundary}) \tag{4.17}$$

These estimations indicate that although stacking faults or twins would form every 4 atomic layer on average, the secondary nucleation with incoherent grain boundaries would occur after the growth of hundreds of millions of atomic layers. This determines the minimum grain size of films contributed by the secondary nucleation without the impingement of the independent nuclei. Even if the errors made in the estimation are considered, the predicted minimum grain size by secondary nucleation is at least larger than a few micrometres.

With this theoretical background on the secondary nucleation, let's examine two microstructures which are commonly observed in thin film processes: one is nanocrystals embedded in an amorphous matrix and the other is nanostructured films.

4.10 Nanocrystalline or Microcrystalline Grains Embedded in an Amorphous Matrix

Nanocrystalline or microcrystalline grains embedded in an amorphous matrix are observed extensively in many thin film processes such as CVD, laser ablation, sputtering and electrodeposition. Figure 4.17 shows a typical example of such structures where silicon nanocrystals are embedded in the amorphous silicon matrix. The sample was prepared by PECVD at the substrate temperature of 200 °C with a gas mixture of dichlorosilane and hydrogen (Monroy et al. 2011).

This microstructure is puzzling in two aspects: one is how the secondary nucleation can occur on the amorphous phase which has a rough interface and the other is how the crystalline silicon can be formed at the substrate temperature as low as 200 °C. The secondary nucleation cannot occur on the amorphous phase because every sites are kinks and the supersaturation cannot build on the rough interface. Besides, it is well known that the diffusion of silicon atoms is so slow below 600 °C that the amorphous silicon is expected when growth occurs at temperatures below 600 °C.

Therefore, all the silicon nanocrystals in Fig. 4.17 cannot have formed by secondary nucleation but must have come from the gas phase. This means that silicon nanocrystals shown in Fig. 4.17 were originally the gas phase nuclei. In relation to

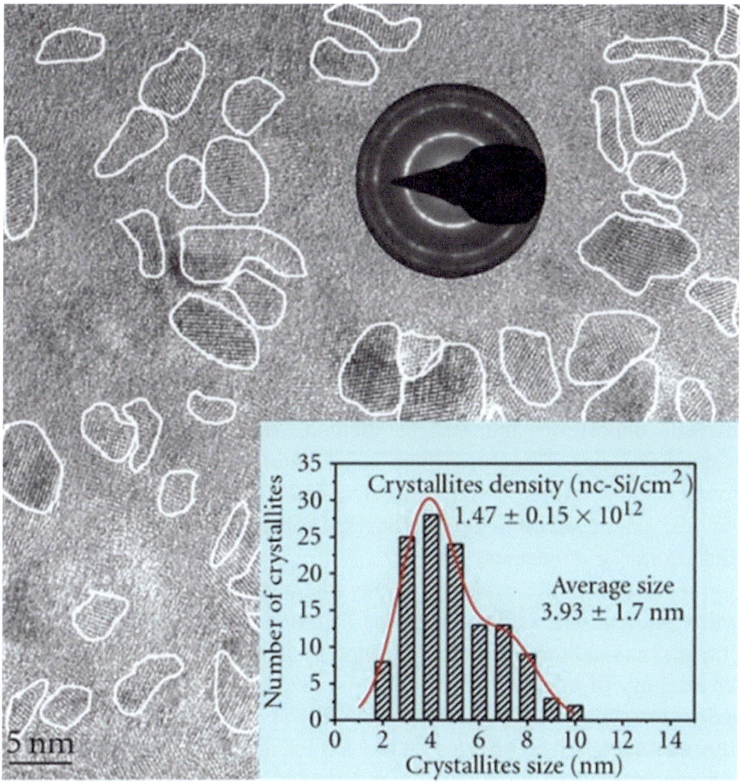

Fig. 4.17 High Resolution TEM (HRTEM) image of silicon nanocrystals embedded in an amorphous matrix deposited by PECVD. The circular insert shows the small area diffraction patterns of the sample and the square insert shows the size distributions and number densities of the silicon nanocrystalline inclusions (Monroy et al. 2011)

this possibility, Cabarrocas and his colleagues (Roca i Cabarrocas 2000, 2002, 2004; Pere Roca i et al. 2004; Johnson et al. 2008; Roca i Cabarrocas et al. 2008) extensively studied the incorporation of the gas phase generated silicon nanocrystals in the growing film and called it a 'polymorphous' structure. Besides, the gas phase nucleation in the PECVD process has also been extensively studied by Shiratani and his colleagues (Shiratani et al. 1994, 1996; Koga et al. 2000; Watanabe et al. 2001; Nunomura et al. 2006).

Even if it is accepted that silicon nanocrystals in Fig. 4.17 were formed in the gas phase, it is difficult to explain their crystallinity because the temperature of the gas phase is not so high as to produce crystalline silicon in the PECVD process. It can be explained if it is assumed that gas phase nuclei are electrically charged because charged nanoparticles (CNPs) are liquid-like as mentioned in Chap 1.

This scenario is not limited to Fig. 4.17 but can be applied to all the microstructures of nanocrystals embedded in an amorphous matrix prepared by CVD, PVD and electrodeposition. Considering that nanocrystals embedded in an amorphous matrix are so commonly observed, the film growth by non-classical crystallization seems to be general.

4.11 Nanocrystalline Films

Nanocrystalline or microcrystalline films are widely fabricated by CVD, laser ablation, sputtering, electro-spraying deposition, aerosol deposition, chemical bath deposition, sol-gel method and electrodeposition. In the electro-spraying deposition, aerosol deposition and sol-gel method, starting materials are nano or micron-sized particles. Therefore, there is no doubt that the building block for the film growth is nano or micron-sized particles. In CVD, laser ablation and sputtering, however, the nanocrystalline or microcrystalline films are believed to grow by individual atoms or molecules.

Figure 4.18 shows the nanocrystalline diamond deposited from an Ar–CH_4 microwave plasma. The lattice image of diamond is clearly shown. Figure 4.19 shows the nanocrystalline NiO film deposited by DC reactive magnetron sputtering. Another category of nanocrystalline films is a cauliflower structure, which is also observed in most thin film processes. Cauliflower structures are characterized by numerous nanometer-sized nodules on the surface as shown in Figs. 4.20 and 4.21.

These nanocrystalline films have been believed to grow by repeated secondary nucleation with the growth unit of individual atoms or molecules. However, as

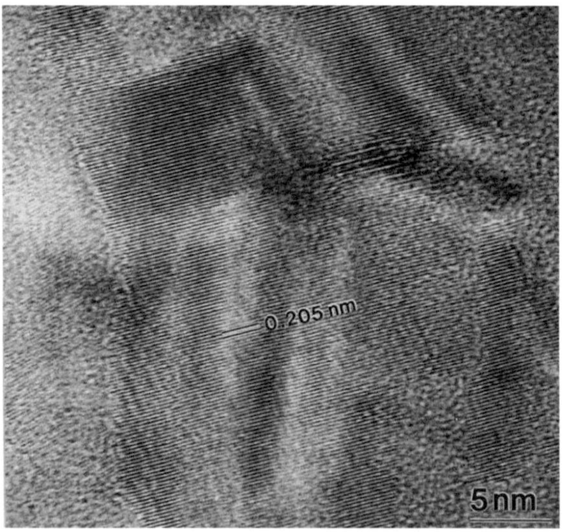

Fig. 4.18 A HRTEM image of the nanocrystalline diamond film produced from an Ar–CH_4 microwave plasma at 100 Torr (Zhou et al. 1998)

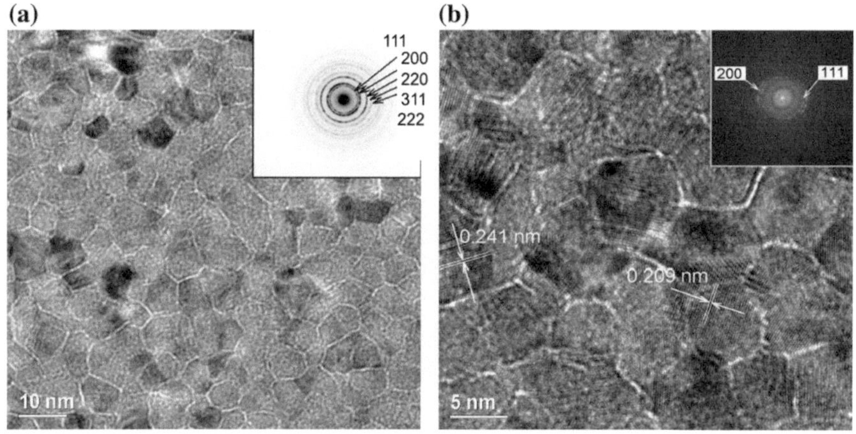

Fig. 4.19 TEM images of NiO films deposited by DC reactive magnetron sputtering from a Ni target. **a** TEM image shows the nanostructure at lower magnification and selected area electron diffraction (SAED) pattern in *inset* and **b** HRTEM image reveals the lattice fringes inside the selected grains of 200 and 111 planes of NiO with fast Fourier transformation (FFT) diffraction pattern in *inset* (Hotový et al. 2007)

Fig. 4.20 SEM image of the cauliflower structure of silicon deposited for 10 min with 1 % SiH_4–1 % HCl–98 % H_2 gas mixture, 1123 K substrate temperature, and 100 torr reactor pressure (Hwang et al. 2000)

analyzed by (4.17), the secondary nucleation cannot occur so frequently as to produce the nanocrystalline structure. Therefore, the nanocrystalline structure should be formed by the impingement of independent nuclei. More specifically the gas phase nuclei should be incorporated into films. Therefore, the cauliflower structure, which is evolved so frequently in thin film processes, provides the indirect evidence that the films grow by non-classical crystallization. It can be argued how the nanoparticles formed in the gas phase nuclei can produce such dense films without voids or pores. Again, this is attributed to the presence of charge in the nanoparticles.

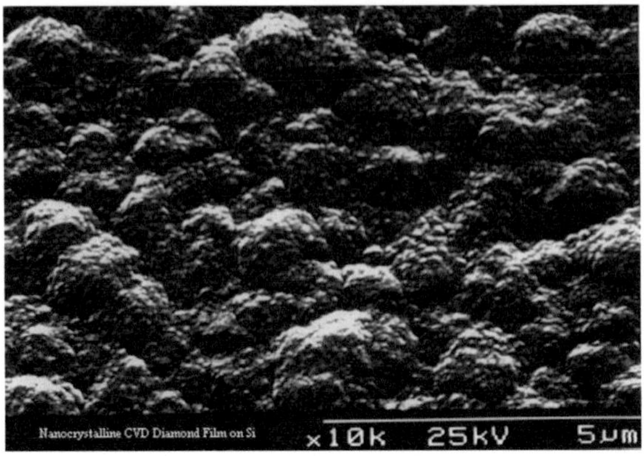

Fig. 4.21 Nanocrystalline film, exhibiting 'cauliflower' morphology, typical of diamond grown under high (>2 %) methane concentrations (May 1995)

References

Berg WF (1938) Crystal growth from solutions, Proc R Soc 164A:79–95. doi:10.1098/rspa.1938.0006.

Burton W-K, Cabrera N, Frank F (1951) The growth of crystals and the equilibrium structure of their surfaces. Phil Trans R Soc A 243(866):299–358.

Flemings M (1974) Solidification processing, McGraw-Hill Series in Material Science and Engineering.

Frank F (1949) On the equations of motion of crystal dislocations. Proc Phys Soc Sect A 62(2):131.

Hartman P (1973) Crystal growth: An introduction. North Holland, Amsterdam.

Heyraud JC, Métois JJ (1987) Equilibrium shape of an ionic crystal in equilibrium with its vapour (NaCl). J Cryst Growth 84(3):503–508. doi:http://dx.doi.org/10.1016/0022-0248(87)90282-X.

Hirth JP, Pound GM (1963) Condensation and evaporation. Pergamon Press, Oxford.

Hotový I, Liday J, Spiess L, Romanus H, Čaplovičová M, Búc D, Sitter H, Bonanni A (2007) TEM investigations of Au–NiO nanocrystalline thin films as gas sensing material. J Electr Eng 58(6):347–350.

Hwang N-M, Jeon I-D, Kim D-Y (2000) Charged cluster model as a new paradigm of crystal growth. J Ceram Proc Res 1:34–44.

Johnson EV, Patriarche G, Roca i Cabarrocas P (2008) Directional growth of Ge on GaAs at 175 °C using plasma-generated nanocrystals. Appl Phys Lett 92(10):103108. doi:10.1063/1.2895636.

Koga K, Matsuoka Y, Tanaka K, Shiratani M, Watanabe Y (2000) In situ observation of nucleation and subsequent growth of clusters in silane radio frequency discharges. Appl Phys Lett 77(2):196–198.

Kossel W (1927) Zur Theorie des Kristallwachstums. Nachrichten von der Gesellschaft der Wissenschaften zu Göttingen, Mathematisch-Physikalische Klasse 1927:135–143.

Kossel V-W (1934) Zur Energetik von Oberflächenvorgängen. Ann Phys 413(5):457–480.

May PW (1995) CVD diamond: a new technology for the future? Endeavour 19(3):101–106. doi:http://dx.doi.org/10.1016/0160-9327(95)97494-S.

Monroy BM, Remolina A, García-Sánchez MF, nchez, Ponce A, Picquart M, Santana G (2011) Structure and optical properties of silicon nanocrystals embedded in amorphous silicon thin films obtained by PECVD. J Nanomater 2011:1–9. doi:10.1155/2011/190632.

Nunomura S, Kita M, Koga K, Shiratani M, Watanabe Y (2006) In situ simple method for measuring size and density of nanoparticles in reactive plasmas. J Appl Phys 99(8):083302.

Pere Roca i C, Chaâbane N, Kharchenko AV, Tchakarov S (2004) Polymorphous silicon thin films produced in dusty plasmas: application to solar cells. Plasma Phys Contr F 46(12B):B235.
Reed-Hill RE (1973) Physical metallurgy principles. D. Van Nostrand Co., New York.
Roca iC, P. (2000) Plasma enhanced chemical vapor deposition of amorphous, polymorphous and microcrystalline silicon films. J Non-Cryst Solids 266–269, Part 1:31–37. doi:http://dx.doi.org/10.1016/S0022-3093(99)00714-0.
Roca i Cabarrocas P (2002) Plasma enhanced chemical vapor deposition of silicon thin films for large area electronics. Curr Opin Solid State Mater Sci 6(5):439–444. doi:http://dx.doi.org/10.1016/S1359-0286(02)00112-2.
Roca i Cabarrocas P (2004) New approaches for the production of nano-, micro-, and polycrystalline silicon thin films. Phys Status Solidi C 1(5):1115–1130. doi:10.1002/pssc.200304328.
Roca i Cabarrocas P, Djeridane Y, Bui VD, Bonnassieux Y, Abramov A (2008) Critical issues in plasma deposition of microcrystalline silicon for thin film transistors. Solid-State Electronics 52(3):422–426. doi:http://dx.doi.org/10.1016/j.sse.2007.10.028.
Samotoin N (2004) Morphology, intergrowths, and growth mechanisms of diamond crystals at different stages of their formation from gases. Geochem Int 42(2):134–144.
Samotoin N (2001) Mechanisms of layered spiral growth of synthetic diamond crystals based on SEM data. In: Doklady Earth Sciences. vol 8. Interperiodica Po box 1831, Birmingham, AL 35201–1831 USA, pp 925–928.
Shiratani M, Kawasaki H, Fukuzawa T, Tsuruoka H, Yoshioka T, Watanabe Y (1994) Study on growth processes of particulates in helium-diluted silane RF plasmas using scanning electron microscopy. Appl Phys Lett 65(15):1900–1902.
Shiratani M, Kawasaki H, Fukuzawa T, Yoshioka T, Ueda Y, Singh S, Watanabe Y (1996) Simultaneous in situ measurements of properties of particulates in RF silane plasmas using a polarization-sensitive laser-light-scattering method. J Appl Phys 79(1):104–109.
Stranski I (1928) Zur theorie des kristallwachstums. Z Phys Chem 136(259):259–278.
Sunagawa I (1987a) Morphology of crystals morphology of crystals, vol PART B : Fine particles, minerals and snow. Terra, Tokyo.
Sunagawa I (1987b) Morphology of minerals. Morphology of crystals 2:511–587.
Sunagawa I (1990) Growth and morphology of diamond crystals under stable and metastable conditions. J Cryst Growth 99(1–4 pt 2):1156–1161. doi: 10.1016/S0022-0248(08)80100-5.
Sunagawa I (1995) The distinction of natural from synthetic diamonds. J Gemmol 24(7):485–499.
van Beijeren H (1977) Exactly solvable model for the roughening transition of a crystal surface. Phys Rev Lett 38(18):993–996.
Van Enckevort W, Janssen G, Vollenberg W, Schermer J, Giling L, Seal M (1993) CVD diamond growth mechanisms as identified by surface topography. Diamond Relat Mater 2(5):997–1003.
Watanabe Y, Shiratani M, Koga K (2001) Formation kinetics and control of dust particles in capacitively-coupled reactive plasmas. Phys Scripta T89:29–32.
Zhou D, McCauley TG, Qin LC, Krauss AR, Gruen DM (1998) Synthesis of nanocrystalline diamond thin films from an Ar-CH4 microwave plasma. J Appl Phys 83(1):540–543.

Chapter 5
Diamond Synthesis at Low Pressure

All synthetic processes of materials are irreversible, which can be approached thermodynamically and kinetically. When we say that we understand some phenomena, it normally means that we understand their thermodynamics and kinetics. The process of diamond CVD is a good example, which shows how thermodynamics and kinetics can be applied to understand the underlying principle of diamond synthesis at low pressure.

5.1 High Pressure Synthesis of Diamond

Since it was established in 1797 that diamond is a crystalline form of carbon, the diamond synthesis began to attract attention. The attempt to synthesize diamond was first made in 1880 by Hannay (1879) but not successful. The second attempt was reported in 1894 by Moissan (1894). The attempts to grow diamond from the gas phase were also made in 1911. The starting material was acetylene in the presence of mercury vapour at 1000 °C. In the 1940s and 1950s, there were a few claims of growing diamond using an acetylene torch or carbon arcs. It should be noted that the acetylene torch and carbon arcs are known to produce diamond films these days.

The first successful diamond synthesis was reported by researchers of General Electric Company Research Laboratory with patents filed in 1954, announced in February of 1955 and published in 1955 (Bundy et al. 1955). Here, diamond was synthesized at high temperature and pressure, where diamond is more stable than graphite. Shortly after, a second success was reported by the Swedish firm of ASEA in 1955 (Liander and Lundblad 1955).

© Springer Science+Business Media Dordrecht 2016

N.M. Hwang, *Non-Classical Crystallization of Thin Films and Nanostructures in CVD and PVD Processes*, Springer Series in Surface Sciences 60,
DOI 10.1007/978-94-017-7616-5_5

5.2 Low Pressure Synthesis of Diamond (Old CVD Process)

Soon after, the diamond was successfully synthesized by gas phase reaction at low pressure by Eversole (1962), who was a researcher of the Union Carbide Corporation. Here, diamond was grown on the pre-existing diamond seed crystal. Therefore, diamond growth occurred as epitaxial thin films on diamond substrates. Eversole's pioneering work on the CVD method was taken up by Angus et al. (1968) at Case Western Reserve University (USA) up to 1970s, who made systematic and sustained efforts. In the presence of a diamond seed, diamond need not nucleate but graphite need to nucleate. Therefore, diamond has only a growth barrier whereas graphite has a nucleation barrier.

Since the theory of nucleation and growth is well formulated now, Eversole's idea of diamond synthesis can be understood more clearly by comparing the barriers of nucleation and growth. Assuming the spherical shape and the isotropic surface energy of the graphite nucleus, the barrier of graphite nucleation on the diamond seed would be given as [see (3.12) in Chap. 3]

$$\Delta G^*_{3D,gra} = \frac{2 - 3\cos\theta + \cos^3\theta}{4}\frac{16\pi\sigma^3}{3(\Delta f^{gas\rightarrow gra})^2} \tag{5.1}$$

where θ is the contact angle of the graphite nucleus with the diamond seed, σ is the surface energy of graphite and $\Delta f^{gas\rightarrow gra}$ is the driving force for the precipitation of graphite from the gas phase.

Assuming the disc shape of the 2-D nucleus of diamond, the growth barrier of diamond would be given as [see (4.8) in Chap. 4]

$$\Delta G^*_{2D,dia} = \frac{\pi\varepsilon^2}{h\Delta f^{gas\rightarrow dia}} \tag{5.2}$$

where ε is the step free energy of diamond, h is the height of the 2-D nucleus and $\Delta f^{gas\rightarrow dia}$ is the driving force for the precipitation of diamond from the gas phase. It should be noted that the step free energy ε in (5.2) is much smaller than the surface energy σ in (5.1) as shown in Fig. 4.10 in Chap. 4. Although the driving force for the diamond precipitation from the gas phase is slightly smaller than that for the graphite precipitation, the growth barrier of diamond in (5.2) is much smaller than the nucleation barrier of graphite on the diamond seed in (5.1).

If the ratio of the 2D to 3D nucleation rate, which is similar to the concept for the frequency of the secondary nucleation, is given by

$$\frac{I_{2D,dia}}{I_{3D,gra}} = \exp\left(\frac{\Delta G^*_{3D,gra} - \Delta G^*_{2D,dia}}{kT}\right) \tag{5.3}$$

Considering that the ratio of 2D to secondary nucleation in (4.17) in Chap. 4 is $\sim 10^8$, the value of (5.3) should not be less than $\sim 10^8$.

Therefore, the diamond growth on a diamond seed would be kinetically more favourable than graphite nucleation at low pressure where diamond is less stable than graphite. Eversole's idea is sound both thermodynamically and kinetically. However, there was a problem in the growth rate, which was only ~ 10 Å/h. This growth rate was too low to be considered for any commercial application. Besides, graphite tends to be co-deposited. Once graphite is co-deposited, the growth rate of graphite is so much faster than that of diamond, the entire surface of diamond seeds would be covered with graphite. Then, diamond cannot be grown any more. For further growth of diamond, the surface layer of graphite should be removed by etching. Therefore, the process was cyclic: growth of diamond and etching of graphite.

Then, why was the growth rate of diamond so low? Why was graphite co-deposited? What was wrong with Eversole's idea? These questions in the old diamond CVD process should be answered to understand the new diamond CVD process using the gas activation such as plasma and hot filament.

5.2.1 Possibility of Gas Phase Nucleation in the Old CVD Diamond Process

Considering (5.3), the diamond growth is expected to be dominant over graphite nucleation. However, this is true only when the reaction takes place only on the surface of the diamond seed. In other words, Eversole's idea works only under the condition that the gas phase nucleation does not take place. If gas phase nucleation takes place, the diamond seed would not provide any beneficial effect and graphite would nucleate dominantly over diamond. If the graphite gas phase nuclei fall on the diamond seed, the graphite will start to grow much faster than diamond, covering the entire surface.

As mentioned in Chap. 3, there are numerous heterogeneous sites such as ions in the gas phase. It should be reminded that ions are continuously generated by cosmic rays and natural radio activity and a steady state concentration of $\sim 10^3$ ion pairs cm^{-3} is maintained in air.

To inhibit the gas phase nucleation, the supersaturation for precipitation of solid carbon should be controlled very carefully. If the supersaturation is too high, an avalanche of gas phase nucleation would take place. If the supersaturation is too low, the growth rate of diamond would be too low. An optimum supersaturation can be made between the two extremes to produce the maximum growth rate of diamond. The optimized growth rate with cyclic growth and etching seemed to be ~ 10 Å/h. This is why this cyclic process of diamond synthesis had been studied only by a few scientists. It should be remembered that in order to inhibit the gas phase nucleation, the film growth rate is impractically low, which also applies to other thin film growth by CVD

5.3 Diamond Synthesis at Low Pressure (New CVD Process Using Gas Activation)

A research group led by Derjaguin at the Physical Chemistry Institute in Moscow also continued the diamond synthesis on diamond seed crystals using hydrocarbon at low pressure (Derjaguin et al. 1968). In order to increase the growth rate of diamond in this cyclic process, Derjaguin and Fedoseev (1977), Varnin et al. (1977) tried to increase the etching rate of graphite using atomic hydrogen. To produce atomic hydrogen, they used hot tungsten wires.

While synthesizing diamond using hot wires, some unexpected results were obtained (Spitsyn et al. 1981). First, they could increase the growth rate of diamond about 1000 times as high as that of the old diamond CVD process. Second, they did not need to use the diamond seeds any more. Faceted diamond crystals could be grown by a chemical transport reaction (CTR) method. They attributed such unexpected changes to atomic hydrogen, which etches graphite much faster than diamond.

The next significant progress in the new CVD diamond process took place in Japan. A research group led by Setaka at the National Institute for Research in Inorganic Materials (NIRIM) in Tsukuba started the diamond synthesis at low pressure since 1972. Initially they tried to decompose methane gas thermally on diamond powders following the method by Derjaguin et al. (1968) and Angus et al. (1968) without much success. Inspired by the Spitsyn et al's synthesis of faceted diamond crystals using atomic hydrogen but instead of using graphite as a carbon source, they used a mixture gas of hydrogen and hydrocarbon expecting that not only atomic hydrogen but also hydrocarbon radicals would be effective for diamond growth. They succeeded in diamond growth at low pressures using hot filaments (Matsumoto et al. 1982a, b).

They also succeeded in synthesizing diamond by using microwave plasma (Kamo et al. 1983). In the case of RF plasma, their initial result with 400 kHz RF was not successful but by increasing the RF frequency to 4 MHz, the result was partially successful (Takagi 1983). Then, by using the RF frequency to 13.56 MHz they could produce faceted diamond crystals (Matsumoto 1985).

After these methods, many other methods such as thermal plasma by RF induction (Matsumoto et al. 1987), DC plasma (Suzuki et al. 1987; Kurihara et al. 1988) combustion flame (Hirose and Mitsuizumi 1988), inductively-coupled plasma (Okada et al. 1999), capacitively-coupled plasma (CCP) (Chattopadhyay and Matsumoto 1995; Asakura et al. 1998) were developed. Thanks to the effort to increase the growth rate of diamond films, the growth rate as high as 1 mm/h was reported (Otake et al. 1990).

5.3.1 CVD Diamond Reactor and Processing Conditions

Figure 5.1 shows the schematic of the HFCVD reactor that has been used to synthesize diamond in our group. A water-cooled substrate is placed less than ~10 mm below the tungsten hot filament. The gas mixture of CH_4 and H_2 is supplied to the hot filament typically with the total flow rate of 100 standard cubic centimetre per minute (SCCM). The concentration of CH_4 typically ranges between 0.5 and 3 %. The reactor pressure is typically 20 Torr. The thermocouple tip is placed near the substrate to measure the substrate temperature, which typically ranges between 800 and 1100 °C. The two-colour pyrometer is used through the window to measure the filament temperature, which typically ranges between 2000 and 2200 °C.

Figure 5.2 shows the scanning electron microscopy (SEM) images of the diamond deposited on the silicon substrate. Figure 5.2a, b show that the film was not continuous after deposition of 2 h but continuous after deposition of 8 h. The film quality can be examined qualitatively by the SEM morphology. Raman spectroscopy, which has a diamond line at 1332 cm^{-1} and a graphite line at 1582 cm^{-1}, is used for quantitative analysis (see Table 5.1).

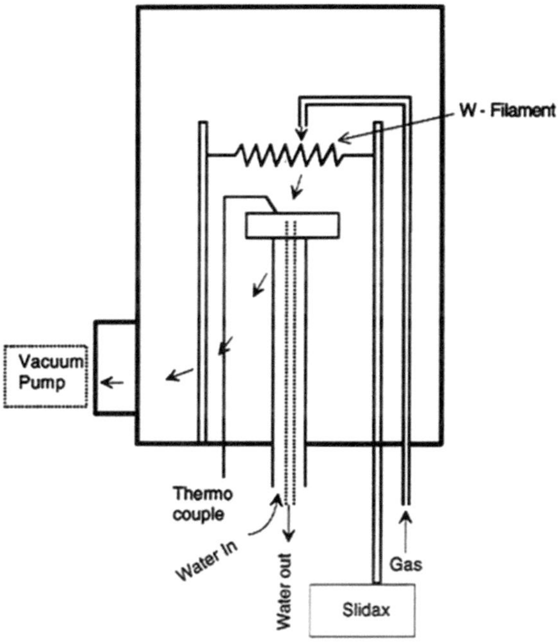

Fig. 5.1 Schematic of the HFCVD diamond reactor

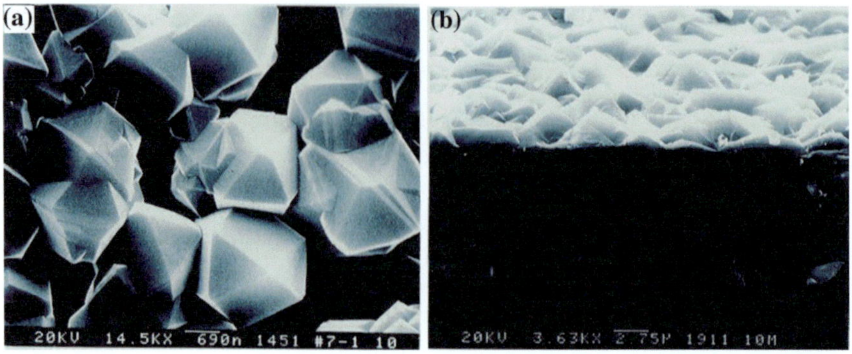

Fig. 5.2 SEM images showing diamond deposited on the silicon substrate at the substrate temperature of 990 °C and the filament temperature of 2100 °C under 20 Torr for the gas mixture of 1 % CH_4–99 % H_2 with the deposition time of **a** 2 h and **b** 8 h

Table 5.1 Peaks expected in the Raman spectrum of CVD diamond or carbon (Prawer and Nemanich 2004)

Position (cm^{-1})	Typical FWHM (cm^{-1})	Assignment
520	3–5	First-order silicon Raman peak
1100–1150	40–80	Most likely transpolyacetylene at grain boundaries; Often observed in nanocrystalline CVD diamond films
1332	5–10	First-order diamond Raman line
1345	250	sp^2 amorphous carbon (the D peak)
1430–1470	80	Most likely transpolyacetylene at grain boundaries
1520–1580	100	sp^2 amorphous carbon (the G peak)

5.3.2 *Phase Diagram of Carbon*

Although the diamond synthesis at low pressure has been studied extensively, its understanding is still lacking. There are many puzzling phenomena occurring in the diamond synthesis. One of them is how diamond can be synthesized at low pressure where graphite is more stable. Figure 5.3 shows the phase diagram of carbon.

In order for diamond to be stable, the pressure should be higher than at least a few GPa. Therefore, at pressure for the synthesis of CVD diamond, definitely graphite is more stable than diamond. Then, here comes a fundamental question. Is the synthesis of CVD diamond contradictory to the thermodynamic and kinetic law or not? One thing for sure is that all the phenomena in the universe follow the thermodynamic and kinetic law without exception. Therefore, the synthesis of CVD diamond cannot be contradictory to the thermodynamic and kinetic law. If it appears to do so, there must be something we are missing and we need to find out what it is. The next section will deal with this fundamental problem.

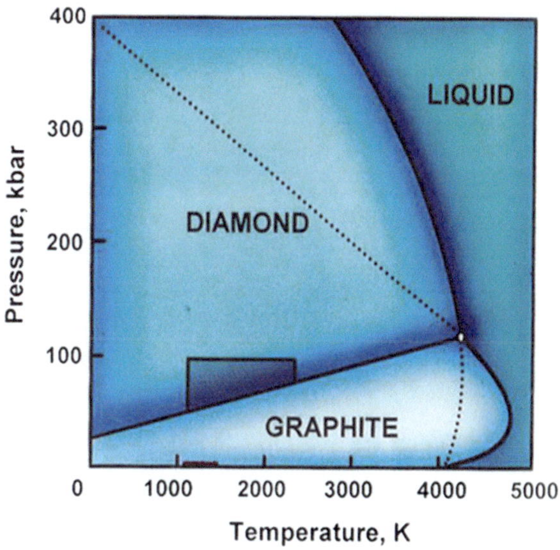

Fig. 5.3 Phase diagram of carbon showing the stable regions of graphite and diamond (Bundy 1980)

5.4 Thermodynamics and Kinetics of the Diamond Synthesis at Low Pressure

In the synthesis of CVD diamond, a mixture of hydrogen and hydrocarbon is used. In the CVD reactor, hydrocarbon would decompose into hydrogen and carbon, which will be the source of diamond. Typically, methane (CH_4) is used for hydrocarbon. We can think of the three states that carbon can exist: in the gas phase decomposed from hydrocarbon, in diamond and in graphite. Carbon in graphite is most stable and carbon in the gas phase is least stable. The stability of carbon in diamond is in the middle. The stability of carbon in each state can be represented by its chemical potential (μ_C). Then the relative stability of carbon in each state can be expressed as

$$\mu_C^{graphite} < \mu_C^{diamond} < \mu_C^{gas} \tag{5.4}$$

Then, the problem is why carbon decomposed from hydrocarbon transforms to diamond rather than to more stable graphite. Since μ_C^{gas} is higher than both $\mu_C^{diamond}$ and $\mu_C^{graphite}$, the second law of thermodynamics says that carbon in the gas phase can transform irreversibly either to graphite or to diamond. In other words, carbon in the gas phase have two choices in its kinetic path: one is the path to graphite and the other is the path to diamond. The two paths are kinetically parallel. When nature has two choices of kinetically parallel paths, which path does she take? According to the kinetic law, nature takes the path of the lower kinetic barrier.

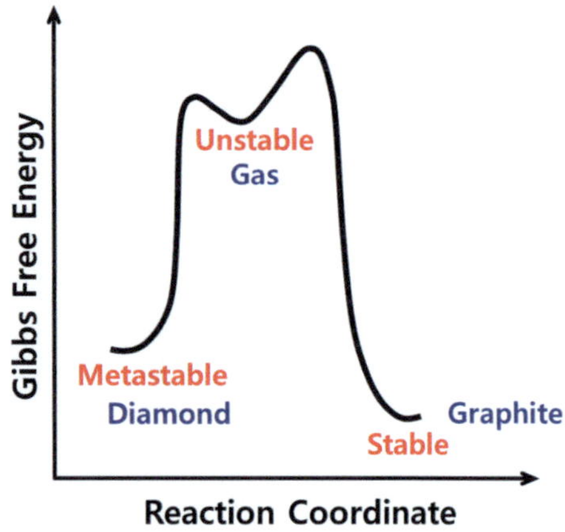

Fig. 5.4 Thermodynamic and kinetic description for the dominant formation of metastable diamond over stable graphite (Hwang and Lee 2010)

Therefore, if the kinetic barrier for diamond formation from the gas phase is lower than that for graphite formation, the less stable diamond can be formed at low pressure. This can be explained by the diagram in Fig. 5.4.

The ordinate in Fig. 5.4 can be Gibbs free energy or chemical potential. Figure 5.4 shows that even though diamond is less stable than graphite, its kinetic barrier from gas is lower than that of graphite. Then, diamond can form dominantly over graphite. The problem that remains to be solved is to check if this condition is satisfied in the CVD diamond process using the gas activation.

It should be noted that there are many examples of metastable phase formation that satisfies the condition of Fig. 5.4 (Mason 1958; Ishihara et al. 1985; Garvie 1965; Hori et al. 1984; Hurle 1993). For example, in the Fe-C system graphite is more stable than cementite (Fe_3C). However, when carbon steel is cooled from ~900 °C, where the austenite phase is stable, the austenite is decomposed into ferrite and cementite. Stable graphite cannot form because the nucleation barrier for graphite is much higher than that for cementite. As another example, when nucleation from supercooled water vapour takes place below the freezing point of water, water is nucleated first and then, after appreciable growth, transformed to ice (Mason 1958).

Such phenomena, which had been observed so commonly in many solutions by Ostwald, were called 'Ostwald stage rule' (Tammann and Mehl 1925). On the other hand, metastable tetragonal zirconia is nucleated dominantly over stable monoclinic zirconia when precipitated from the gas phase or the solution (Ishihara et al. 1985; Garvie 1965). In these cases, a clear answer was made as to why the metastable phase is formed dominantly over the stable one. The formation of the metastable phase is explained by the capillary effect of the small particles, particularly in the nucleation stage (Ishihara et al. 1985; Garvie 1965). The capillary pressure built up in the nuclei can be so large that the high-pressure phase of a smaller molar volume tends to become more stable than the low pressure phase of a larger molar volume.

If the capillary effect is roughly applied to the low-pressure synthesis of diamond with the spherical particle with a radius (r) of 1 nm and an isotropic diamond surface energy (σ) of 3.7 J/M^2 (Belokurov 1960), the pressure (P) built up inside the particle given by the Laplace equation ($\Delta P = 2\ \sigma/r$) becomes 7400 MPa. Under such a high capillary pressure, the stability of diamond may be comparable to that of graphite. This analysis implies that we need to analyse in more detail the possibility of the dominant formation of metastable diamond.

Usually, a kinetic barrier in Fig. 5.4 corresponds to the nucleation barrier. Therefore, the nucleation barrier between diamond and graphite should be compared. Before the nucleation barrier is compared, the thermodynamic stability between diamond and graphite in their nucleation stage needs to be compared.

5.5 Comparison of Stability Between Diamond and Graphite in the Nucleation Stage

In order to compare the stability between diamond and graphite in their nucleation stage, their nucleation curves should be compared. The free energy of nucleation consists of the driving force for precipitation and the surface energy of the nucleus. For rough comparison, we will use the reported values of 3.7 and 3.1 J/M^2, respectively, for isotropic surface energies of diamond (Belokurov 1960) and graphite (Fedosayev et al. 1989). It should be noted that the surface energy of diamond is higher than that of graphite. And the driving force for the precipitation of diamond from the gas phase is smaller than that of graphite. Therefore, the free energy of diamond is higher than that of graphite in all ranges of radius as shown in Fig. 5.5.

It appears that diamond becomes increasingly less stable than graphite as the size gets smaller. Judging only from Fig. 5.5, the diamond formation cannot be explained by the conventional approach of metastable phase formation. However, graphite and diamond nuclei of the same radius have different number of atoms: the diamond nucleus has a larger number of atoms than graphite because molar volume

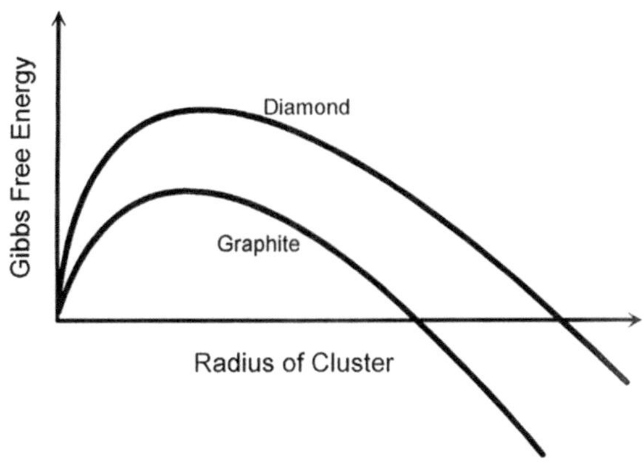

Fig. 5.5 Dependence of Gibbs free energy on radius for diamond and graphite during precipitation from the gas phase (Hwang and Lee 2010)

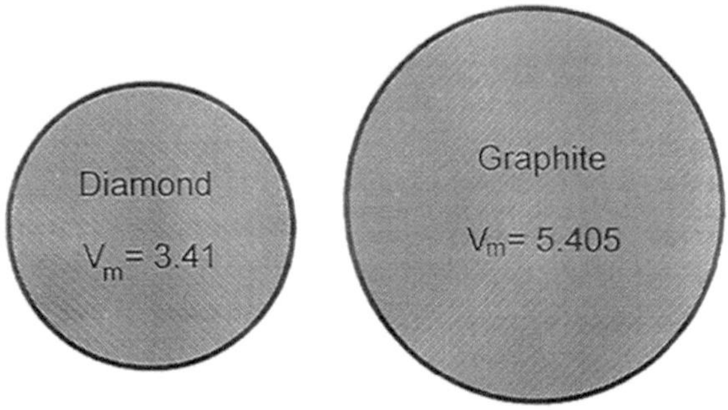

Fig. 5.6 Diamond and graphite nuclei with the same number of atoms showing that the surface area of diamond is much smaller than that of graphite, which results in $V_{dia}^{2/3}\sigma_{dia}(8.38 \times 10^{-4}) < V_{gra}^{2/3}\sigma_{gra}(9.55 \times 10^{-4})$

of diamond is smaller than that of graphite. Therefore, Fig. 5.5 can be misleading and the comparison of the free energy between diamond and graphite should not be made with respect to the radius but be made with respect to the number of atoms.

It should be noted that even though the surface energy of diamond (3.7 J/M^2) is larger than that of graphite (3.1 J/M^2), the molar volume of diamond (3.41 cm^3/mol) is much smaller than that of graphite (5.405 cm^3/mol). Therefore, if diamond and graphite nuclei with the same number of atoms are compared, the diamond nucleus is smaller than the graphite nucleus as shown in Fig. 5.6. The surface area of diamond is much smaller than that of graphite. Since the surface energy is given by the specific surface energy multiplied by the area, the quantity $V^{2/3}\sigma$, where V is a molar volume, should be compared between diamond and graphite. $V_{dia}^{2/3}\sigma_{dia}$ and $V_{gra}^{2/3}\sigma_{gra}$ are respectively 8.38×10^{-4} and 9.55×10^{-4} J. Therefore, in the nucleation stage, the surface energy contribution of diamond is smaller than that of graphite.

The Gibbs free energy of diamond and graphite nuclei can be expressed in terms of number of atoms respectively as

$$\Delta G_{dia} = n\Delta\mu^{gas\rightarrow dia} + \eta_{dia}\sigma_{dia}n^{2/3} \tag{5.5}$$

$$\Delta G_{gra} = n\Delta\mu^{gas\rightarrow gra} + \eta_{gra}\sigma_{gra}n^{2/3} \tag{5.6}$$

where η is a shape factor and for a sphere is given as

$$\eta = (4\pi)^{1/3}(3\Omega)^{2/3} \tag{5.7}$$

where Ω is an atomic volume, which is obtained by dividing the molar volume V by Avogadro's number.

If (5.5) and (5.6) are equated and solved for n, it is expressed after rearrangement as follows

$$n^* = 36\pi\left(\frac{\Omega_{gra}^{2/3}\sigma_{gra} - \Omega_{dia}^{2/3}\sigma_{dia}}{\Delta\mu^{gra\rightarrow dia}}\right)^3 \tag{5.8}$$

In order for metastable diamond to nucleate dominantly over graphite, it is necessary that n* should be positive. Since graphite is more stable than diamond at low pressure, $\Delta\mu^{gra\rightarrow dia}$ in (5.8) is positive. In order for n* to be positive, the numerator should be positive. It requires $\Omega_{gra}^{2/3}\sigma_{gra} > \Omega_{dia}^{2/3}\sigma_{dia}$, which is identical to $V_{gra}^{2/3}\sigma_{gra} > V_{dia}^{2/3}\sigma_{dia}$. This inequality is shown to be satisfied by the numerical evaluation above.

It should be noted that the positive value of n* is not a sufficient condition for dominant nucleation of diamond. In order for diamond to nucleate dominantly over graphite, its nucleation barrier should be lower than that of graphite. For evaluation of the nucleation barrier, $\Delta\mu^{gas\rightarrow dia}$ in (5.5) and $\Delta\mu^{gas\rightarrow gra}$ in (5.6) should be determined. In the paper published in 1992 (Hwang et al. 1992), Hwang et al. derived (5.8) but could not properly evaluate the nucleation barrier because $\Delta\mu^{gas\rightarrow dia}$ and $\Delta\mu^{gas\rightarrow gra}$ could not be determined at that time. This is why Hwang et al. (1994), Hwang (1994), Hwang and Yoon (1994) tried to formulate the driving force for deposition in the CVD process. The thermodynamic formulation for this is described in detail in Chap. 2.

The driving forces for deposition of diamond and graphite were also calculated in Chap. 2. Under the condition of T = 1200 K and P = 2700 Pa for the gas mixture of 1 % CH_4–99 % H_2, the calculation shows that $\alpha^{gas\rightarrow gra} = 23.3$ and $\Delta\mu^{gas\rightarrow gra} = -31{,}413$ J/mole, where α and Δμ are respectively the supersaturation ratio and driving force for graphite precipitation from the gas phase. The calculation further shows that $\alpha^{gas\rightarrow dia} = 11.7$, $\Delta\mu^{gas\rightarrow dia} = -24{,}540$ and $\Delta\mu^{dia\rightarrow gra} = -6873$ J/mole. If these values are used together with the reported surface energy ($\sigma_{dia} = 3.7$ J/M^2 and $\sigma_{gra} = 3.1$ J/M^2) and molar volume ($V_{dia} = 3.41$ cm^3/mol and $V_{gra} = 5.405$ cm^3/mol), the nucleation curves for diamond and graphite can be estimated as shown in Fig. 5.7.

Figure 5.7 is quantitative to a degree although there are uncertainties in the reported values of surface energy of diamond and graphite and in the assumption of a spherical shape. n*, at which the two curves intersect, is 351. This means that for the nucleus containing less than 351 atoms, diamond is more stable than graphite. This result is similar to the calculation of Badziag et al. (1990); The title of their paper is “Nanometer sized diamond is more stable than graphite”. The main purpose of their study was to explain the formation of interstellar diamonds, which turned out to exist in abundance in the universe.

Fig. 5.7 Dependence of Gibbs free energy of diamond and graphite on the number of atoms. *Solid* and *dashed lines* are respectively for diamond and graphite (Choi et al. 1997)

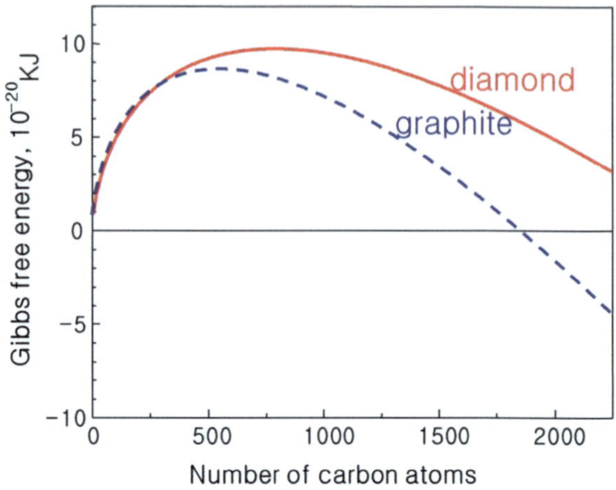

5.6 Capillary Pressure of Diamond and Graphite in the Nucleation Stage

The nucleus is under high capillary pressure given by Laplace equation, which is given for a spherical nucleus of radius r as

$$P_2 - P_1 = \frac{2\sigma}{r} \tag{5.9}$$

where P_1 and P_2 are the pressure respectively outside and inside the nucleus and σ the interface energy. This capillary pressure increases the chemical potential of the nucleus, which is expressed as

$$\Delta\mu_r = \frac{2\sigma}{r} V_m, \tag{5.10}$$

where V_m is the molar volume of the nucleus. Since the stability of diamond with respect to that of graphite increases with increasing pressure, the capillary pressure in the nucleation stage may stabilize diamond over graphite.

The partial derivative of chemical potential (or molar Gibbs free energy) with respect to pressure at constant temperature is equal to the molar volume, which is expressed as

$$\left(\frac{\partial\mu}{\partial P}\right)_T = V_m. \tag{5.11}$$

When the chemical potential is evaluated per atom, V_m in (5.11) becomes the atomic volume, Ω.

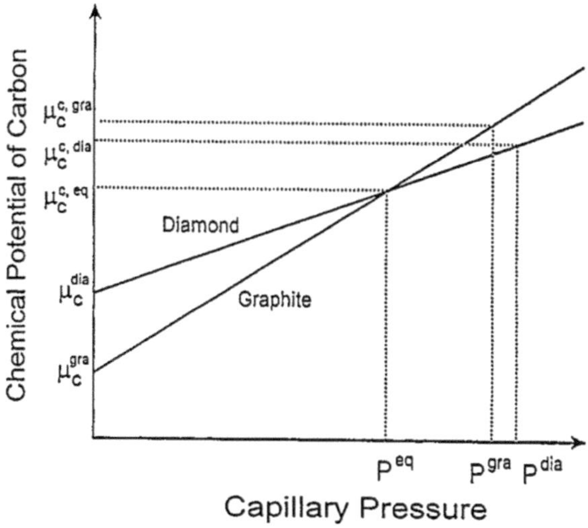

Fig. 5.8 Dependence of chemical potentials of carbon in diamond and graphite on capillary pressure (Hwang et al. 1996b)

The chemical potentials per atom in diamond and graphite nuclei are expressed as

$$\mu_C^{c,dia} = \mu_C^{dia} + \Omega^{dia}\left(2\sigma^{dia}/r^{dia}\right), \tag{5.12}$$

and

$$\mu_C^{c,gra} = \mu_C^{gra} + \Omega^{gra}(2\sigma^{gra}/r^{gra}), \tag{5.13}$$

where $\mu_C^{c,dia}$ and μ_C^{dia} represent the chemical potentials of carbon in diamond with and without the capillary effect, respectively, and likewise for graphite.

The capillary pressure on the stability of diamond and graphite is illustrated in Fig. 5.8. The slopes of the two lines in Fig. 5.8 are the atomic volumes of diamond and graphite, respectively. It should be noted that in analysing the stability between the nuclei of diamond and graphite the number of carbon atoms in each nucleus should be the same, and thus, the radius of each nucleus is different from each other. The radius of the diamond nucleus is smaller than that of the graphite nucleus for the same number of atoms.

It is assumed that the molar volume does not vary with pressure. Since graphite has a larger atomic volume than diamond, its slope is steeper. P^{gra} and P^{dia} in Fig. 5.8 are the pressure inside the graphite and diamond nuclei, respectively. These capillary pressures are determined by the surface energy and the radius of the nucleus as given by (5.9). The chemical potential of carbon in the graphite nucleus becomes higher than that of the diamond nucleus when the pressure inside the nuclei becomes sufficiently high by decreasing the size of the nuclei. The pressure at the intersection, P^{eq}, in Fig. 5.8, where the stability between graphite and diamond changes, corresponds to the phase boundary between graphite and diamond in the phase diagram of carbon.

The number of atoms that makes the chemical potentials of graphite and diamond nuclei the same is determined by equating (5.12) and (5.13):

$$\Omega^{gra}\left(2\sigma^{gra}/r^{gra}\right) - \Omega^{dia}\left(2\sigma^{dia}/r^{dia}\right) = \mu_C^{dia} - \mu_C^{gra}. \tag{5.14}$$

The right-hand size of (5.14) is the chemical potential difference $\Delta\mu_C^{gra\to dia}$ between diamond and graphite without the capillary effect, which will appear in the final derivation. The radius r can be expressed in terms of the number of atoms, n as

$$r = \left(\frac{3\Omega}{4\pi}\right)^{1/3} n^{1/3}, \tag{5.15}$$

Substituting (5.15) into (5.14) and solving for n lead to

$$n^{\dagger} = \frac{32\pi}{3}\left[\frac{\sigma^{gra}\left(\Omega^{gra}\right)^{2/3} - \sigma^{dia}\left(\Omega^{dia}\right)^{2/3}}{\Delta\mu_C^{gra\to dia}}\right]^3. \tag{5.16}$$

It should be noted that the expression for $n^{\dagger}$ in (5.16) is slightly different from that for n^* in (5.8). It will be a good exercise to try to understand why (5.8) and (5.16) are different.

n^* in (5.8) is derived under the condition where the free energy of a diamond nucleus is the same as that of a graphite nucleus whereas $n^{\dagger}$ in (5.16) is derived under the condition where the chemical potential of a diamond nucleus is the same as that of a graphite nucleus. Differentiating (5.5) and (5.6) with respect to 'n' produces respectively the chemical potentials of diamond and graphite nuclei.

$$\frac{\partial \Delta G_{dia}}{\partial n} = \Delta\mu^{gas\to dia} + \frac{2}{3}\eta_{dia}\sigma_{dia}n^{-1/3} \tag{5.17}$$

$$\frac{\partial \Delta G_{gra}}{\partial n} = \Delta\mu^{gas\to gra} + \frac{2}{3}\eta_{gra}\sigma_{gra}n^{-1/3} \tag{5.18}$$

Equating (5.17) and (5.18) and solving for n produce the same expression as (5.16).

5.6.1 *Relative Nucleation Rate of Diamond and Graphite*

The dominant nucleation of diamond can be approached more quantitatively by estimating the nucleation rate. The activation barrier for nucleation is the free energy change associated with forming an embryo of a critical size. Since the nucleation rate depends on the activation barrier, capillarity has a significant effect.

According to the classical nucleation theory by Becker and Döring, the nucleation rate for the solid from the vapor is expressed as

$$I = q_o O_c \left(\frac{\Delta G_c}{kT} \right) \exp\left(-\frac{\Delta G_c}{kT} \right), \tag{5.19}$$

where I is the number of stable nuclei created per second, q_o is the probability per unit time per unit area of capturing a vapor atom, O_c is the area of the critical embryo and ΔG_c the free energy of formation of the critical embryo, which is the activation energy for nucleation.

The ratio of the nucleation rate for diamond and graphite becomes

$$\frac{I^{dia}}{I^{gra}} = A \exp\left(\frac{\Delta G_c^{gra} - \Delta G_c^{dia}}{kT} \right), \tag{5.20}$$

where the pre-exponential term, A, is the ratio of q_o, O_c, and ΔG_c between diamond and graphite. When this nucleation ratio is greater than unity, the nucleation of diamond is dominant over that of graphite and vice versa. The pre-exponential term of (5.20) would not be much different from unity. Thus, the intensity ratio would be affected mainly by the exponential term.

The difference in activation energy between diamond and graphite is

$$\Delta G_c^{gra} - \Delta G_c^{dia} = \frac{16\pi (\Omega^{gra})^2 (\sigma^{gra})^3}{3(\Delta\mu^{gas\to gra})^2} - \frac{16\pi (\Omega^{dia})^2 (\sigma^{dia})^3}{3(\Delta\mu^{gas\to dia})^2}. \tag{5.21}$$

This equation shows that the nucleation intensity ratio will depend on the driving force of precipitation of solid carbon from the gas phase. Since in the previous section, it was shown that $\Omega_{gra}^{2/3}\sigma_{gra} > \Omega_{dia}^{2/3}\sigma_{dia}$, $(\Omega^{gra})^2(\sigma^{gra})^3$ is larger than $(\Omega^{dia})^2(\sigma^{dia})^3$ and then $\Delta G_c^{gra} - \Delta G_c^{dia}$ increases from negative to positive with increasing driving force; a positive value means the dominant nucleation of diamond over graphite. With further increasing the driving force, the difference reaches a maximum and then decreases.

By assuming that the number of carbon atoms of the critical nucleus is 100, the ratio of the nucleation rate can be estimated. Based on the surface energies of diamond, 3.7 J/M^2 and graphite, 3.1 J/M^2, the ratio of the diamond to graphite nucleation rate is estimated to be $\sim 1.3 \times 10^{10}$ at 1200 K. This ratio is very sensitive to the variation of the surface energy. For the surface energies of diamond, 4.0 and 4.1 J/M^2, the ratios are 0.033 and 1.8×10^{-6}, respectively. These estimations show that within the error of the surface energy of diamond and graphite, the dominance in nucleation is reversed.

When a solid carbon nucleates on the substrate with contact angle θ, the activation energy for heterogeneous nucleation is given by that for homogeneous nucleation multiplied by the heterogeneous factor $S(\theta) = (2 - 3\cos\theta + \cos^3\theta)/4$

as explained in Chap. 3. Thus, in the case of nucleation on the substrate, (5.21) becomes

$$\Delta G_c^{gra} - \Delta G_c^{dia} = \frac{16\pi(\Omega^{gra})^2(\sigma^{gra})^3}{3(\Delta\mu^{gas\to gra})^2} S(\theta) - \frac{16\pi(\Omega^{dia})^2(\sigma^{dia})^3}{3(\Delta\mu^{gas\to dia})^2} S(\phi), \quad (5.22)$$

where θ and ϕ are the contact angles with the substrate respectively for graphite and diamond. Equation (5.22) indicates that when the interfacial energy between diamond and the substrate is smaller than that between graphite and the substrate, the dominant nucleation is favoured and vice versa.

5.7 Diamonds in the Sky

Diamonds in the interstellar medium were first proposed by Saslaw and Gaustad (1969) and were experimentally found in unshocked chondrite meteorites (Lewis et al. 1987). Diamonds found in meteorites are extremely small just 2–5 nm in diameter. How these interstellar diamonds are formed has not yet been known. Lewis et al. (1987) suggested that these diamonds were formed in a circumstellar environment as metastable condensates. Then, the question similar to that in CVD diamond arises: how are these diamond formed? Nuth (1987) tried to explain the formation of metastable diamond by thermodynamics of small-particles. Interestingly, he emphasized that since diamond has a small molar volume than graphite, diamond can be more stable than graphite for a sufficiently small size even though its surface energy may be larger than that of graphite. He suggested the following equation to determine the size where diamond and graphite has the same stability.

$$(area \times surface\,energy)_{dia} + \Delta G_{dia} = (area \times surface\,energy)_{gra} + \Delta G_{gra} \quad (5.23)$$

Note that ΔG_{dia} and ΔG_{gra} are the volume free energy. If they are converted to the free energy per atom, (5.23) is identical to equating (5.5) and (5.6). Therefore the solution of (5.23) would be identical to (5.8). Believing that the reported surface energy data of diamond and graphite are too uncertain, Nuth tried to estimate the surface energy from (5.23) based on the assumption that diamond smaller than 5 nm is more stable than graphite. This assumption comes from the size of diamond found in unshocked chondrite meteorites.

The size dependence of the stability between diamond and graphite has been studied experimentally by Fedosayev et al. (1989) and Bundy et al. (1996). They observed that the nanodiamond that initially formed with a grain size below 3 nm transforms into graphite at a larger size. This problem was approached theoretically by many scientists (Nuth 1987; Badziag et al. 1990; Gamarnik 1996; Ree et al.

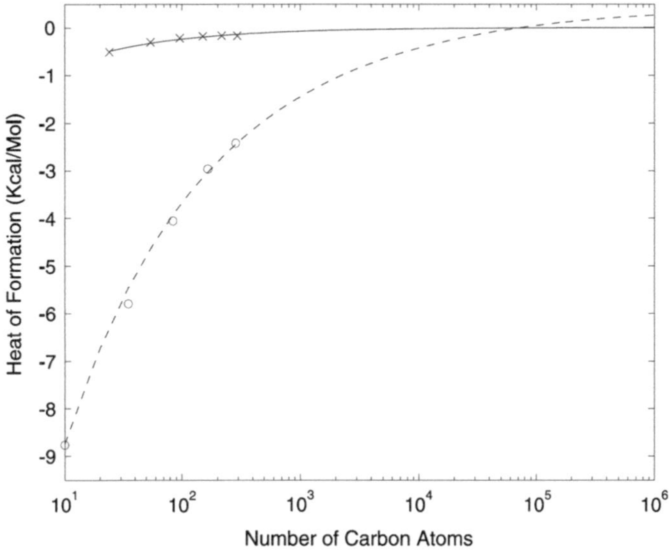

Fig. 5.9 Comparison of the cluster size dependence of $\Delta H_f(sp^3)$ and $\Delta H_f(sp^2)$ determined by the PM3 HF method. The fits to the sp^3 (plotted as o's) and sp^2 (plotted as x's) respectively (Winter and Ree 1998)

1999; Barnard et al. 2003). There is a big difference between calculations in size, where the stability between diamond and graphite is the same. One example is the calculation by Winter and Ree (1998), which is shown in Fig. 5.9.

In Fig. 5.9, the carbon cluster heat of formation as a function of size was determined using the experimental cohesive energies of diamond and graphite along with the calculated C–H bond energies of the hydrogen capped clusters. These results predict that diamond clusters smaller than 30,000–70,000 atoms are more stable than graphite clusters.

In contrast, the calculation of Badziag et al. (1990) predicted the stability change between diamond and graphite at much smaller size. For surface bonds terminated with hydrogen atoms, diamonds smaller than ~3 nm in diameter were energetically favoured over polycyclic aromatics (the precursors to graphite), without requiring the high pressures or extreme kinetic conditions such as detonations usually associated with diamonds. In this size range, metastable diamond is not metastable any more but it is more stable than graphite. Therefore, pressure is not necessary to form ultra-small diamonds in detonations or in cosmic environments. Besides, extreme non-equilibrium kinetic conditions such as gas activation are not necessary to justify nucleation in initial stages of the vapour growth of CVD diamond.

Considering the two results by Winter and Ree (1998) and Badziag et al. (1990), the results may vary in a wide range, depending on the calculation methods and the parameters used. Even the number of carbon atoms calculated by Badziag et al.

(1990) for the stability change between diamond and graphite is much larger than $n^* = 351$ in Fig. 5.7, which is evaluated by thermodynamic data. These calculations and evaluations might have some errors but one thing clear is that the stability of diamond relative to that of graphite increases with decreasing size.

5.8 Comparison of Nucleation Barrier Between Diamond and Graphite

Let's go back to experimental facts about the CVD diamond. If the gas activation such as hot filament or plasma is not adopted with other conditions being the same as the CVD diamond process, graphite or non-diamond carbon is formed. These facts do not agree with calculations by Winter and Ree (1998) and Badziag et al. (1990) but perhaps accidentally agree with Fig. 5.7, where the nucleation barrier of diamond is slightly higher than that of graphite.

Then, what would happen if the nucleation barrier of diamond is slightly lower than that of graphite not like Fig. 5.7? Diamond synthesis in the gas phase would be much easier than now. For example, oxygen-deficient combustion of oxy-hydrocarbon materials such as fuel and wood could have produced diamond particles instead of graphitic soot.

In reality, the gas activation such as hot filament and plasma is essential to the successful synthesis of diamond at low pressure. From this fact, it can be inferred that the role of the gas activation would be to increase the stability of diamond relative to that of graphite. Then, the gas activation is expected to increase n^* of (5.8). To identify the possible role of the gas activation, let's examine the parameters in (5.8). $\Delta\mu^{gra \rightarrow dia}$ and the atomic volume Ω would not be affected by the gas activation. The only parameter that can be affected by the gas activation would be the surface energy of diamond and graphite.

If we assume that the gas activation reduces the surface energy of diamond, the nucleation of diamond can be more dominant than that of graphite. For example, if the surface energy of diamond is reduced by 10 and 20 %, n^* in (5.8) becomes respectively 1784 and 5092 at 1200 K. Then, the nucleation barrier of diamond becomes lower than that of graphite as shown in Fig. 5.10.

Therefore, if the gas activation reduces the surface energy of diamond relative to that of graphite, the diamond formation in the new CVD process can be understood. In the next stage, it should be understood how the gas activation reduces the surface energy of diamond. Two possibilities were suggested. One is the adsorption of the atomic hydrogen suggested by Badziag et al. (1990). The other is an electrical double layer on the surface formed by charge (Hwang et al. 1996a). The reason why Hwang et al. (1996a) suggested this is based on the experimental observation that a diamond film was deposited on a silicon substrate whereas a porous skeletal soot structure was evolved on an iron substrate. This aspect will be explained in more detail in Chap. 6.

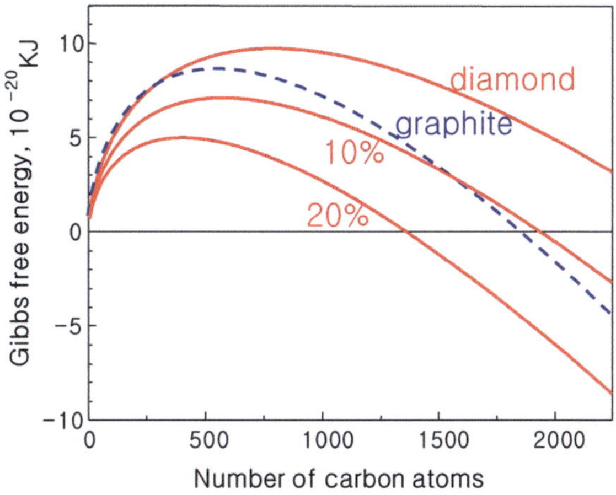

Fig. 5.10 Plots of the Gibbs free energy versus the number of atoms. *Solid lines* denote the free energies of diamond for 0, 10 and 20 % reduction of the surface energy and the *dashed line* denoted the free energy of graphite (Choi et al. 1997)

Equation (5.23) is almost identical to the equation related to the melting point depression of nanoparticles, which will be dealt with in Chap. 13.

5.9 Atomic Hydrogen Hypothesis

As mentioned earlier, the old CVD diamond process, where graphite tended to be co-deposited, was a cyclic process of diamond deposition and graphite etching. If atomic hydrogen is used, the rate of graphite etching is enhanced. Therefore, the time taken for growth of the same amount of diamonds can be reduced. Based on this idea, Derjaguin and Fedoseev (1975, 1977) tried to produce atomic hydrogen by hot tungsten filaments, which led to the discovery of the new diamond CVD process. They attributed such unexpected changes to atomic hydrogen, which etches graphite preferentially over diamond. Based on these results, they suggested that atomic hydrogen acts as a means of gasifying or etching graphite or graphitic carbon. This suggestion of atomic hydrogen hypothesis is equivalent to saying that atomic hydrogen etches graphite much faster than diamond and therefore less stable diamond can be deposited in the presence of atomic hydrogen. In other words, in the presence of atomic hydrogen, diamond is kinetically stable relative to graphite (Sommer et al. 1989).

5.9.1 Distinction Between Reversible and Irreversible Fluxes

In dealing with the atomic hydrogen hypothesis, where graphite etches and diamond deposits simultaneously, reversible etching and deposition should be distinguished from irreversible etching and deposition. For example, when water is in equilibrium with its vapour at 70 °C in a closed vessel, H_2O molecules continue to evaporate and condensate simultaneously. In this case of dynamic equilibrium, we say that evaporation and condensation occur simultaneously. This case corresponds to reversible evaporation and condensation. In equilibrium, the rate of evaporation is equal to that of condensation. In other words, the flux of reversible evaporation is cancelled out by the opposite flux of reversible condensation. As a result, the net flux is zero.

However, irreversible etching or deposition is accompanied by a net flux. Therefore, reversible etching and deposition can occur simultaneously whereas irreversible etching and deposition cannot. Either irreversible etching or irreversible deposition can occur. Reversible etching or deposition is not driven by the chemical potential difference and does not produce a net flux whereas irreversible etching or deposition is driven by the chemical potential difference and produces a net flux.

5.9.2 Criticism of Atomic Hydrogen Hypothesis

Although the atomic hydrogen hypothesis has been the most popular explanation for the synthesis of the new CVD diamond process using gas activation, it has a critical drawback in that the irreversible transfer of carbon atoms violates the second law of thermodynamics (Yarbrough 1992; Hwang and Yoon 1996). The CVD process is thermodynamically an open system in that the deposited film which is a system, exchanges a material with the gas phase, which is a surrounding. The criterion for the irreversible transfer of atoms between the system and the surrounding is the chemical potential. The chemical potential of an atom is the partial derivative of the Gibbs free energy of the system with respect to the number of corresponding atoms under a constant temperature and pressure (Gaskell 1995). As atoms transfer from the region with a high chemical potential to the region with a low chemical potential, the total Gibbs free energy is decreased.

Since graphite is more stable than diamond under the condition of gas-activated CVD according to the phase diagram of carbon (Whittaker 1978), the chemical potential of carbon in diamond is higher than that in graphite, which can be expressed as

$$\mu_C^{dia} > \mu_C^{gra}, \tag{5.24}$$

where μ is a chemical potential, the subscript 'C' represents carbon and the superscripts 'dia' and 'gra' represent diamond and graphite, respectively.

It should be noted that reversible etching and deposition can occur simultaneously whereas irreversible etching and deposition cannot. Etching of a condensed phase into a gas phase means that the chemical potential of the element in a condensed phase is higher than that in a gas phase. Therefore, the irreversible graphite etching can be written as

$$\mu_C^{gra} > \mu_C^{gas}, \tag{5.25}$$

where the superscript 'gas' represents the gas phase. Similarly, the irreversible diamond deposition can be written as

$$\mu_C^{gas} > \mu_C^{dia}. \tag{5.26}$$

Here, etching and deposition refer to the irreversible process. Hereafter, the terms of etching and deposition will be used without the adjective 'irreversible'.

According to the atomic hydrogen hypothesis, inequalities (5.25) and (5.26) should be simultaneously satisfied and can be written as

$$\mu_C^{gra} > \mu_C^{gas} > \mu_C^{dia}. \tag{5.27}$$

Inequality (5.27) produces the inequality of

$$\mu_C^{gra} > \mu_C^{dia}, \tag{5.28}$$

which says that diamond is more stable than graphite, being definitely contradictory to inequality (5.24).

The contradiction would be clearer if the inequality is shown graphically. The inequality given by thermodynamics or phase diagram of carbon can be drawn as Fig. 5.11a, which is identical to Fig. 5.4. However, the inequality imposed by the atomic hydrogen hypothesis can be drawn as Fig. 5.11b.

According to the second law of thermodynamics, it is quite clear that if stable graphite should be etched, less stable diamond should be etched also. But this clear statement is contradictory to inequality (5.27). The atomic hydrogen hypothesis implicitly assumes that the unbalanced etching rate between graphite and diamond can change the stability between graphite and diamond; it neglects the fact that the thermodynamic stability, which depends only on the initial and final states, is not affected by kinetics. Simple thermodynamic analyses described above show that the atomic hydrogen hypothesis has a critical weak point.

Fig. 5.11 Comparison of chemical potential of carbon for graphite, diamond and gas based on **a** phase diagram of carbon and **b** atomic hydrogen hypothesis

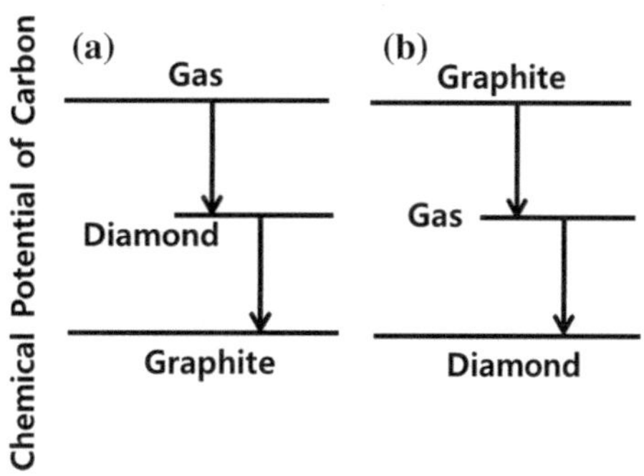

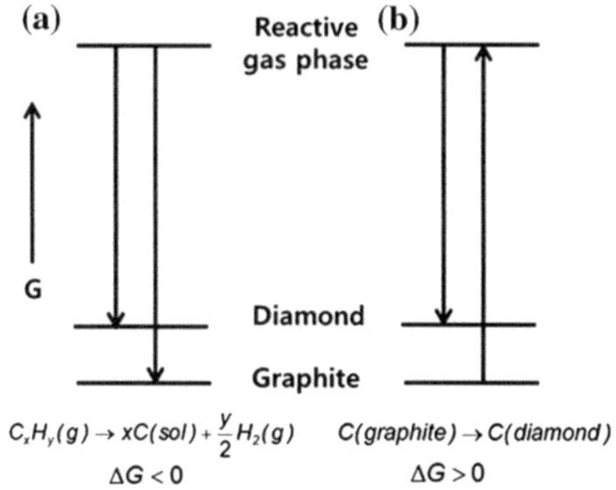

Fig. 5.12 **a** CVD of solid carbon from a gaseous hydrocarbon where the gaseous hydrocarbon is metastable with respect to solid carbon and molecular hydrogen. Using accepted values for the bulk solids, diamond has only a slightly higher free energy than graphite (~0.5 kcal/mol at 298 K), and the free energy of reaction is negative, regardless of the form of solid carbon deposited, **b** steady-state gasification of graphite and deposition of diamond. i.e., the chemical vapor transport of carbon from graphite to diamond through the formation of a gaseous hydrocarbon intermediate. This is believed by most to be dependent on the presence of atomic hydrogen. Free energy of the gaseous hydrocarbon is drawn consistent with (**a**) for the sake of simplicity (Yarbrough 1992)

Yarbrough (1992) made the same conclusion based on his thermodynamic analysis as shown in Fig. 5.12.

Therefore, it is quite clear that the atomic hydrogen hypothesis violates the second law of thermodynamics. Then, why is it so popular in the diamond CVD community? The reason is that the atomic hydrogen hypothesis is supported by experimental observations as described in the following section.

5.10 Diamond Deposition with Simultaneous Graphite Etching

Vakil (1989) measured the total solid carbon mass including the graphite substrate in gas-activated diamond CVD. The diamond content continuously increased while the total solid carbon mass continuously decreased by etching of the graphite substrate. Vakil's finding of diamond deposition with simultaneous graphite etching was reproduced by many other scientists. For example, Badzian et al. (1988) showed that diamond deposited on the etch pit of the graphite substrate as shown in Fig. 5.13. The etch pit tells that the graphite substrate was being etched, indicating diamond deposition with simultaneous graphite etching.

Salvadori et al. (1992) measured the weight change of graphite ring with deposition time under the condition of 0.5 % CH_4–H_2, 80 Torr. And T_{sub} = 870 °C as shown in Fig. 5.14. The mass decreased by etching for the first 1 h. Then, as the

Fig. 5.13 Diamond deposition on the etch pit of graphite substrate (Badzian et al. 1988)

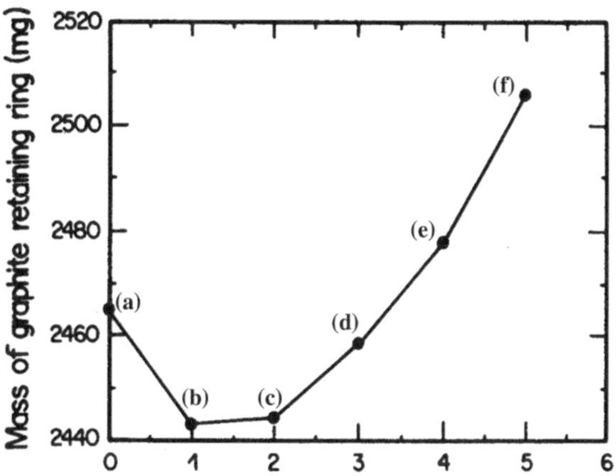

Fig. 5.14 Weight change of graphite ring with deposition time (Salvadori et al. 1992)

diamond film started to deposit on the graphite ring between 1 and 2 h, the mass decrease stopped. After deposition for 2 h, the mass started to increase because the graphite ring was covered with the diamond film.

The atomic hydrogen hypothesis was strongly supported by the experimental fact that diamond is synthesized with simultaneous etching of graphite (Vakil 1989; Badzian et al. 1988; Salvadori et al. 1992). These well-organized experiments seem to be regarded as the evidence for the atomic hydrogen hypothesis. Considering such experimental observations, some people say that the atomic hydrogen hypothesis is not a hypothesis but an experimental fact.

Figure 5.15 shows the microstructure of diamonds, which were deposited on a graphite substrate for 2 h under 2700 Pa at a wire temperature of 2100 °C and a substrate temperature of 1050 °C with a gas mixture of 1 % CH_4–99 % H_2. Comparing the mass of the graphite substrate before and after deposition, a mass of

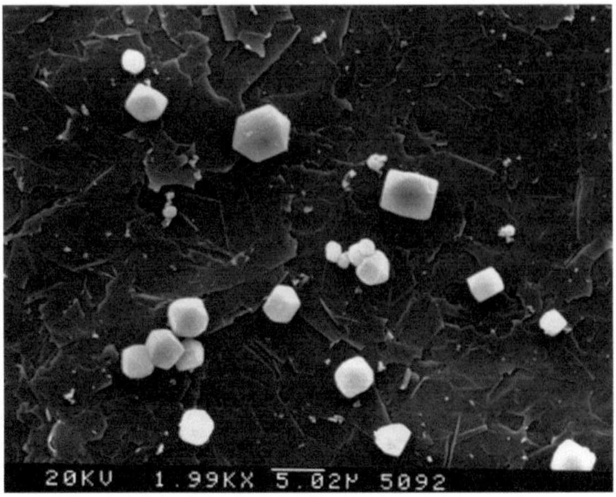

Fig. 5.15 Microstructure showing diamond deposition with simultaneous graphite etching. Diamond was deposited on the graphite substrate at a hot wire temperature of 2100 °C and a substrate temperature of 1050 °C. The mass of the graphite substrate was 43.21 mg before deposition and 37.06 mg after deposition, being decreased by 6.15 mg although the diamond crystals were deposited on the graphite (Hwang and Lee 2010)

6.15 mg was decreased during deposition although the diamond crystals were deposited on the graphite. Therefore, Fig. 5.15 reproduced the well-known result of diamond deposition with simultaneous graphite etching.

Figures 5.13, 5.14 and 5.15 clearly show that diamonds deposit on graphite, which is being etched simultaneously. These experimental observations are puzzling not only thermodynamically but also kinetically. It should be noted in the old CVD diamond process that a small debris of graphitic carbon on diamond seed would soon cover the entire surface and the graphitic carbon should be etched before diamond can be further grown. Therefore, the presence of graphite is a poison to the growth of diamond. Therefore, graphite should be the worst substrate for growth of metastable diamond from the conventional concept of crystal growth. In the new CVD process, however, almost all kinds of carbon including graphite are relatively good for initiation and growth of diamond.

5.11 Experimental Fact Appearing to Violate the Second Law of Thermodynamics

In order to see clearly the direction of an irreversible transfer of carbon atoms allowed by the second law of thermodynamics, the chemical potential diagrams are drawn in Fig. 5.16 for the three possible stabilities among diamond, graphite and gas under a low pressure.

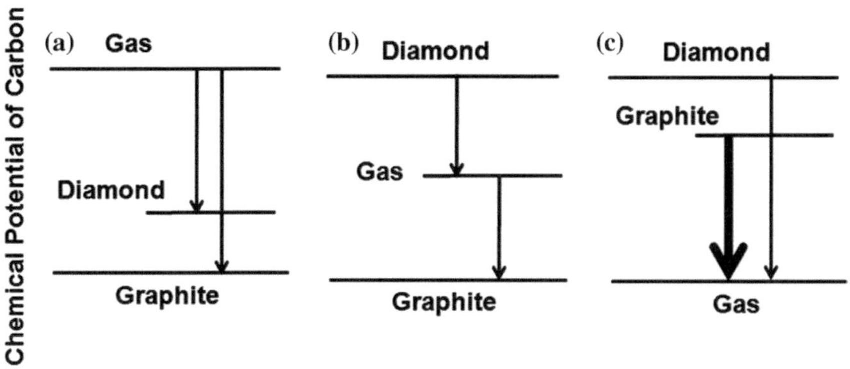

Fig. 5.16 The direction of carbon atoms allowed by the second law of thermodynamics when the chemical potential of carbon in gas is highest (**a**), between diamond and graphite (**b**) and lowest (**c**) (Hwang and Lee 2010)

Figure 5.16a–c show the direction of the carbon flux allowed by the second law of thermodynamics when the chemical potential of carbon in gas is highest, between diamond and graphite and lowest, respectively. Figure 5.16a shows that if less stable diamond should deposit, stable graphite should deposit also. Figure 5.16b also shows that graphite deposition with simultaneous diamond etching is possible, which is opposite to the experimental observation of diamond deposition with simultaneous graphite etching. Finally, Fig. 5.16c shows that if stable graphite should etch, less stable diamond should also etch.

The etching rate of graphite is much higher than that of diamond. This kinetic effect is considered in Fig. 5.16c by drawing the line of graphite etching thicker. It should be reminded that kinetics cannot override thermodynamics whose mistake is made in the atomic hydrogen hypothesis. Instead, kinetics should be pursued under the condition where thermodynamics is satisfied as in Fig. 5.16c.

Any case of Fig. 5.16a–c does not allow for diamond deposition with simultaneous graphite etching. Therefore, it is quite obvious that the experimental observation appears to be contradictory with the second law of thermodynamics. In approaching such a puzzling problem, something, which appears to be relatively absurd or strange, should be identified. In the experimental observation of diamond deposition with simultaneous graphite etching, one strange thing is that in the gas-activated diamond CVD, we supply methane concentration high enough to deposit the condensed phase of carbon, which can be diamond, graphite or other carbon allotropes such as amorphous carbon, C60 or carbon nanotubes. Then, why does graphite etch away into the gas phase under the condition that the thermodynamic driving force is clearly for deposition of graphite?

In order to understand this situation, the CVD phase diagram of the C–H system shown in Fig. 5.17a should be examined (Hwang and Yoon 1996). In Fig. 5.17a, the diamond and graphite lines indicate the solubility of carbon in the gas phase in equilibrium with diamond and graphite, respectively. The metastability of diamond

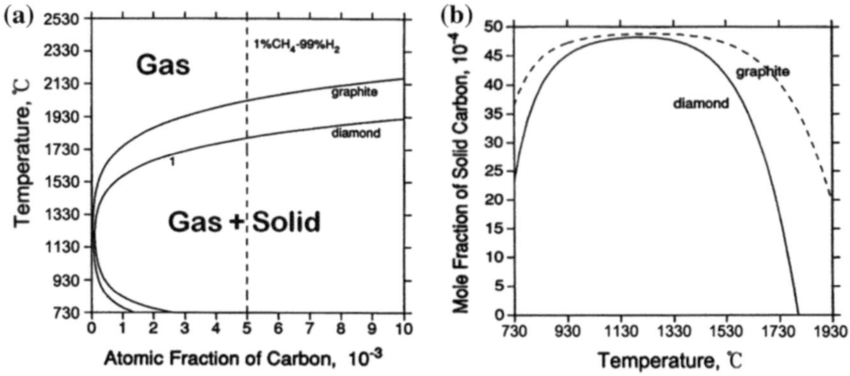

Fig. 5.17 **a** The CVD phase diagram of the C–H system under 2700 Pa. The *graphite* and *diamond lines* indicate the maximum solubility of carbon in the gas phase. Outside this region, the single phase of gas is stable and inside this region, the two phases of gas + solid are stable. The vertical *dashed line* indicates the composition of the gas mixture of 1 % CH_4–99 % H_2, which is typically used in the gas-activated diamond CVD process. **b** The calculated temperature dependence of the equilibrium mole fraction of solid carbon under the same condition as (**a**) (Hwang and Yoon 1996)

with respect to graphite is represented by the fact that the diamond line is inside the graphite line.

For example, outside the graphite line, the single phase of gas is stable and inside this line, the two phases of gas and graphite are stable with diamond being metastable inside the diamond line. The composition of 1 % CH_4–99 % H_2, which corresponds to the atomic fraction of carbon, 0.00493, falls in the two phase region of gas + solid below ~1730 and ~2030 °C, respectively, for diamond and graphite. This composition is shown as the vertical line in Fig. 5.17a. The methane concentration range, typically used in the CVD diamond process, is 0.5–3 %. In this composition range, the driving force is for deposition of both graphite and diamond in the substrate temperature range 730–1070 °C. The notable feature in the CVD phase diagram of the C–H system in Fig. 5.17a is that the solubility of carbon in the gas phase is minimal around 1230 °C and then increases with decreasing temperature. The increase in carbon solubility with decreasing temperature below ~1230 °C indicates that the equilibrium amount of solid carbon in the gas phase decreases with decreasing temperature. This aspect can also be revealed by the temperature dependence of the equilibrium amount of precipitation of solid carbon from the gas mixture of 1 % CH_4–99 % H_2 at 2700 Pa as shown in Fig. 5.17b (Hwang and Yoon 1996). The equilibrium amount of solid carbon, which is maximal at ~1230 °C, decreases with decreasing temperature below ~1230 °C.

Considering Fig. 5.17, the experimental fact of graphite etching under the condition of the driving force for deposition can only be explained by assuming that gas-phase nucleation occurs in the diamond CVD reactor. If gas-phase nucleation does not occur with the gas mixture of 1 % CH_4–99 % H_2, the carbon concentration in the gas phase would be represented by the vertical line in Fig. 5.17a. Under this condition carbon in the gas phase is supersaturated and thus the driving force is for the precipitation of solid carbon. If gas-phase nucleation occurs, however, the carbon concentration in the gas phase would follow the solubility line in Fig. 5.17a.

Let us examine what would happen when gas-phase nucleation occurs in the CVD diamond reactor. In the reactor, there exists a steep temperature gradient between the hot wire and the substrate, which are only ~1 cm away from each other.

In this steep temperature gradient, if the gas phase drifted from the high-temperature region near the wire towards the low temperature region near the substrate, the gas phase should etch solid carbon. The reason is that when the gas with a minimal solubility of carbon at ~1230 °C reaches the substrate at a lower temperature, the gas becomes deficient of carbon because its equilibrium solubility of carbon is much higher at a lower temperature as shown in Fig. 5.17a. Similarly, if the gas-phase nuclei move from the high-temperature region near the wire towards the low-temperature region near the substrate, some of them should be etched because their equilibrium amount of precipitation is maximal at ~1230 °C and becomes much lower at lower temperatures as shown in Fig. 5.17b.

Therefore, the driving force becomes for etching or for deposition, depending on whether gas-phase nucleation occurs or not. This aspect can be understood more clearly by the chemical potential diagram. If gas-phase nucleation does not occur, the chemical potential diagram would be Fig. 5.16a, where the driving force is for deposition of both diamond and graphite. If gas-phase nucleation occurs, however, the chemical potential diagram would be Fig. 5.16c, where the driving force is for etching of both diamond and graphite. Besides, in order to explain the experimental observation of diamond deposition with simultaneous graphite etching, the gas phase nuclei should be a building block for deposition of diamond films. In other words, diamond films should grow by non-classical crystallization.

In summary, in order for the experimental observation of diamond deposition with simultaneous graphite etching not to violate the second law of thermodynamics, the diamond nanoparticles should be generated in the gas phase and deposit as diamond films. Therefore, the experimental verification of gas-phase nucleation in the gas activated diamond CVD process is critical and will be shown in Chap. 6.

References

Angus JC, Will HA, Stanko WS (1968) Growth of diamond seed crystals by vapor deposition. J Appl Phys 39(6):2915–2922. doi:http://dx.doi.org/10.1063/1.1656693

Asakura Y, Chattopadhyay K, Matsumoto S, Hirakuri K (1998) Diamond synthesis in capacitively coupled 13.56 MHz radio frequency plasma using parallel plate electrodes with the addition of direct current power. J Vac Sci Technol A 16(6):3185–3189. doi:http://dx.doi.org/10.1116/1.581518

Badziag P, Verwoerd WS, Ellis WP, Greiner NR (1990) Nanometre-sized diamonds are more stable than graphite. Nature 343(6255):244–245

Badzian AR, Badzian T, Roy R, Messier R, Spear KE (1988) Crystallization of diamond crystals and films by microwave assisted CVD (Part II). Mater Res Bull 23(4):531–548. doi:http://dx.doi.org/10.1016/0025-5408(88)90161-4

Barnard AS, Russo SP, Snook IK (2003) Coexistence of bucky diamond with nanodiamond and fullerene carbon phases. Phys Rev B 68(7):073406

Belokurov B (1960) Concerning the surface tension of germanium, silicon and diamond. Zh Fiz Khim+ 34(2):4400–4443

Bundy FP (1980) The P, T phase and reaction diagram for elemental carbon, 1979. J Geophys Res-Sol Ea 85(B12):6930–6936. doi:10.1029/JB085iB12p06930

Bundy FP, Hall HT, Strong HM, Wentorf RH (1955) Man-made diamonds. Nature 176(4471):51–55

Bundy FP, Bassett WA, Weathers MS, Hemley RJ, Mao HU, Goncharov AF (1996) The pressure-temperature phase and transformation diagram for carbon; updated through 1994. Carbon 34(2):141–153. doi:http://dx.doi.org/10.1016/0008-6223(96)00170-4

Chattopadhyay K, Matsumoto S (1995) Diamond synthesis by capacitively coupled radio frequency plasma with the addition of direct current power. Appl Phys Lett 67(26):3972–3974. doi:http://dx.doi.org/10.1063/1.114422

Choi K, Kang S-JL, Jang HM, Hwang NM (1997) Nucleation behavior in the presence of charge in the CVD diamond process. J Cryst Growth 172(3–4):416–425. doi:http://dx.doi.org/10.1016/S0022-0248(96)00759-2

Derjaguin BV, Fedosayev DV (1975) The synthesis of diamond at low pressure. Sci Am 233: 102–109. doi:10.1038/scientificamerican1175-102

Derjaguin BV, Fedoseev DV (1977) The growth of diamond and graphite from the gas phase. Nauka, Moscow

Derjaguin BV, Fedoseev DV, Lukyanovich VM, Spitzin BV, Ryabov VA, Lavrentyev AV (1968) Filamentary diamond crystals. J Cryst Growth 2(6):380–384. doi:http://dx.doi.org/10.1016/0022-0248(68)90033-X

Eversole W (1962) Synthesis of diamond. US Patent 3030188, 17 Apr 1962

Fedosayev DV, Deryagin BV, Varasavskaja IG (1989) Special issue on diamond growth and films the crystallization of diamond. Surf Coat Tech 38(1):1–122. doi:http://dx.doi.org/10.1016/0257-8972(89)90127-8

Gamarnik MY (1996) Size-related stabilization of diamond nanoparticles. Nanostruct Mater 7(6): 651–658. doi:http://dx.doi.org/10.1016/0965-9773(96)00034-7

Garvie RC (1965) The occurrence of metastable tetragonal zirconia as a crystallite size effect. J Phys Chem 69(4):1238–1243. doi:10.1021/j100888a024

Gaskell DR (1995) Introduction to the thermodynamics of materials. Taylor & Francis, Washington, DC

Hannay JB (1879) On the artificial formation of the diamond. Proc R Soc 30:188–189. doi:10.1098/rspl.1879.0105

Hirose Y, Mitsuizumi M (1988) Diamond synthesis in air using a flame. New Diamond 4(3):34–35

Hori S, Yoshimura M, Somiya S, Takahashi R (1984) Al_2O_3-ZrO_2 ceramics prepared from CVD powders in Advances in ceramics, vol. 12, Science and technology of zirconia II. American Ceramic Society, Inc, United States

Hurle DT (1993) Handbook of crystal growth. North-Holland, Amsterdam

Hwang NM (1994) Thermodynamic analysis of the chemical vapor deposition of diamond in the C-H, C-O and C-H-O systems. J Cryst Growth 135(1):165–171. doi:http://dx.doi.org/10.1016/0022-0248(94)90738-2

Hwang N-M, Lee D-K (2010) Charged nanoparticles in thin film and nanostructure growth by chemical vapour deposition. J Phys D Appl Phys 43(48):483001

Hwang NM, Yoon DY (1994) Thermodynamic approach to the chemical vapor deposition process. J Cryst Growth 143(1–2):103–109. doi:10.1016/0022-0248(94)90372-7

Hwang NM, Yoon DY (1996) Thermodynamic approach to the paradox of diamond formation with simultaneous graphite etching in the low pressure synthesis of diamond. J Cryst Growth 160(1–2):98–103. doi:http://dx.doi.org/10.1016/0022-0248(95)00549-8

Hwang NM, Bahng GW, Yoon DN (1992) Thermodynamics and kinetics for nucleation of diamond in the chemical vapor deposition process. Diam Relat Mater 1(2):191–194
Hwang NM, Hahn JH, Bahng GW (1994) Thermodynamic approach to the C•H•O deposition diagram in the diamond chemical vapour deposition process. Diam Relat Mater 3(1–2):163–167. doi:http://dx.doi.org/10.1016/0925-9635(94)90051-5
Hwang NM, Hahn JH, Yoon DY (1996a) Charged cluster model in the low pressure synthesis of diamond. J Cryst Growth 162(1):55–68
Hwang NM, Hahn JH, Yoon DY (1996b) Chemical potential of carbon in the low pressure synthesis of diamond. J Cryst Growth 160(1–2):87–97. doi:http://dx.doi.org/10.1016/0022-0248(95)00548-X
Ishihara KN, Maeda M, Shingu PH (1985) The nucleation of metastable phases from undercooled liquids. Acta Metall Mater 33(12):2113–2117. doi:http://dx.doi.org/10.1016/0001-6160(85)90172-5
Kamo M, Sato Y, Matsumoto S, Setaka N (1983) Diamond synthesis from gas phase in microwave plasma. J Cryst Growth 62(3):642–644. doi:http://dx.doi.org/10.1016/0022-0248(83)90411-6
Kurihara K, Sasaki K, Kawarada M, Koshino N (1988) High rate synthesis of diamond by dc plasma jet chemical vapor deposition. Appl Phys Lett 52(6):437
Lewis RS, Ming T, Wacker JF, Anders E, Steel E (1987) Interstellar diamonds in meteorites. Nature 326(6109):160–162
Liander H, Lundblad E (1955) Artificial diamonds. Asea J 28(5–6):97–98
Mason BJ (1958) The supercooling and nucleation of water. Adv Phys 7(26):221–234. doi:10.1080/00018735800101237
Matsumoto S (1985) Chemical vapour deposition of diamond in RF glow discharge. J Mater Sci Lett 4(5):600–602. doi:10.1007/BF00720043
Matsumoto S, Sato Y, Kamo M, Setaka N (1982a) Vapor deposition of diamond particles from methane. Jpn J Appl Phys 21(4A):L183
Matsumoto S, Sato Y, Tsutsumi M, Setaka N (1982b) Growth of diamond particles from methane-hydrogen gas. J Mater Sci 17(11):3106–3112. doi:10.1007/BF01203472
Matsumoto S, Hino M, Kobayashi T (1987) Synthesis of diamond films in a rf induction thermal plasma. Appl Phys Lett 51(10):737
Moissan H (1894) Nouvelles expériences sur la reproduction du diamant. Cr Hebd Acad Aci 118:320
Nuth JA (1987) Small-particle physics and interstellar diamonds. Nature 329(6140):589–589
Okada K, Komatsu S, Matsumoto S (1999) Preparation of microcrystalline diamond in a low pressure inductively coupled plasma. J Mater Res 14(2):578–583. doi:10.1557/JMR.1999.0082
Otake N, Tokura H, Yoshikawa M, Yang C (1990) Deposition of diamond film on a sintered diamond surface and its application to a cutting tool. In S. Saito, O. Fukunaga, M. Yoshikawa (ed) Science and technology of new diamond, Terrapub, Tokyo, pp 139–143
Prawer S, Nemanich RJ (2004) Raman spectroscopy of diamond and doped diamond. Phil Trans R Soc A 362(1824):2537–2565
Ree FH, Winter NW, N. Glosli J, Viecelli JA (1999) Kinetics and thermodynamic behavior of carbon clusters under high pressure and high temperature. Physica B 265(1–4):223–229. doi:http://dx.doi.org/10.1016/S0921-4526(98)01380-5
Salvadori M, Brewer M, Ager J, Krishnan K, Brown I (1992) The effect of a graphite holder on diamond synthesis by microwave plasma chemical vapor deposition. J Electrochem Soc 139(2): 558–560. doi:10.1149/1.2069256
Saslaw WC, Gaustad JE (1969) Interstellar dust and diamonds. Nature 221:160–162. doi:10.1038/221160b0
Sommer M, Mui K, Smith FW (1989) Thermodynamic analysis of the chemical vapor deposition of diamond films. Solid State Commun 69(7):775–778. doi:http://dx.doi.org/10.1016/0038-1098(89)90829-6
Spitsyn BV, Bouilov LL, Derjaguin BV (1981) Vapor growth of diamond on diamond and other surfaces. J Cryst Growth 52:219–226. doi:http://dx.doi.org/10.1016/0022-0248(81)90197-4

Suzuki K, Sawabe A, Yasuda H, Inuzuka T (1987) Growth of diamond thin films by DC plasma chemical vapor deposition. Appl Phys Lett 50(12):728–729

Takagi T (1983) Proceedings of the international ion engineering congress: the 7th symposium (1983 international) on ion sources and ion assisted technology, ISIAT 83, and the 4th international conference on ion and plasma assisted techniques, IPAT '83, 12–16 September, 1983, Kyoto Internatioal Conference Hall, Sakyo-ku, Kyoto, Japan

Tammann G, Mehl RF (1925) The states of aggregation. Nan Nostrand Company, New York

Vakil HB (1989) Thermodynamic and kinetic aspects of CVD diamond growth. Paper presented at the 1st annual diamond technology workshop, Wayne State University, Detroit, Michigan, 18 Sep 1989

Varnin V, Deryagin B, Fedoseev D (1977) Growth of polycrystalline diamond films from the gas phase. Sov Phys Crystallogr 22(4):513–515

Vladimirov SV, Ostrikov K (2004) Dynamic self-organization phenomena in complex ionized gas systems: new paradigms and technological aspects. Phys Rep 393(3–6):175–380. doi:http://dx.doi.org/10.1016/j.physrep.2003.12.003

Whittaker AG (1978) Carbon: a new view of its high-temperature behavior. Science 200(4343):763–764. doi:10.1126/science.200.4343.763

Winter N, Ree F (1998) Carbon particle phase stability as a function of size. J Compu-Aided Mater Design 5(2–3):279–294. doi:10.1023/A:1008674712837

Yarbrough WA (1992) Vapor-phase-deposited diamond—problems and potential. J Am Ceram Soc 75(12):3179–3200. doi:10.1111/j.1151-2916.1992.tb04411.x

Chapter 6
Growth Mechanism of CVD Diamond

The dominant formation of metastable diamond over stable graphite could be explained by comparing the nucleation barrier between diamond and graphite in Chap. 5 based on the classical nucleation theory. However, very puzzling deposition behaviour was observed on an iron substrate under the condition where diamonds deposit on a silicon substrate. This puzzling deposition behaviour was a starting point for the endeavour to investigate the growth mechanism of diamond.

6.1 Deposition Behaviour Between Silicon and Iron Substrates

When silicon, iron and palladium substrates were placed side by side under the same processing condition of HFCVD, diamonds grew on a silicon substrate, highly porous skeletal graphitic soot particles grew on an iron substrate and carbon nanotubes grew on a palladium substrate as shown in Fig. 6.1 (Hwang et al. 1996; Huh et al. 2005). This result implies that the growth mechanisms of diamond, soot and carbon nanotubes should be closely related to one another. Therefore, if the growth mechanism of soot is understood, such understanding would provide a great insight as to the growth mechanism of diamond and carbon nanotubes. Besides, the microstructure of soot is very peculiar and cannot be explained by any conventional crystal growth theory based on the atomic or molecular unit.

Analysis of the Raman spectra shows that soot is graphitic. Soot is porous and skeletal, being very fragile and weakly connected. It can be very easily rubbed by fingers, which means that the bonding between soot particles is not atomic nor molecular bonding but weak van der Waals bonding between particles. The morphology of soot is like the powder compacts formed by landing of particles formed in the gas phase in the conventional CVD process (Pierson 1992).

© Springer Science+Business Media Dordrecht 2016

N.M. Hwang, *Non-Classical Crystallization of Thin Films and Nanostructures in CVD and PVD Processes*, Springer Series in Surface Sciences 60,
DOI 10.1007/978-94-017-7616-5_6

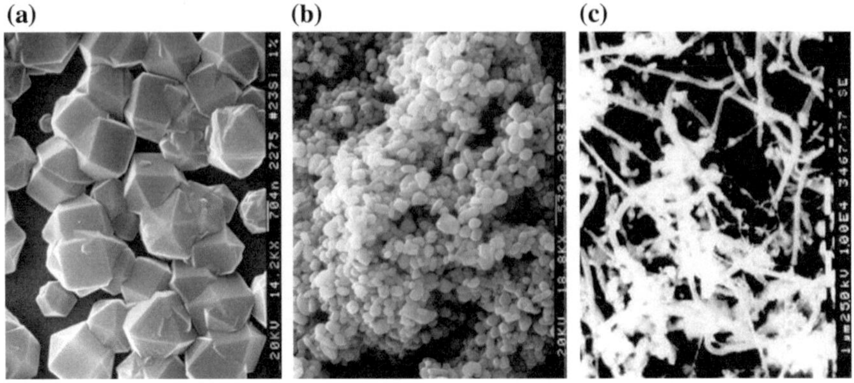

Fig. 6.1 SEM images of **a** diamond deposited on a silicon substrate, **b** soot deposited on an iron substrate and **c** carbon nanotubes (CNTs) on a palladium substrate at a flow rate, 100 sccm, of 1 % CH_4–99 % H_2 gas mixture for 2 h under 2700 Pa at a hot filament temperature of 2200 °C and a substrate temperature of 990 °C. The substrates were placed side by side during hot wire diamond CVD (Hwang et al. 1996; Huh et al. 2005)

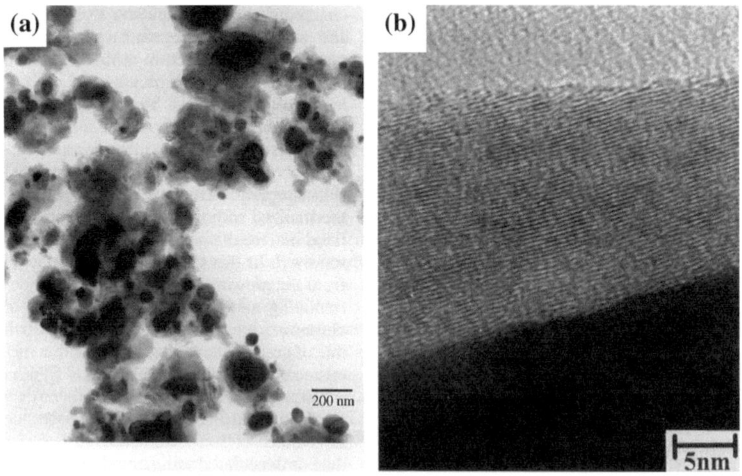

Fig. 6.2 TEM images of soot on the iron substrate at **a** low and **b** high magnifications (Huh et al. 2005)

The growth rate of soot on the iron substrate in Fig. 6.1b is higher than that of diamond on the silicon substrate by more than 10 times. The high growth rate of soot is difficult to interpret by the atomic hydrogen hypothesis because the graphite should be etched by the atomic hydrogen. If the soot is formed by the conversion from diamond by the catalytic effect of the iron substrate, the growth rate of soot should not be higher than that of diamond.

When soot particles are dispersed in ethanol, they are attracted to the magnet, implying that soot particles should carry iron nanoparticles inside. In order to confirm this, soot nanoparticles were observed by TEM. Figure 6.2a, b shows the TEM images of soot respectively in low and high magnification. With the iron particles at the center, the graphitic layer grow, making an onion structure. Iron particles must have originated from the iron substrate presumably by diffusion, which will be more enhanced at high temperature. The formation of iron particles on the growing surface would be retarded as the soot layer gets thicker. This might be why diamonds start to deposit on soot if the deposition time is long enough.

6.2 Formation Mechanism of Soot

Considering that soot is commonly formed during combustion, its formation mechanism is expected to have been studied in the field of combustion and flame. Indeed, the formation mechanism of soot has been extensively studied in the field of combustion and flame (Calcote 1981; Harris and Weiner 1985; Gerhardt and Homann 1990a, b). In this field, soot is regarded as aggregates of fine carbon nanoparticles nucleated and grown in the gas phase.

Since the soot formation is known to be affected by the applied electric bias (Gerhardt and Homann 1990a, b), it is explained by the ionic mechanism, where soot is aggregates of charged carbon nanoparticles after losing their charges (Calcote 1981; Harris and Weiner 1985; Gerhardt and Homann 1990a, b). Homann and his colleagues (Gerhardt et al. 1987; Homann and Traube 1987; Gerhardt and Homann 1990a, b; Weilmünster et al. 1999) made an extensive study to measure the size distribution of charged soot particles under varying combustion conditions. Considering that diamonds can be synthesized by hydrocarbon/oxygen flame (Hirose et al. 1990; Hirabayashi et al. 1993; Cappelli and Paul 1990; Kapil et al. 1996), the study made by Homman and his colleagues is worth mentioning in some detail. Actually, the conditions of hydrocarbon/oxygen flame that were studied by Homann and his colleagues include the diamond synthesis condition although they did not seem to pay attention to the diamond synthesis.

When Homann et al. (Homann and Traube 1987; Gerhardt and Homann 1990a, b) measured the mass distribution, they extracted the charged carbon particles through an orifice from a low vacuum chamber into a high vacuum chamber for the analysis by energy analyzer or Wien filter. The extracted charged particles form a directed beam with a supersonic velocity, which depends on the mass of the particles. Figure 6.3 shows the relation between the average velocities of charged soot particles in the beam as a function of their mass.

The abscissa in Fig. 6.3 is the particle mass in terms of atomic mass unit (amu). Since the amu of carbon is 12, C_{60} corresponds to 720 amu. It should be noted that most of these particles are far smaller than the wavelength of visible light (300 nm), they are invisible to the naked eye.

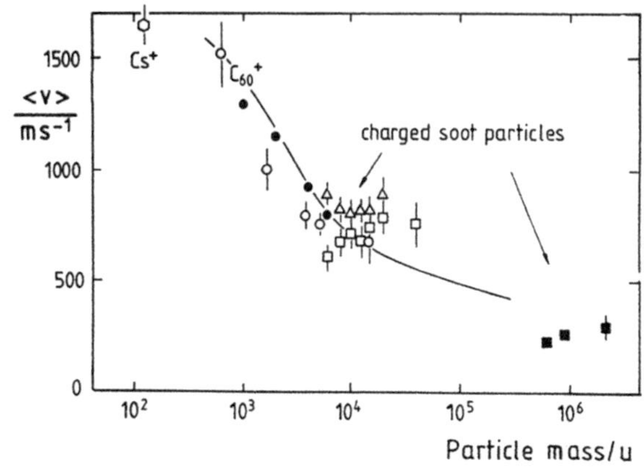

Fig. 6.3 Average velocities of ions in the beam as a function of their mass. The values were obtained by different methods. Cs^+ energy analysis (CsCl-seeded flame), C_{60}^+ and ○ electric deflection with TOF analysis; ● Wien-filter experiments; △ ion draw pulse-method, f = 10 kHz; □ ion draw pulse-method, f = 5 kHz; ■ electric deflection with electron micrographic size determination; p = 2.70 kPa (Gerhardt and Homann 1990b)

6.3 Preferential Formation of Soot at the Corner of Iron Substrates

By repeated experiments on soot formation on the iron substrate in the HWCVD reactor, we observed that soot grows preferentially at the corner of the substrate as shown in Fig. 6.4. As a result, the pinnacle of soot is developed at the corner. This preferential growth at the corner gives the impression that the flux making soot might be electrically charged. The sharp point such as the corner provides the high electric field gradient, which would attract the charged species. In this respect, the tendency of soot to be formed preferentially on the corner might be similar to that of diamond to be formed preferentially on the convex edge as described in Chap. 4.

6.4 Current Measurements in the CVD Diamond Reactor

In order to check whether the flux that produces soot should be electrically charged or not, the current between the substrate and the ground was measured by the electrometer with varying filament temperature at the conditions of 100 sccm of 1 % CH_4–99 % H_2 gas mixture, the reactor pressure of 2700 Pa and the distance of 10 mm between filament and substrate. For current measurements, a stainless steel block was used as a substrate. Figure 6.5 shows the dependence of current data on filament temperature. The current is negative and its magnitude increases with increasing filament temperature.

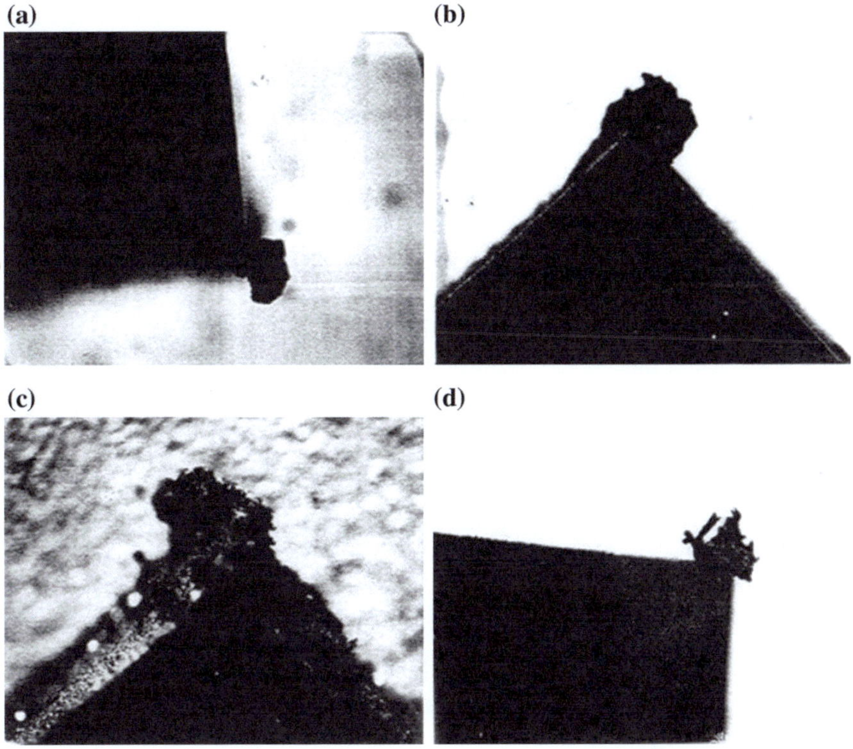

Fig. 6.4 Stereoscopic images of the soot showing the preferential growth at the corner of the iron substrate

Fig. 6.5 The current measured between the substrate and the ground with varying filament temperature

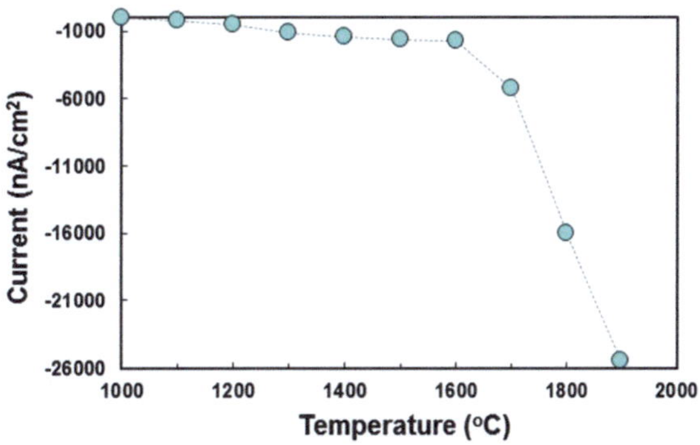

At the typical filament temperature for diamond synthesis, the negative current of more than 25,000 nA/cm^2 is measured. Since one ampere corresponds to $\sim 10^{19}$ charges per s, 25,000 nA/cm^2 corresponds to $\sim 2.5 \times 10^{14}$ charges per s, which would be in the steady state. So a huge amount of charges is generated in the

HFCVD diamond reactor. The source of these negative charges would be electrons emitted from the hot filament. However, it is not certain whether a part of these charges should be negative ions or negatively-charged nanoparticles.

6.5 Ion-Induced Nucleation

As described in Chap. 3, an electric charge acts as a strong heterogeneous nucleation site, triggering nucleation in the gas phase. This phenomenon is 'ion-induced nucleation', which is well known in Wilson's cloud chamber (Wilson 2014; Gupta and Ghosh 1946) and bubble chamber (Peyrou 1967) experiments. It should be reminded of the historical fact that homogeneous nucleation could not be obtained even in room temperature air, where no artificially-ionizing source exists, because ion-induced nucleation could not be avoided due to continuous generation of ion pairs by cosmic rays and natural radio activity. Therefore, in a HFCVD diamond reactor where negative charges exist so abundantly, charge-induced nucleation would occur. As a result it is expected that there exist negatively-charged carbon nuclei in the gas phase.

6.6 Which Property of Substrates Is Related with Soot Formation?

It is reasonable to assume that the formation mechanism of porous skeletal soot on the iron substrate in Figs. 6.1b and 6.4 would be similar to that formed in the field of combustion and flame. Therefore, negatively-charged carbon nuclei seem to produce the soot after being neutralized on the iron substrate. Then, which property of the substrate would be related with the soot formation? The soot formation is not related with the conductivity of the substrate because diamond can be deposited on copper and gold.

In the electrochemistry, the hydrogen H^+ ions migrate to the cathode, receive their missing electrons, and bubble up through the solution as hydrogen gas as follows,

$$2H^{+} + 2e^{-} \rightarrow H_2 \tag{6.1}$$

This reaction of hydrogen evolution tends to be the slowest and a rate-determining step in overall reactions of electrochemistry. The reason why platinum is used as an electrode is that platinum has a high rate of hydrogen evolution. The rate of hydrogen evolution depends on metals as shown in Table 6.1. Palladium has the highest rate and platinum has the second highest rate. This means that Pd and Pt have the high rate to give electrons to hydrogen ions. Therefore, the rate of hydrogen evolution is the rate of charge transfer. Pd, Pt, Rh,

Table 6.1 The rate of the hydrogen evolution reaction on a series of metals at the equilibrium potential (Bockris and Reddy 1977)

Metal	log rate (A/cm)	Metal	log rate (A/cm)
Pd	−3.0	Ag	−5.1
Pt	−3.1	Nb	−5.4
Rh	−3.2	Mo	−5.5
Ir	−3.7	Cu	−6.7
Ni	−5.2	Ta	−7.0
Fe	−5.2	Bi	−8.0
Au	−5.7	Al	−8.1
W	−5.9	Ti	−8.2

Ir, Ni and Fe with a high rate of hydrogen evolution are transition metals which have incompletely filled d orbitals. Transition metals are often good catalysts.

Hwang et al. (1996) and Huh et al. (2005) found out that there exists almost perfect correlation between diamond or soot deposition and the hydrogen evolution rate of the materials. According to Hwang et al. (1996), the substrate materials making soot in the CVD diamond process correspond to the materials having a high charge transfer rate: Pd (Huh et al. 2005), Pt (Huh et al. 2005; Belton and Schmieg 1992), Rh (Huh et al. 2005), Ir (Huh et al. 2005), Ni (Belton and Schmieg 1992) and Fe (Hwang et al. 1996) produce soot or a non-diamond phase whereas Au, W, Mo, Cu, Ta, Ag, Nb (Spitsyn et al. 1981; Huh et al. 2005; Kobashi et al. 1988), Al (Eddy et al. 1992) and Ti (Mori et al. 1992) produce diamond. Bi has too low a melting point to be used as a substrate.

Therefore, it can be said that when negatively-charged carbon nanoparticles formed in the gas phase approach iron or substrates of a high rate of hydrogen evolution, they lose charge to the substrate and then produce soot after being neutralized like the soot formation mechanism in the field of flame and combustion. This way, the formation of porous skeletal structure of soot can be explained. However, the evolution of dense diamond films on silicon or substrates of a low rate of hydrogen evolution cannot be explained. From common sense, gas phase nuclei of negatively-charged carbon nanoparticles are expected to be critically harmful to the growth of high quality diamond films. It is generally believed that the gas phase nucleation should be avoided to produce high quality films.

It is difficult to understand that under the condition where charged nuclei are produced so abundantly in the gas phase, the high quality diamond film can grow as shown in Fig. 6.1a. A question arises as to why charged gas phase nuclei do not spoil the diamond growth. Why don't charged gas phase nuclei land on the diamond growing surface? It should be reminded that the growth rate of soot is higher than that of diamond by more than 10 times. How could charged gas phase nuclei avoid landing on the diamond growing surface although they produce soot at such a high rate on the iron substrate placed next to the silicon substrate?

Charged carbon nanoparticles suspended in the gas phase would be like colloids. In order to understand their behavior in the HWCVD reactor, the behavior of colloidal particles needs to be understood although the size of colloidal particles is

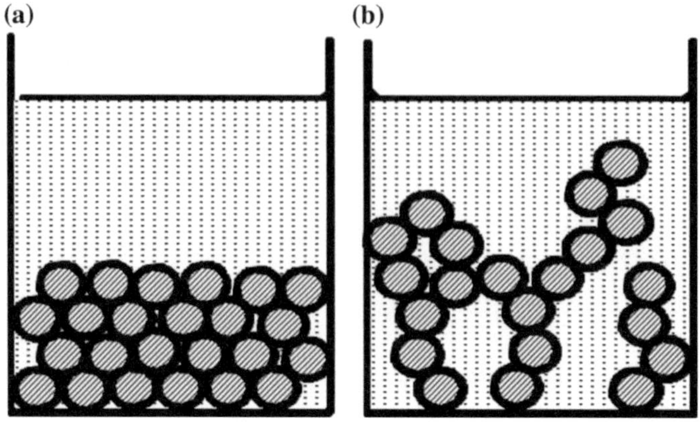

Fig. 6.6 Sedimentation behavior for **a** deflocculation and **b** flocculation of colloidal particles (Shaw and Costello 1993)

much larger than that of charged carbon particles. There are two forces involved in colloids: van der Waals attraction and Coulomb repulsion. When the Coulomb repulsion is dominant, the colloidal particles can be stable and suspended in solution for an extended time. The colloidal particles settle relatively slowly but in a highly regular way undergoing a self-assembly like Fig. 6.6a. This case is called deflocculation. When van der Waals attraction is dominant, colloidal particles are unstable but undergo random Brownian coagulation each other and the aggregated particles fall down and produce a porous skeletal structure such as a fractal structure by diffusion-limited aggregation (DLA) (Meakin 1992) as shown in Fig. 6.6b. This case is called flocculation.

If we compare Fig. 6.6a, b respectively with Fig. 6.1a, b, it appears that they can be matched to a degree. By matching Fig. 6.6a, b respectively to Fig. 6.1a, b, the following scenario can be made. The negatively-charged carbon nanoparticles have a diamond structure. When they approach a silicon substrate, they maintain their charge. They undergo repulsion-dominant slow sedimentation in a highly regular way like Fig. 6.6a. In other words, charged diamond nanoparticles should be the building block. However, Fig. 6.1a is quite different from Fig. 6.6a in that each diamond grain neither consists of nanoparticles nor contains nanometer-sized voids. Therefore, in order to match Fig. 6.6a to Fig. 6.1a, it should be assumed that charged diamond nanoparticles be liquid-like at the substrate temperature, undergoing epitaxial recrystallization without leaving nanometer-sized voids.

When the negatively-charged diamond nanoparticles approach the iron substrate, they seem to lose charge to the substrate just before landing by a tunneling process. After losing charge, the diamond structure changes to the graphite structure, implying that the negative charge stabilizes diamond. The neutral graphite nanoparticles undergo random Brownian coagulation each other and produce a porous skeletal structure. Once the iron substrate is covered with graphite particles, the charge transfer rate would be no longer so high as the iron surface. However,

Fig. 6.2 shows that the iron nanoparticles are supplied from the iron substrate to the growing surface. As far as these iron nanoparticles are supplied, they would continue to convert charged diamond nanoparticles to graphite nanoparticles.

6.7 Comparison of Deposition Behaviour Between Floating and Grounded Iron Substrates

This scenario explains not only the diamond deposition on a silicon substrate and the soot deposition on an iron substrate but also the seemingly-paradoxical phenomena of diamond deposition with simultaneous graphite etching. According to this scenario, the negative charges should stabilize the diamond structure of nanoparticles. In order to confirm the possibility that the electric charge should stabilize diamond, Hwang et al. (1996) compared the deposition behavior between grounded and floating iron substrates. One iron substrate was made grounded by being placed on a steel block and the other was made floating by being placed on a quartz. The iron substrate on an insulating quartz plate is electrically floating and the charge will build up whereas the iron substrate on a steel block is electrically grounded.

Under the same configuration of floating and grounded substrates, the floating substrate has a higher temperature than the grounded substrate because the quartz additionally has a thermal insulation effect. Since the high substrate temperature is favorable for diamond deposition, the substrate temperature was made the same at 990 °C between the floating and grounded substrates by setting the filament temperatures of 2200 and 2050 °C respectively for grounded and floating substrates. It should be noted that the higher filament temperature is favorable for diamond formation.

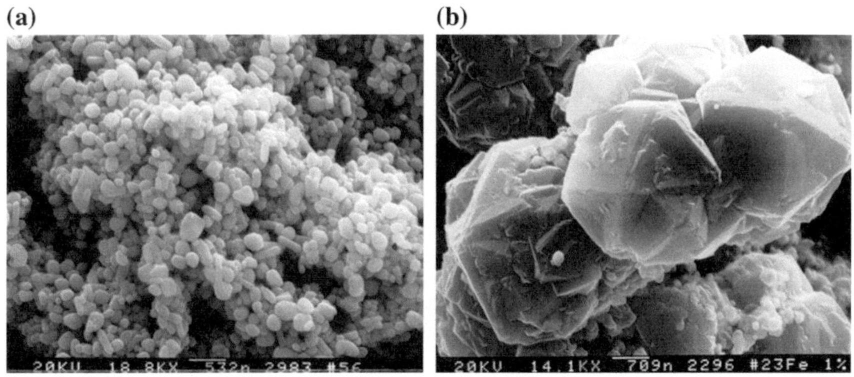

Fig. 6.7 SEM images showing the evolution of **a** soot on a grounded iron substrate and **b** diamond on an initially-formed soot on a floating iron substrate. The other deposition conditions were the same as those of Fig. 6.1 (Hwang et al. 1996)

After 2 h of deposition, soot continued to grow on the grounded iron substrate whereas diamond grew on the initially-formed soot on the floating iron substrate as shown in Fig. 6.7a, b respectively. These results suggest that the electric charge stabilizes diamond over graphite.

6.8 Charged Cluster Model

Based on these results, Hwang et al. (1996) suggested a charged cluster model as a growth mechanism of the low pressure synthesis of diamond using the gas activation. The charged cluster model can be summarized as follows:

1. Negatively charged clusters are formed in the gas phase of HFCVD diamond process.
2. Their size is of a few nanometers and invisible, suspended in the gas phase like colloidal particles.
3. These clusters have the diamond structure. The diamond stability comes from the two factors: capillary pressure and electrical double layer by charge.
4. If they lose charge, they will transform to graphite.
5. When charged clusters approach the Fe substrate of the high charge transfer rate, they lose charge just before landing.
6. After losing charge, they lose the diamond stability coming from the electrical double layer and instantly transform into graphite clusters.
7. These neutral graphite clusters undergo random Brownian coagulation, producing soot on Fe.
8. When charged clusters approach the Si substrate, they retain charge and maintain the diamond structure.
9. They undergo repulsion dominant sedimentation and result in the highly ordered packing and self-assembly combined with epitaxial recrystallization, growing into diamond crystals.
10. Small clusters are liquid-like, undergo epitaxial recrystallization and deposit as high quality diamond crystals.
11. Medium-size clusters made twins or stacking faults.
12. Large clusters deposit as a cauliflower structure.
13. The high substrate temperature enhances epitaxial recrystallization and is favorable for the deposition of high quality films.
14. The low substrate temperature enhances non-epitaxial recrystallization and is favorable for deposition of nano-diamond structure or cauliflower structure.

Although the charged cluster model can uniquely explain not only the diamond deposition with simultaneous graphite etching without violating the second law of thermodynamics but also the microstructure evolution of diamond on a silicon substrate and soot on an iron substrate, it is a revolutionary proposal as to the growth mechanism of diamond at low pressure using a gas activation technique. From the viewpoint of the firmly-believed crystal growth based on an atomic and molecular unit, the idea of diamond growth by charged nanoparticles is unbelievable. Its validity would critically depend on the experimental verification of the generation of the hypothetical negatively-charged diamond nanoparticles in the gas phase during the CVD diamond process.

6.9 Experimental Verification of Hypothetical Diamond Nanoparticles in the Gas Phase

Following the method of Homann's group (Homann and Traube 1987; Gerhardt and Homann 1990b) to use an energy analyzer and Wien filter to determine the energy and mass distribution of charged soot particles in the hydrocarbon/oxygen flame, Jeon et al. (2000, 2001) tried to verify the existence of charged nanoparticles in the gas phase during the HWCVD diamond process and to determine their size distribution. A three-chamber system with two-stage differential pumping, as shown in Fig. 6.8, was used in study. With this system, the hot filament reaction chamber pressure was 800 Pa (6 torr) while the measuring chamber was maintained at less than 0.0013 Pa (10^{-5} torr). The gas adjacent to the substrate for diamond deposition in the hot filament reactor was extracted through a sampling orifice of 1.2 mm diameter to the second chamber and then through a skimmer of 2.0 mm diameter to the measuring chamber. The orifice was made of aluminum, which was connected

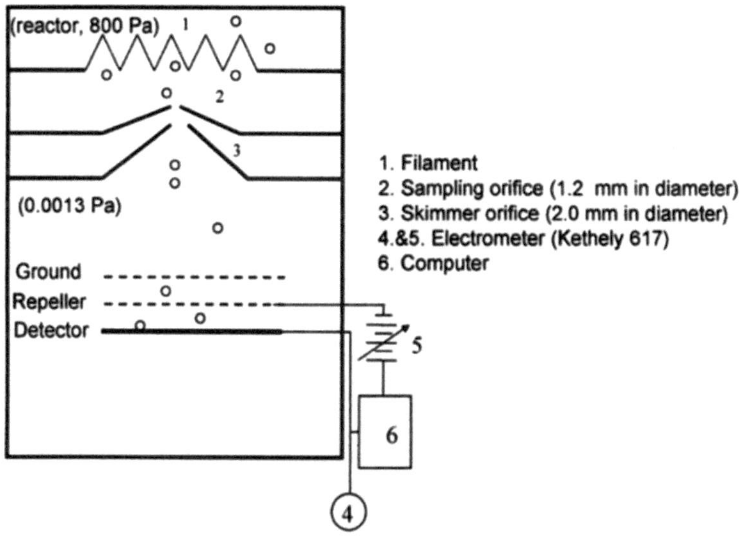

Fig. 6.8 Schematic of the ion mobility analyzer used to measure the energy distribution of charged clusters. The system consists of three chambers. The top chamber is the hot filament reactor. The measuring chamber is at the bottom (Jeon et al. 2001)

to a water-cooled brass plate. The distance between the filament and the orifice was 5 mm. The filament temperature was 2100 °C.

Because of the high thermal conductivity of aluminum, the temperature of the orifice could be kept below 350 °C even when the orifice was only 5 mm away from the hot filament. However, when a 1 mm thick molybdenum plate was placed over the orifice, the temperature increased to 750 °C and there well-crystalline diamonds were deposited.

The kinetic energy of the charged clusters ejected from the reactor was measured by an ion energy analyzer. In the measuring chamber, a repelling voltage was applied to the grid above the current detector. The current variation was measured on the detector with concurrent scanning of the repelling voltage. An energy distribution of the charged clusters was obtained by differentiating the current with respect to the repelling voltage.

During these measurements, diamonds tended to coat near the hole of the aluminum orifice in the reactor. Presumably due to the Coulomb repulsion between the coated diamond at the orifice and the charged clusters, the current measured in the measuring chamber decreased markedly with processing time and the data was not reproducible. In order to avoid this, platinum was coated on the orifice. Another factor affecting the current signal on the detector is the negative electric bias on the filament with respect to the chamber. A stable current was obtained at a bias of −20 V, which was not high enough to trigger a DC glow discharge between the filament and the orifice.

Figure 6.9 shows the current measured on the current detector with repelling voltage for various methane concentrations. When the repelling voltage is −1 V,

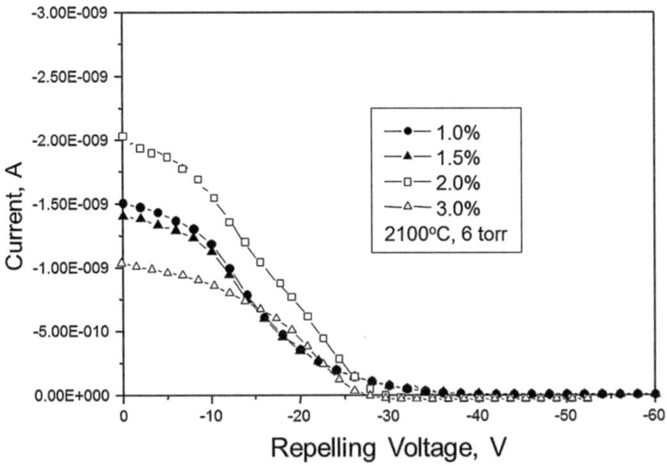

Fig. 6.9 The current measured on the detector in the third chamber with repelling voltage for methane concentrations of 1.0, 1.5, 2.0 and 3.0 % in hydrogen (Jeon et al. 2001)

the negatively charged clusters with energy less than 1 eV will be repelled. Therefore, the current would decrease with increasing repelling voltage.

If the current data in Fig. 6.9 is differentiated with respect to repelling voltage, the energy distribution of charged clusters is obtained as shown in Fig. 6.10. The ordinate in Fig. 6.10 represents the number of charged nanoparticles for the corresponding energy in the abscissa. The number is larger at 2100 °C than at 1900 °C. At the filament temperature of 1900 °C in Fig. 6.10a, energy distributions tend to have two peaks, implying the bimodal distribution of charged nanoparticles. At 2100 °C in Fig. 6.10b, 1 and 1.5 % CH_4 have peaks around −15 V whereas 3 % CH_4 has peaks around −27 V, indicating that clusters at 3 % CH_4 tend to have higher energy than those at 1 and 1.5 % CH_4. On the other hand, 2 % CH_4 in Fig. 6.10b has double peaks, implying the bimodal size distribution.

The extracted gas or clusters have supersonic velocities. The energy of gas or charged clusters through the orifice is given by $\boldsymbol{E} = \frac{1}{2}\boldsymbol{m}\boldsymbol{v}^2$.

In order to determine the mass distribution, the velocity of charged clusters needs to be determined. This is why a Wien filter, which is known as a velocity selector, is used. The schematic of the system installed with a Wien filter is shown in Fig. 6.11. Inside the Wien filter, magnetic field and electric filed are applied in a vertical direction. Charged clusters of a specific velocity that pass through the Wien filter do not deflect but go straight.

By adjusting the electric field of the Wien filter, charged clusters only with a certain velocity can pass through the Wien filter and are measured as a current on the detector. Appreciable current was measured in the velocity range of 500–1000 m/s. Mass distributions for various selected velocities were measured. Four of them are shown in Fig. 6.12a. The total charged cluster mass distribution is the sum of the data for all selected velocities, which is shown in Fig. 6.12b. Since the lower filtering limit of our Wien filter is ~2000 amu, data below this value are not shown.

Nearly all the clusters were negatively charged and the number of positively charged clusters was negligible in the hot filament diamond CVD reactor used in

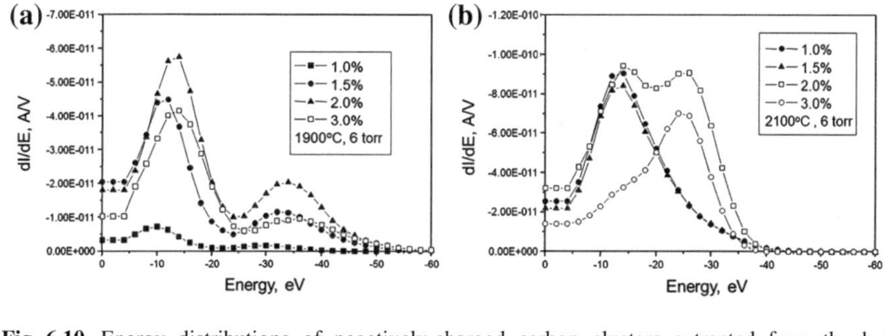

Fig. 6.10 Energy distributions of negatively-charged carbon clusters extracted from the hot filament reactor using gas mixtures of 1 % CH_4–99 % H_2, 1.5 % CH_4–98.5 % H_2, 2 % CH_4–98 % H_2 and 3 % CH_4–97 % H_2 at filament temperatures of **a** 1900 °C and **b** 2100 °C (Jeon et al. 2001)

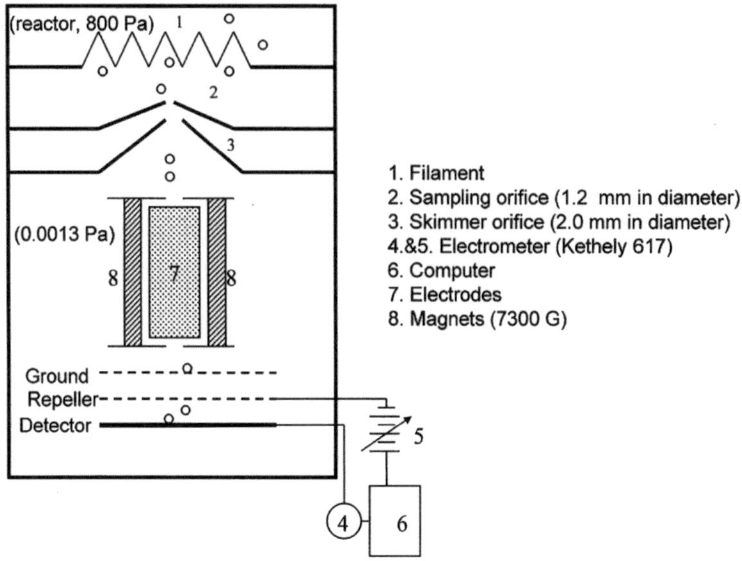

Fig. 6.11 Schematic of the HFCVD system with a Wien filter and an ion mobility analyzer (Jeon et al. 2000)

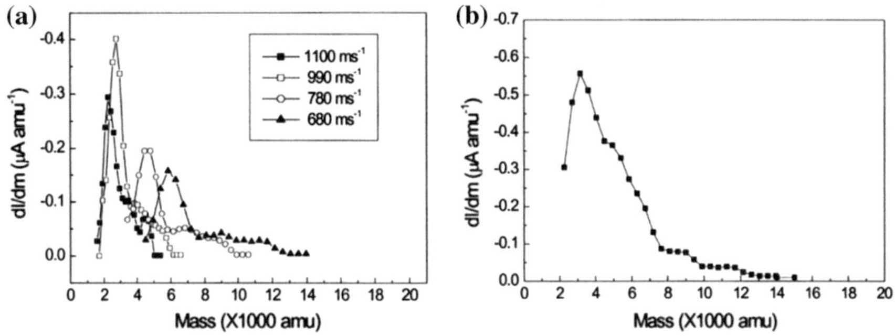

Fig. 6.12 Mass distribution of negatively charged carbon clusters extracted from the hot lament reactor **a** for four selected velocities by a Wien filter and **b** for the sum of the mass distributions for all measured velocities. A gas mixture of 1.5 % CH_4–98.5 % H_2 was supplied to the reactor at a rate of 20 sccm (Jeon et al. 2000)

this study. This fact indicates that the source of charge must be electrons emitted from the hot filament. The distribution peak in Fig. 6.12b is ~3000 amu. This mass would correspond to ~250 carbon atoms if the clusters consist solely of carbon.

The hypothetical charged clusters, whose existence in the diamond CVD reactor was suggested by the charged cluster model, were experimentally confirmed. Such extensive gas phase nucleation together with Fig. 5.17a, b would cause not only graphite but also diamond to be etched at the substrate temperature. In order to avoid the thermodynamic paradox of diamond deposition and simultaneous graphite etching, these gas phase nuclei should be the building block of diamond growth. Besides, atomic contribution to the diamond growth is negative because diamonds are being etched. Therefore, the CVD diamonds grow 100 % by charged clusters.

Gas phase nucleation of diamond powder particles was already reported in an RF plasma by Mitura (1987) and in DC arc plasma by Chonan et al. (1989). The powder particles that they observed are expected to be the product of appreciable growth from charged clusters and to be much larger than individual nanometer-sized charged clusters. Frenklach et al. (1989, 1991) and Howard et al. (1990) also reported the gas phase nucleation of diamond.

6.10 Experimental Evidences for Diamond Etching During the CVD Diamond Process

If diamonds are etched by the atomic unit while growing by nanoparticles, its evidence might be found during the CVD diamond process especially when its growth is retarded for some reason. Figure 6.13 shows the diamond morphology, indicating that diamonds are being etched. The diamonds were deposited for 4 h 30 min under conditions of 100 sccm of 1.6 % CH_4–98.4 % H_2, the distance between the filament and the Si substrate being 7 mm at the filament temperature of 1900 °C. Under this condition, the continuous diamond film was not formed but some isolated diamond particles were formed along the scratch of the Si substrate. Also diamonds, which had grown initially, did not grow appreciably even when the processing time was extended to 4 h 30 min. Instead, the edges are grooved and corners are caved in, indicating they are being etched. It should be noted that corners and edges are the preferential site for etching.

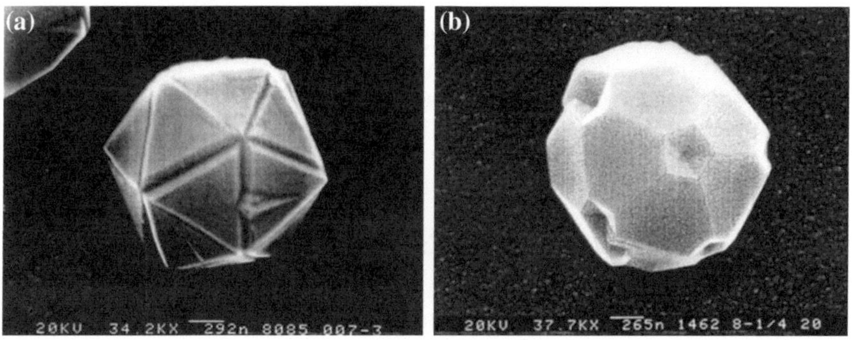

Fig. 6.13 SEM morphology of diamonds evolved after being deposited for 4 h 30 min under the conditions of the reactor pressure of 2700 Pa, the 100 sccm flow rate of 1.6 % CH_4–98.4 % H_2, the distance between hot filament and Si substrate being 7 mm and the filament temperature of 1900 °C. **a** Edges are grooved and **b** corners are caved in

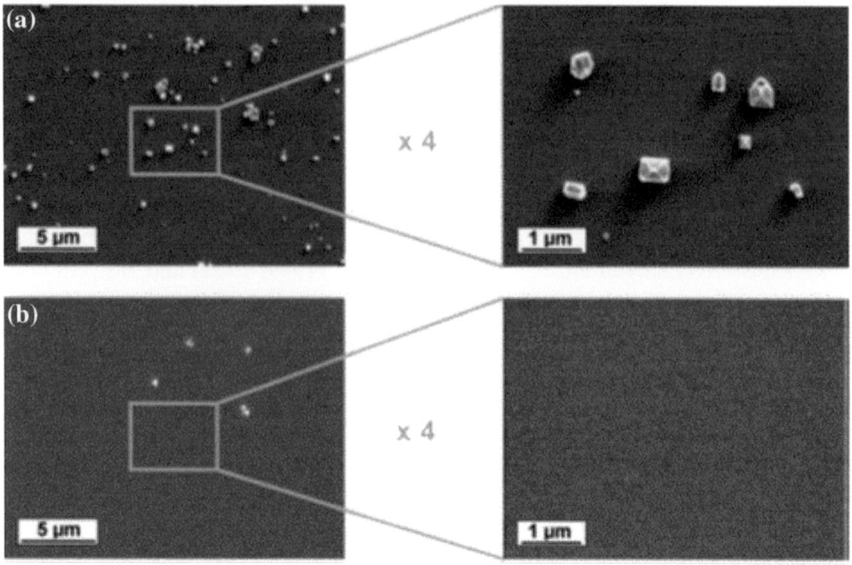

Fig. 6.14 SEM images of a diamond film with very low nucleation density **a** before, and **b** after exposure to 60-min BEN conditions. Edges of image parallel to [110] and $[1\bar{1}0]$ (Hörmann et al. 2001)

Also diamond etching during bias-enhanced nucleation (BEN) treatment was reported by Hörmann et al. (2001). While studying the BEN treatment for nucleation enhancement or epitaxy improvement, their experiments implied that diamond may not grow under conditions to induce diamond nucleation by BEN. In order to check this hypothesis, they conducted a further experiment using thin diamond films with very low nucleation density. Figure 6.14 shows the sample sparsely covered with diamond crystallites of 100–400 nm size before the experiment and after the 60-min treatment step under standard BEN conditions, respectively. Identical positions on the sample are depicted. Figure 6.14a, b clearly show that most of the crystallites at the surface have disappeared through etching. Not only the number of diamond crystals but also their size decreased drastically. Thus, diamond is etched under the standard BEN conditions.

6.11 Ex-situ Observation of Diamond Clusters

Ahn et al. (2002) tried to capture the diamond clusters on a membrane during the oxyacetylene flame diamond CVD process and observe them by TEM. A Mo mesh grid coated with an amorphous silica membrane for TEM was used. The substrate

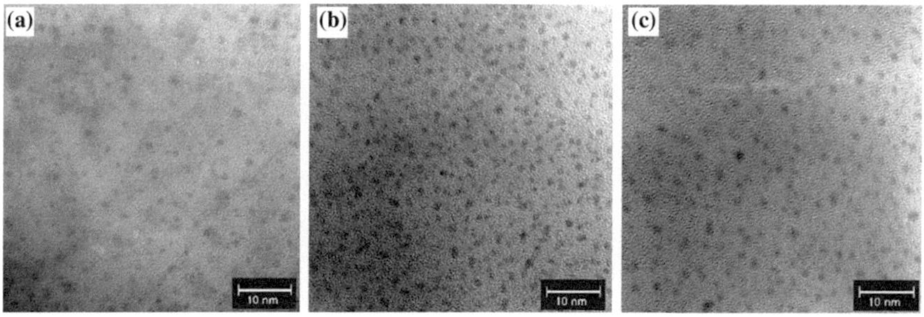

Fig. 6.15 TEM images of carbon clusters captured on the silica membrane for **a** 10 s, **b** 30 s and **c** 60 s with a gas ratio of C_2H_2 to O_2 of 1.04 during the flame deposition of diamond (Ahn et al. 2002)

was placed on the water-cooled copper plate. Diamond crystals were deposited in the flame 16 mm away from the nozzle for the gas ratio of acetylene to oxygen of 1.04. The surface temperature of the water-cooled copper plate was measured to be 600 °C by the thermocouple buried in the copper plate. The actual grid temperature was estimated to be 700–800 °C or even higher, considering that the silica membrane tended to melt with the prolonged exposure in the flame. Figure 6.15a–c show carbon clusters captured for 10, 30 and 60 s, respectively.

Individual clusters are shown as dark spots. Isolated clusters with an average size of 1.5 nm are distributed uniformly on an amorphous silica membrane. The number density of clusters was $2.1 \times 10^{11}/mm^2$ after 10 s, increased to $3.2 \times 10^{11}/mm^2$ after 30 s but decreased to $2.0 \times 10^{11}/mm^2$ after 60 s. The decrease in the number density between 30 and 60 s was repeatedly reproduced. This means that the clusters are being etched away after some initial period under the identical processing conditions.

This phenomenon of cluster deposition with simultaneous atomic etching is similar to that observed in the diamond CVD using the C–H system (Hwang and Yoon 1996). Unlike the C–H system, however, the gas phase nucleation does not change the driving force for etching on the substrate in the C–H–O system using the gas ratio of acetylene to oxygen of 1.04 (Hwang 1994; Gueroudji and Hwang 2000). However, the oxyacetylene torch CVD process is done in open air and the outer flame must be oxygen rich due to the supply of oxygen from the air. This excess oxygen coming from the air appears to make the driving force for etching on the substrate.

In the initial stage before 30 s, the cluster flux landing on the membrane must have exceeded the atomic flux of etching, which results in the increase of the number density of carbon clusters with deposition time. As the charged clusters continue to land on the insulating membrane, the electric charge will build up. This built-up charge will exert the Coulomb repulsion on the incoming charged clusters

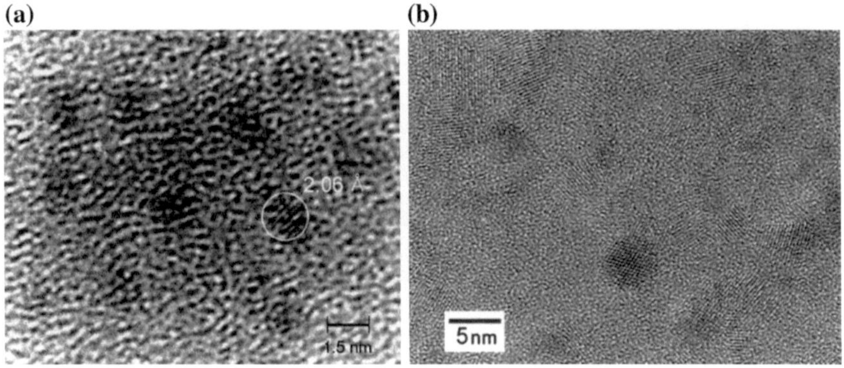

Fig. 6.16 HRTEM images of carbon clusters captured on the silica membrane for **a** 60 s with a ratio of $C_2H_2:O_2$ = 1.04 and **b** 10 s with a ratio of $C_2H_2:O_2$ = 1.09 during the flame deposition of diamond (Ahn et al. 2002)

especially when they have the same sign of the charge. Thus, the cluster flux landing on the membrane will decrease with processing time. If the flux for atomic etching exceeds the flux for cluster landing, the number density of clusters on the membrane will decrease. This picture explains the decrease in the number density of clusters after 60 s in Fig. 6.15c. Most of clusters shown in Fig. 6.15 appear to etch away during prolonged deposition, judging from the final diamond grain size of a few microns evolved after prolonged deposition.

Although most clusters in Fig. 6.15 were amorphous, a few of them have a lattice fringe as shown in Fig. 6.16a, which is a high-resolution image of Fig. 6.15c. Lattice spacing was 2.06 Å, which indicates the spacing of (111) plane of diamond.

The amorphous clusters in Figs. 6.15 and 6.16 are not necessarily amorphous in the gas phase. Because of the liquid-like property of charged clusters, their structure is greatly influenced by the surface they land on. Charged clusters have a strong tendency to accommodate the structure of the surface on which they land. Even if the clusters have the diamond structure in the gas phase, they might change to the amorphous structure on the amorphous membrane. This tendency was observed in gold clusters captured on the amorphous silica membrane (Barnes et al. 2000). Small clusters (2.4 nm) were amorphous while large ones (8 nm) had a gold lattice. This tendency was also observed in the silicon HWCVD process, as will be shown in Chap. 14.

Figure 6.16b shows the HRTEM image of the carbon clusters captured for 10 s at 12.5 mm away from the nozzle for the gas ratio of 1.09. Compared with the clusters in Fig. 6.15, the clusters in Fig. 6.16b are much larger (>5 nm) and have a lattice fringe. Most clusters have a graphite lattice and only a minor fraction of them has a diamond lattice.

6.12 Effect of Methane Concentration on the Size of Charged Clusters

If diamond deposition occurs by the building block of charged clusters, one of the important parameters would be the size of charged clusters. Jeon et al. (2001) examined how the size distribution of charged clusters are varied with methane concentration by measuring their energy distribution as shown in Fig. 6.17. The condition for measurements of the energy distribution are the same as that of Fig. 6.9.

In this case also, most clusters were negatively charged with a negligible amount of positively charged clusters in the hot filament diamond CVD reactor. In the diamond synthesis by the oxy-acetylene frame, however, both positive and negative clusters were measured (Ahn et al. 2002). Figure 6.17a shows that clusters formed at

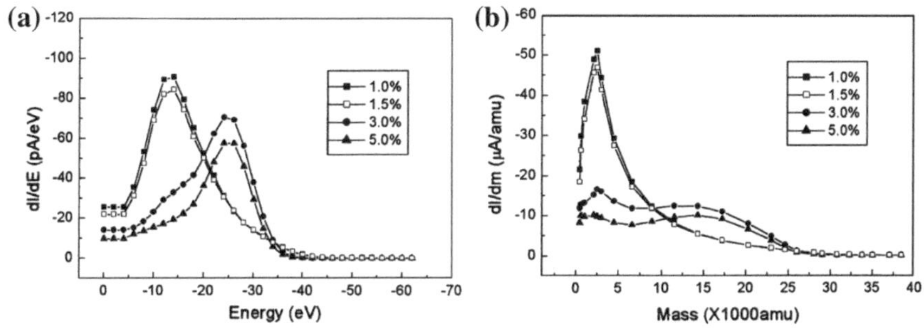

Fig. 6.17 **a** Energy and **b** mass distributions of negatively charged carbon clusters extracted from the hot filament reactor using gas mixtures of 1 % CH_4–99 % H_2, 1.5 % CH_4–98.5 % H_2, 3 % CH_4–97 % H_2, and 5 % CH_4–95 % H_2 (Jeon et al. 2001)

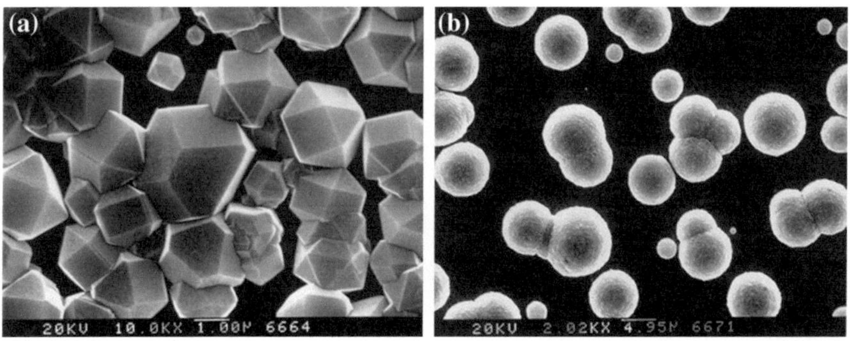

Fig. 6.18 SEM images of diamond films deposited in situ for 1 h during the measurement of energy distribution of charged clusters at the filament temperature of 2100 °C and the reactor pressure of 800 Pa with **a** 1 % CH_4–99 % H_2 and **b** 3 % CH_4–97 % H_2 (Jeon et al. 2001)

low CH_4 concentrations (1 % CH_4–99 % H_2 and 1.5 % CH_4–98.5 % H_2) have low energy while those at high CH_4 concentrations (3 and 5 % CH_4) have higher energy.

Following the scheme suggested by Gerhardt and Homann (1990b), the mass distributions in Fig. 6.17b were obtained from the energy distribution data in Fig. 6.17a based on the measured relations between the velocity and the cluster mass. Clusters of ~3000 amu were dominant for low methane concentrations (1 and 1.5 % CH_4) while an appreciable number of large clusters, approximately ~18,000 amu, existed at higher methane concentrations (3 and 5 % CH_4). If the negatively charged clusters measured in Fig. 6.17b consist of pure carbon, the clusters of ~3000 and ~18,000 amu contain ~250 and ~1500 carbon atoms, respectively.

During the energy distribution measurements, diamonds were deposited in situ on a Mo substrate placed near the orifice with a substrate temperature of 750 °C. The diamond films deposited at 1 and 1.5 % CH_4 showed good crystalline quality, while those at 3 and 5 % CH_4 showed a ball-like or cauliflower-shape structure, as has been well established in the diamond CVD process (Angus and Hayman 1988). Figure 6.18a, b show the SEM images after deposition for 1 h at 1 and 3 % CH_4, respectively.

It is difficult to believe that these faceted diamond crystals in Fig. 6.18a grow by charged clusters. Although it is almost indistinguishable whether they grow by individual atoms or by clusters, the diamond crystals have developed the (100) surface. It should be reminded from Chap. 4 that the growth shape of diamond crystals should be an octahedron enclosed by (111) surfaces according to the periodic bonding chain (PBC) analysis if the diamond growth occurs by individual atoms. According to Sunagawa (1987, 1990), the development of (100) surfaces implied that these diamond crystals should not grow by individual atoms but grow by clusters.

The growth into faceted crystals means that the growth rate is anisotropic or differs among the surfaces, indicating that the attachment rate of charged clusters differs among the surfaces. Since the crystal's surfaces consist mainly of (111) and (100), the attachment rate is low for (111) and (100) surfaces compared with other faces.

Considering that diamond crystals in Fig. 6.18a grow by individual clusters of ~1 nm, a diamond crystal of ~1 μm must have grown by epitaxial recrystallization roughly of billion clusters. Under this condition, secondary nucleation, which is identical to non-epitaxial landing, hardly occurs and initially nucleated diamonds would grow continuously. If a single grain exists initially, this grain would grow as a single crystal. However, in a real growth condition, the substrate is pretreated for enhanced nucleation. If the initial nucleation density on the substrate is high, the grains with the highest vertical growth velocity has the highest probability of survival, grains that grew normal to the substrate survived and grains that grew off normal to the substrate will be terminated by the evolutionary selection based on the Van der Drift model (Van der Drift 1967). In this situation, the film would have a columnar grain structure. The resulting polycrystalline diamond structure is normally of high quality with a low value of full width at half maximum (FWHM) of a Raman peak for diamond.

These diamonds are called microcrystalline diamond (MCD) in comparison with nanocrystalline diamonds (NCD), which consist of the grains of 10–100 nm as shown in Fig. 6.18b. Under the growth condition where the methane concentration is relatively high, the size of the charged nanoparticles is larger as shown in Fig. 6.17b. The diamond morphology is ball-like or spherical in Fig. 6.18b, which is drastically different from cuboctahedrons with well-defined facets in Fig. 6.18a. This means that in contrast with the anisotropic attachment rate of small charged clusters, large clusters have the isotropic attachment rate, resulting in an isotropic growth rate and producing a spherical shape. If the spherical diamond particle of Fig. 6.18b is observed with higher magnification, the surface is covered with numerous nanometre-sized nodules as shown in Fig. 6.19. This morphology is called a cauliflower structure.

In this cauliflower structure, some charged nanoparticles undergo epitaxial recrystallization and others do not. As a result, twins and grains are formed. In order to explain the cauliflower structure such as Fig. 6.19 based on a classical concept of crystal growth by atomic growth, secondary nucleation should occur in such a high rate that numerous secondary nuclei must be formed. As described in Chap. 4, since the secondary nucleation barrier is much higher than the growth barrier, growth is always dominant over secondary nucleation and the secondary nucleation cannot occur in such a high rate as to produce the cauliflower structure. By investigating the structure of globular and cauliflower-like diamond particles prepared by HFCVD from methane-hydrogen gas mixture, Melnikova (2005) suggested that clusters are formed in the vapor phase and then deposit on the surface of diamond particles through coalescence.

As the size of charged clusters increases, the tendency of being liquid-like would decrease. Therefore, large clusters have a tendency to land non-epitaxially on the growing surface, producing secondary nucleation. The minimum grain size would

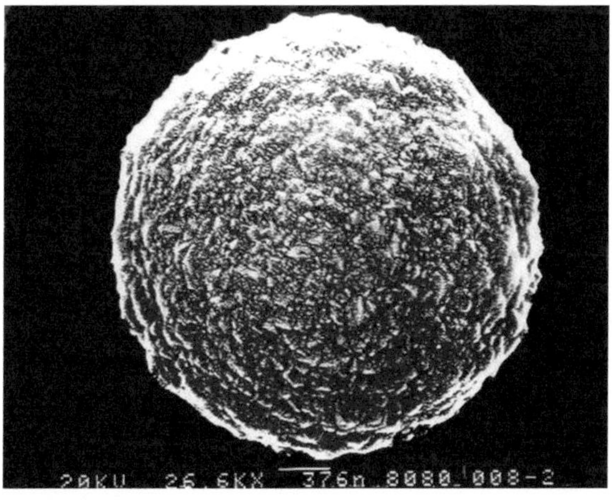

Fig. 6.19 Ball-like or cauliflower diamond with numerous nanonodules on the surface

be obtained when clusters in the gas phase hardly undergo epitaxial landing but deposit as secondary nuclei. Such a case would be nanodiamonds where the diamond morphology consists of equiaxed grains of 3–5 nm (Butler and Sumant 2008). Gruen et al. (1994), Zhou et al. (1998) and Gruen (1999) reported the synthesis of ultra-nanocrystalline diamond films using Ar-C60 (fullerene) or Ar-CH_4 microwave plasma.

6.13 Diamond Synthesis Without Hydrogen at Low Pressure

Although the presence of atomic hydrogen has been considered essential to the growth of the diamond crystals (Badziag et al. 1990; Lambrecht et al. 1993; Butler et al. 1993; Angus et al. 1993; Bachmann 1996). Gruen et al. claimed that they grew diamond films in a hydrogen-free environment (Gruen et al. 1994). Similar claims were made by Mistry et al. (1996) and Badzian et al. (1997). A critical experimental result was reported in 1999 by Palnichenko et al. (1999), who synthesized diamond particles without hydrogen. Instead, they used a high-current arc pulse >1000 A with a 60-ms duration between two 0.5-mm-diameter graphite rods as shown in Fig. 6.20. A high-current arc pulse would produce plasma, generating an abundant amount of electric charges.

Yoshimoto et al. (1999) also reported the nucleation and growth of diamond by pulsed laser deposition of graphite in a hydrogen-free, pure oxygen environment. Beside, their diamonds were heteroepitaxially aligned on a single-crystal sapphire substrate as shown in Fig. 6.21. These diamond syntheses without hydrogen

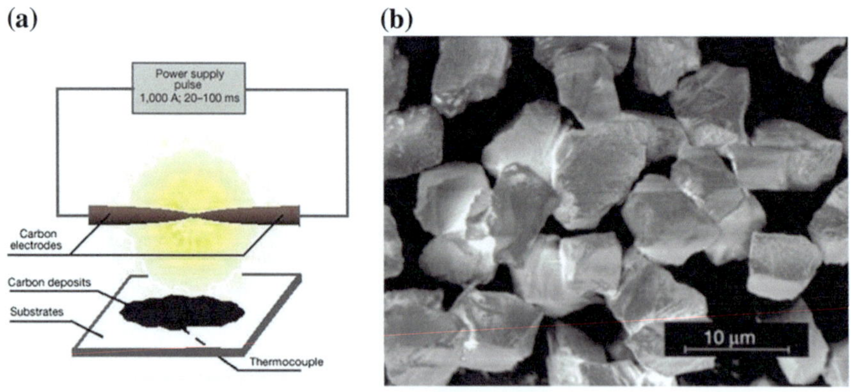

Fig. 6.20 **a** Schematic representation of the thermal activation arc-discharge technique used to synthesize diamond microcrystals and **b** SEM image of diamond microparticles synthesized within a dense carbon-vapour process (Palnichenko et al. 1999)

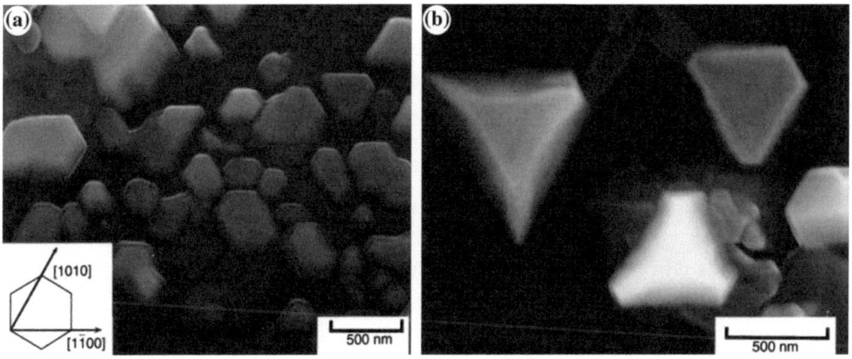

Fig. 6.21 SEM micrographs of the films deposited at 550 °C on the ultrasmooth sapphire (0001) substrate. KrF laser ablation of graphite was employed, in 0.15 torr of oxygen. The hexagonal and well-faceted morphology of the diamond crystals are seen clearly in (**b**). The *inset* in (**a**) shows the orientation of sapphire (0001) substrate, which has a hexagonal structure (Yoshimoto et al. 1999)

Fig. 6.22 Total energy of diamond terminated and graphite terminated cases with respect to various charging states. The abscissa is the excess charge in electron numbers and the ordinate is total energy with respect to that of charge-neutral graphite terminated case (Park et al. 2004)

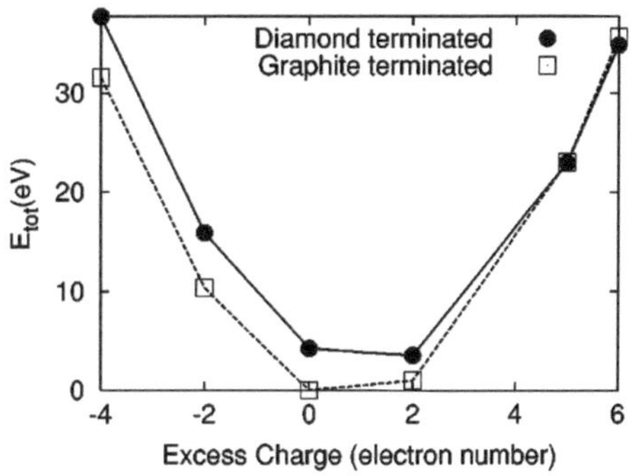

contradict previously held views as to the role of hydrogen in the low pressure synthesis of diamond. Therefore, the role of gas activation techniques such as hot filament or plasma should be reconsidered.

Based on a density functional calculation, Park et al. (2004) studied the effect of charge on the stability of a diamond nanocluster surface and showed that a negative charge stabilized diamond over graphite and a positive charge stabilized the hydrogenated surface. As shown in Fig. 6.22, they showed that excess electrons stabilized diamond terminated surface over graphite terminated one.

Although more elaborate calculations are needed for the quantitative estimation of the charge effect on the stability of diamond clusters, both experiments and theoretical calculation indicate that negative charge should stabilize diamonds. This new understanding suggests that efforts should be made to generate abundant electric charges and the size of carbon clusters should be made small enough to stabilize diamond by the capillary pressure in the presence of charge.

Although numerous experimental results suggested that atomic hydrogen has a beneficial effect on the diamond synthesis, it is not certain whether the role of atomic hydrogen to reduce the surface energy of diamond by adsorption is additive to the diamond stabilization or not. Homann's analysis (Gerhardt et al. 1987; Gerhardt and Homann 1990a) implies that negatively-charged carbon clusters are not hydrogenated but pure carbon although positively-charged carbon clusters are hydrogenated. Besides, if hydrogenated diamond clusters deposit as a building block, the hydrogen content in diamond films is expected to be rather high. However, the hydrogen content in the CVD diamond is not so high. In order to clarify whether negatively-charged carbon clusters generated in the CVD diamond process are hydrogenated or not, these clusters should be extracted to the high vacuum chamber and analyzed by the accurate quadruple or time-of-flight mass analyzer. Taking all experimental results into consideration, it may be that the role of atomic hydrogen is simply to etch carbon clusters to reduce their size, which would maintain high capillary pressure and thereby be beneficial to the diamond stability. In the case of laser ablation of graphite to synthesize a diamond film by Yoshimoto et al. (1999) oxygen would play a role of etching.

6.14 Growth Rate With and Without Gas Phase Nucleation

In Chap. 5, in the old CVD process, where a diamond seed was used as a substrate, a graphitic debris was formed even at the very low growth rate of diamond. Once a graphite debris was formed on the growing diamond surface, the graphite grew much faster than diamond and covered the entire surface of diamond. In order to grow diamond further, the graphite had to be etched. Nevertheless, the growth rate was so low ($\sim$ 10 Å/h).

This growth rate would represent the maximum rate that can be achieved by atomic growth without gas phase nucleation. Such a limited growth rate comes from the fact that the onset of gas phase nucleation can occur at relatively low supersaturation ratio because of strong heterogeneous sites such as ions in the gas phase. In the new CVD diamond process, however, the growth rate is higher than that of old CVD diamond by more than thousands times. This means that such the high growth rate of the new CVD diamond process is possible because gas phase nucleation occurs and the gas phase nuclei contribute to diamond growth. By classical crystallization, the growth rate of films would be limited by gas phase nucleation. If the films grow by non-classical crystallization, the limit of the growth rate imposed by the gas phase nucleation can be increased by more than thousands times.

6.15 Deposition Behaviour of Charged Clusters

If it is understood that diamond films grow by charged clusters, many previous results can be explained based on the deposition behaviour of charged clusters. Because of the electric charge carried by the diamond clusters, very unique deposition behaviour can occur which cannot be explained by neutral fluxes such as atoms and molecules in classical crystallization.

6.15.1 Preferential Diamond Formation on the Edge

In the CVD diamond process, it is well known that diamonds preferentially nucleate along scratches on the substrate or on the edge as mentioned in Chap. 3. The preferential formation of soot at the corners of substrates shown in Fig. 6.5 also represents these aspects. These results can be understood by considering that charged diamond clusters preferentially deposit on the area of the high electric field gradient.

Thinking that it is possible to have an enhancement of electrical fields at the scratch of the substrate, Zhang and Buck (2000) prepared sharp tips by etching in a graphite substrate. Figure 6.23 shows clearly that nucleation occurs at the top of the tips. Buck (2008) suggested Fig. 6.23 as the evidence for charged or polar precursors in diamond nucleation. This result can also be understood by considering that charged clusters are attracted to the tip of the high field gradient.

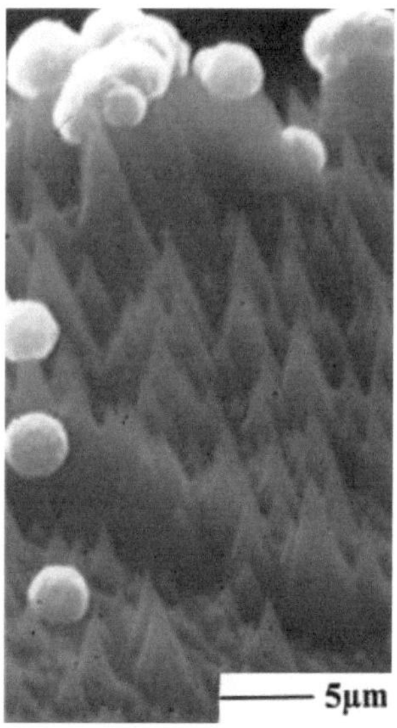

Fig. 6.23 Nucleation of diamonds on top of graphite tips (Zhang and Buck 2000)

Another result related with the deposition behaviour of charged clusters is the bias experiment. The bias effect was studied extensively with lots of experimental results being available.

6.15.2 Bias Effect in the Hot Filament CVD Process

Sawabe and Inuzuka (1985, 1986) first applied the positive bias to the Mo substrate holder, on which the SiC substrate was placed, in the HFCVD diamond process with the intention of bombarding electrons emitted from the hot tungsten filament. They called the process electron-assisted CVD (EACVD). The applied voltage was 150 V with the filament temperature of 1900 °C and the substrate temperature of 600–850 °C. 0.5–2 % CH_4 in hydrogen was supplied at the flow rate of 50 sccm. They could measure the electron current density of 10 mA/cm^2 at the Mo substrate holder. They compared the rate of diamond growth of EACVD with that of the normal HFCVD. The positive bias increased the rate by about 10 times. However, when they applied a negative voltage to the substrate holder, the growth of diamond could not be detected and the growth of graphitic carbon was found over a wide range of experimental conditions.

According to our experiments, the positive bias tends to induce DC glow discharge between the hot filament and the substrate when the bias voltage is higher than ~30 V. The occurrence of glow discharge can be judged most easily by measuring the current. Before the glow discharge occurs, the current density on the substrate is typically 10–100 μA/cm^2, which depends on the filament temperature and other parameters. With the onset of the glow discharge, the current density is increased higher than ~mA/cm^2. Therefore, the current density of 10 mA/cm^2 measured by Sawabe and Inuzuka (1985) indicates that the glow discharge occurred.

The diamond quality is degraded abruptly when the glow discharge occurs. This aspect was mentioned by Ma et al. (1990), who found that a positive surface bias of 150 V with respect to the filament was not favorable to the growth of high quality, low defect density diamond films. Using energy dispersive X-ray spectrometry (EDXS), they showed the substrate surface was contaminated by tungsten, which must have come from the erosion of the tungsten filament by the glow discharge. Therefore, the bias voltage that can be applied without inducing the glow discharge is below ~30 V. Even in this range, the enhancement of diamond formation is observed. This enhancement would come from the attraction of negatively-charged diamond clusters to the substrate rather than the electron bombardment effect.

However, when the bias voltage is negative, the glow discharge does not occur even for ~200 V. The negative bias to the substrate would repel the negatively-charged diamond clusters but would attract the positively-charged hydrocarbon or ions. This would be why Sawabe and Inuzuka (1986) could not detect the diamond growth and found the growth of graphitic carbon over a wide range of experimental conditions. The effect of the negative bias will be treated in the bias-enhanced nucleation (BEN) section. Motivated by the beneficial effect of the bias, which induced the DC glow discharge, Suzuki et al. (1987) tried the DC plasma CVD to synthesize diamond.

6.15.3 *Bias Effect in the DC Plasma Process*

Using the DC plasma CVD at 200 Torr, Suzuki et al. (1987) could produce a nucleation density of $\sim 10^8/cm^2$ on mirror-polished Si wafers without surface scratching by diamond or c-BN powder. This result is in contrast with a very low nucleation density of $\sim 10^4/cm^2$ in the HFCVD process when the substrate is not treated for diamond nucleation. Besides, the growth rate was ~ 20 μm/h. If they placed the substrate on the cathode, however, the diamond growth could not be detected and amorphous or graphitic carbon was found over a wide range of experimental conditions.

Together with the bias experiment in the HFCVD with a DC glow discharge, the DC plasma CVD results provide valuable information as to the characters of negatively and positively charged clusters. The experimental fact is that diamond deposits on the positively-biased substrate or the anode and non-diamond or graphite deposits on the negatively-biased substrate or the cathode. These results imply that negatively-charged clusters should be diamond and positively-charged clusters should be hydrogenated carbon clusters, which agrees with the Homann's analysis of charged clusters in the acetylene flame (Gerhardt et al. 1987).

Using the DC plasma CVD, Koizumi et al. (1990) reported the epitaxial growth of diamond thin films on cubic boron nitride (111) surfaces. They used the mixture of 0.5 % CH_4–99.5 % H_2 under 180 torr. During deposition, the discharge voltage and the current density were maintained at 700 V and 1.8 A/cm^2, respectively, with the substrate temperature of 900 °C. The epitaxial relation was confirmed to be $(111)_{diamond}//(111)_{c\text{-}BN}$ and $[110]_{diamond}//[110]_{c\text{-}BN}$.

In order to understand how the epitaxial growth can be made by the deposition of charged clusters, it is necessary to introduce the concept of 'ionized cluster beam deposition (ICBD) and epitaxy' developed by Tagaki et al. (1976). In the ICBD process, clusters are generated by adiabatic expansion from the crucible, ionized by electron impact, accelerated by electric field and then deposited on the substrate. This process is famous for deposition of epitaxial films which cannot be obtained by conventional methods.

From the ICBD process, it can be concluded that if charged clusters are accelerated on to the substrate, they have the tendency to deposit epitaxially. In relation to this tendency, the bias effect on the homo-epitaxial growth of Si is shown in Fig. 14.4 in Chap. 14. Therefore, considering the deposition behaviour of charged clusters, the DC plasma CVD process naturally provides the situation, where negatively-charged diamond clusters are generated and these clusters are accelerated toward the substrate by the discharge voltage. Therefore, the epitaxial growth of diamond on the BN substrate, which has a very small lattice mismatch with diamond, is not so surprising.

6.15.4 Bias Enhanced Nucleation (BEN)

In the process called 'bias-enhanced nucleation' (BEN), pioneered by Yugo et al. (1990), the negative bias of 50–300 V is applied to the substrate with respect to the hot filament or microwave plasma CVD chamber for a short period (typically <30 min) as a pretreatment prior to actual deposition. Typically, higher methane concentration than the normal process is used in the range of 1–30 % CH_4/H_2. The reactor pressure is 5–40 torr. Then, the bias is turned off and the process changes to normal growth condition with lower methane concentration of 0.5–2 % in plasma. The BEN treatment produces the extremely high nucleation density of 10^9–10^{11}/cm^2 in contrast with 10^7/cm^2 on scratched Si wafers and 10^3–10^5/cm^2 on untreated wafers.

If the bias is left on throughout the entire deposition, the resulting film will have much poorer quality than if the bias is off and the normal growth condition using methane concentration <3 % is allowed to begin. The BEN process not only increases the nucleation density but also produces highly oriented diamond on silicon (100) or (0001) SiC, which tended to form as an interlayer between silicon and diamond (Williams and Glass 1989; Meilunas et al. 1989). Stoner and Glass (1992) produced oriented diamond nuclei on a β-silicon carbide substrate with orientation relationship of $(001)_D//(001)_{SiC}$ and $[110]_D//[110]_{SiC}$. Then, it was shown by Jiang et al. (1993) that BEN produced oriented diamond films on silicon.

If the BEN effect is explained by the charged cluster model, the positively-charged ions or clusters, which contain hydrogenated non-diamond phase as well as charged diamond nanoparticles, are accelerated toward the substrate. Among these, positively charged diamond nanoparticles are accelerated to the substrate and tend to result in oriented diamond formation like the ICBD process. However, diamond nanoparticles are embedded in a large amount of non-diamond carbon. When the bias is off and the normal growth condition begins, the driving force is for etching at the substrate temperature as explained earlier. Under this condition, non-diamond carbon will be etched away and disappear, leaving numerous diamond nanoparticles on the substrate. Diamond nanoparticles left on the substrate would act as seed, increasing the nucleation density. Therefore, the BEN effect of enhanced oriented growth and nucleation density can be explained.

References

Ahn H-S, Park H-M, Kim D-Y, Hwang NM (2002) Observation of carbon clusters of a few nanometers in the oxyacetylene diamond CVD process. J Cryst Growth 234(2–3):399–403. doi:http://dx.doi.org/10.1016/S0022-0248(01)01748-1

Angus JC, Hayman CC (1988) Low-pressure, metastable growth of diamond and "diamondlike" phases. Science 241(4868):913–921. doi:10.2307/1702445

Angus JC, Argoitia A, Gat R, Li Z, Sunkara M, Wang L, Wang Y (1993) Chemical vapour deposition of diamond. Phil Trans R Soc A 342(1664):195–208. doi:10.1098/rsta.1993.0014

Bachmann P (1996) Plasma chemical vapor deposition of diamond films. In: A. Paoletti, A. Tucciarone (ed) The Physics of Diamond, vol 135. IOS Press, Varenna, p 45–71. doi:10.3254/978-1-61499-220-2-195
Badziag P, Verwoerd WS, Ellis WP, Greiner NR (1990) Nanometre-sized diamonds are more stable than graphite. Nature 343(6255):244–245
Badzian AR, Roy R, Mistry P, Turchan M (1997) Laser approaches for diamond synthesis: The QQC process. In: A. Paoletti, A. Tucciarone (ed) The Physics of Diamond, vol 135. IOS Press, Varenna, p 195–208. doi:10.3254/978-1-61499-220-2-195
Barnes MC, Kim D-Y, Ahn HS, Lee CO, Hwang NM (2000) Deposition mechanism of gold by thermal evaporation: approach by charged cluster model. J Cryst Growth 213(1–2):83–92. doi: http://dx.doi.org/10.1016/S0022-0248(00)00359-6
Belton DN, Schmieg SJ (1992) Nucleation of chemically vapor deposited diamond on platinum and nickel substrates. Thin Solid Films 212(1):68–80.
Bockris JO, Reddy AKN (1977) Modern electrochemistry: An introduction to an interdisciplinary area, vol 5. Plenum Press, New York
Buck V (2008) Evidence for charged or polar precursors in diamond nucleation. J Optoelectron Adv Mater 10(1):85–90
Butler JE, Sumant AV (2008) The CVD of nanodiamond materials. Chem Vapor Depos 14(7–8): 145–160
Butler JE, Woodin RL, Brown LM, Fallon P (1993) Thin film diamond growth mechanisms [and Comment]. Phil Trans R Soc A 342(1664):209–224. doi:10.1098/rsta.1993.0015
Calcote HF (1981) Mechanisms of soot nucleation in flames—A critical review. Combust Flame 42:215–242. doi:http://dx.doi.org/10.1016/0010-2180(81)90159-0
Cappelli MA, Paul P (1990) An investigation of diamond film deposition in a premixed oxyacetylene flame. J Appl Phys 67(5):2596–2602
Chonan T, Uenura M, Futaki S, Nishi S (1989) The preparation and characterization of diamond powder by DC arc plasma. Jpn J Appl Phys 28(6A):L1058-L1060
Eddy C, Youchison D, Sartwell B, Grabowski K (1992) Deposition of diamond onto aluminum by electron cyclotron resonance microwave plasma-assisted CVD. J Mater Res 7(12):3255–3259
Frenklach M, Kematick R, Huang D, Howard W, Spear K, Phelps AW, Koba R (1989) Homogeneous nucleation of diamond powder in the gas phase. J Appl Phys 66(1):395–399. doi:http://dx.doi.org/10.1063/1.343890
Frenklach M, Howard W, Huang D, Yuan J, Spear K, Koba R (1991) Induced nucleation of diamond powder. Appl Phys Lett 59(5):546–548. doi:http://dx.doi.org/10.1063/1.105434
Gerhardt P, Homann KH (1990a) Ions and charged soot particles in hydrocarbon flames. 2. positive aliphatic and aromatic ions in ethyne/oxygen flames. J Phys Chem 94(13):5381–5391. doi:10.1021/j100376a039
Gerhardt P, Löffler S, Homann KH (1987) Polyhedral carbon ions in hydrocarbon flames. Chem Phys Lett 137(4):306–310. doi:http://dx.doi.org/10.1016/0009-2614(87)80889-8
Gerhardt PH, Homann KH (1990b) Ions and charged soot particles in hydrocarbon flames I. Nozzle beam sampling: Velocity, energy, and mass analysis; total ion concentrations. Combust Flame 81(3–4):289–303. doi:http://dx.doi.org/10.1016/0010-2180(90)90026-N
Gruen DM (1999) Nanocrystalline diamond films 1. Annu Rev Mater Sci 29(1):211–259
Gruen DM, Liu S, Krauss AR, Luo J, Pan X (1994) Fullerenes as precursors for diamond film growth without hydrogen or oxygen additions. Appl Phys Lett 64(12):1502–1504. doi:http://dx.doi.org/10.1063/1.111872
Gueroudji L, Hwang NM (2000) Thermodynamic limits for the substrate temperature in the CVD diamond process. Diam Relat Mater 9(2):205–211. doi:http://dx.doi.org/10.1016/S0925-9635(00)00232-6
Gupta ND, Ghosh S (1946) A report on the wilson cloud chamber and its applications in physics. Rev Mod Phys 18(2):225

Hörmann F, Schreck M, Stritzker B (2001) First stages of diamond nucleation on iridium buffer layers. Diam Relat Mater 10(9–10):1617–1621. doi:http://dx.doi.org/10.1016/S0925-9635(01)00431-9
Harris SJ, Weiner AM (1985) Chemical kinetics of soot particle growth. Ann Rev Phys Chem 36 (1):31–52. doi:10.1146/annurev.pc.36.100185.000335
Hirabayashi K, Kimura T, Hirose Y (1993) Morphology of flattened diamond crystals synthesized by the oxy-acetylene flame method. Appl Phys Lett 62(4):354–356
Hirose Y, Amanuma S, Komaki K (1990) The synthesis of high-quality diamond in combustion flames. J Appl Phys 68(12):6401–6405
Homann KH, Traube J (1987) Use of a wien filter for mass and velocity analysis of charged soot particles in a nozzle beam. Berich Bunsen Gesell 91(8):828–833. doi:10.1002/bbpc.19870910814
Howard W, Huang D, Yuan J, Frenklach M, Spear K, Koba R, Phelps A (1990) Synthesis of diamond powder in acetylene oxygen plasma. J Appl Phys 68(3):1247–1251. doi:http://dx.doi.org/10.1063/1.346725
Huh J-M, Yoon D-Y, Kim D-Y, Hwang N-M (2005) Effect of substrate materials in the low-pressure synthesis of diamond: approach by theory of charged clusters. Z Metallkd 96 (3):225–232. doi:10.3139/146.101024
Hwang NM (1994) Thermodynamic analysis of the chemical vapor deposition of diamond in the C-H, C-O and C-H-O systems. J Cryst Growth 135(1):165–171. doi:http://dx.doi.org/10.1016/0022-0248(94)90738-2
Hwang NM, Yoon DY (1996) Thermodynamic approach to the paradox of diamond formation with simultaneous graphite etching in the low pressure synthesis of diamond. J Cryst Growth 160(1–2):98–103. doi:http://dx.doi.org/10.1016/0022-0248(95)00549-8
Hwang NM, Hahn JH, Yoon DY (1996) Charged cluster model in the low pressure synthesis of diamond. J Cryst Growth 162(1):55–68
Jeon I-D, Park CJ, Kim D-Y, Hwang NM (2001) Effect of methane concentration on size of charged clusters in the hot filament diamond CVD process. J Cryst Growth 223(1):6–14
Jeon J-D, Park CJ, Kim D-Y, Hwang NM (2000) Experimental confirmation of charged carbon clusters in the hot filament diamond reactor. J Cryst Growth 213(1):79–82
Jiang X, Schiffmann K, Westphal A, Klages CP (1993) Atomic-force-microscopic study of heteroepitaxial diamond nucleation on (100) silicon. Appl Phys Lett 63(9):1203–1205. doi:http://dx.doi.org/10.1063/1.109771
Kapil R, Mehta B, Vankar V (1996) Synthesis of 15R polytype of diamond in oxy-acetylene flame grown diamond thin films. Appl Phys Lett 68(18):2520–2522
Kobashi K, Nishimura K, Kawate Y, Horiuchi T (1988) Synthesis of diamonds by use of microwave plasma chemical-vapor deposition: Morphology and growth of diamond films. Phys Rev B 38(6):4067
Koizumi S, Murakami T, Inuzuka T, Suzuki K (1990) Epitaxial growth of diamond thin films on cubic boron nitride {111} surfaces by dc plasma chemical vapor deposition. Appl Phys Lett 57 (6):563–565. doi:http://dx.doi.org/10.1063/1.103647
Lambrecht WRL, Lee CH, Segall B, Angus JC, Li Z, Sunkara M (1993) Diamond nucleation by hydrogenation of the edges of graphitic precursors. Nature 364(6438):607–610
Ma G-HM, Lee YH, Glass JT (1990) Electron microscopic characterization of diamond films grown on Si by bias-controlled chemical vapor deposition. J Mater Res 5(11):2367–2377. doi:10.1557/JMR.1990.2367
Meakin P (1992) Aggregation kinetics. Phys Scripta 46(4):295
Meilunas R, Wong M, Sheng K, Chang R, Van Duyne R (1989) Early stages of plasma synthesis of diamond films. Appl Phys Lett 54(22):2204–2206. doi:http://dx.doi.org/10.1063/1.101124
Melnikova V (2005) The cluster growth mechanism of nanostructured diamond. In: Veziroglu TN, Yu. Zaginaichenko S, Schur D, Baranowski B, Shpak A, Skorokhod V (eds) Hydrogen Materials Science and Chemistry of Carbon Nanomaterials, vol 172. NATO Science Series II: Mathematics, Physics and Chemistry. Springer, Netherlands, pp 557–562. doi:10.1007/1-4020-2669-2_64

Mistry P, Turchan M, Liu S, Granse G, Baurmann T, Shara M (1996) A revolutionary diamond synthesis technique: the QQC materials deposition process. Innov Mater Res 1(2):193–207
Mitura Sw (1987) Nucleation of diamond powder particles in an RF methane plasma. J Cryst Growth 80(2):417–424. doi:http://dx.doi.org/10.1016/0022-0248(87)90090-X
Mori Y, Deguchi M, Kitabatake M, Yagi H, Hatta A, Ito T, Hirao T, Hiraki A Ion-induced fabrication of high-quality diamond particles on various substrates. In: San Diego'92, 1992. SPIE, pp 18–24
Palnichenko AV, Jonas AM, Charlier JC, Aronin AS, Issi JP (1999) Diamond formation by thermal activation of graphite. Nature 402(6758):162–165
Park N, Park S, Hwang N-M, Ihm J, Tejima S, Nakamura H (2004) First-principles study of the effect of charge on the stability of a diamond nanocluster surface. Phys Rev B 69(19):195411
Peyrou C (1967) Bubble chamber principles. In: Bubble and Spark Chambers: Volume II; Principles and Usc, Academic Press, New York, p 20
Pierson HO (1992) Handbook of chemical vapor deposition: principles, technology and applications. Noyes Publications, New York
Sawabe A, Inuzuka T (1985) Growth of diamond thin films by electron assisted chemical vapor deposition. Appl Phys Lett 46(2):146–147. doi:http://dx.doi.org/10.1063/1.95715
Sawabe A, Inuzuka T (1986) Growth of diamond thin films by electron-assisted chemical vapour deposition and their characterization. Thin Solid Films 137(1):89–99. doi:http://dx.doi.org/10.1016/0040-6090(86)90197-5
Shaw DJ, Costello B (1993) Introduction to colloid and surface chemistry. Butterworth-Heinemann, Oxford
Spitsyn BV, Bouilov LL, Derjaguin BV (1981) Vapor growth of diamond on diamond and other surfaces. J Cryst Growth 52:219–226. doi:http://dx.doi.org/10.1016/0022-0248(81)90197-4
Stoner B, Glass J (1992) Textured diamond growth on (100) β-SiC via microwave plasma chemical vapor deposition. Appl Phys Lett 60(6):698–700. doi:http://dx.doi.org/10.1063/1.106541
Sunagawa I (1987) Morphology of minerals. Morphology of crystals, vol 2. Terra Scientific Publishing Co, Tokyo, pp 511–587
Sunagawa I (1990) Growth and morphology of diamond crystals under stable and metastable conditions. J Cryst Growth 99(1–4 pt 2):1156–1161. doi:10.1016/S0022-0248(08)80100-5
Suzuki K, Sawabe A, Yasuda H, Inuzuka T (1987) Growth of diamond thin films by dc plasma chemical vapor deposition. Appl Phys Lett 50(12):728
Takagi T, Yamada I, Sasaki A (1976) An evaluation of metal and semiconductor films formed by ionized-cluster beam deposition. Thin Solid Films 39:207–217. doi:http://dx.doi.org/10.1016/0040-6090(76)90638-6
Van der Drift A (1967) Evolutionary selection, a principle governing growth orientation in vapour-deposited layers. Philips Res Rep 22(3):267–288
Weilmünster P, Keller A, Homann KH (1999) Large molecules, radicals, ions, and small soot particles in fuel-rich hydrocarbon flames: Part I: positive ions of polycyclic aromatic hydrocarbons(PAH) in low-pressure premixed flames of acetylene and oxygen. Combust Flame 116(1–2):62–83. doi:http://dx.doi.org/10.1016/S0010-2180(98)00049-2
Williams BE, Glass JT (1989) Characterization of diamond thin films: Diamond phase identification, surface morphology, and defect structures. J Mater Res 4(02):373–384. doi:10.1557/JMR.1989.0373
Wilson JG (2014) The principles of cloud-chamber technique. Cambridge University Press
Yoshimoto M, Yoshida K, Maruta H, Hishitani Y, Koinuma H, Nishio S, Kakihana M, Tachibana T (1999) Epitaxial diamond growth on sapphire in an oxidizing environment. Nature 399(6734):340–342
Yugo S, Kimura T, Muto T (1990) Effects of electric field on the growth of diamond by microwave plasma CVD. Vacuum 41(4):1364–1367

Zhang GF, Buck V (2000) Influence of geometry factors of in situ dc glow discharge on the diamond nucleation in a hot-filament chemical vapor deposition system. Surf Coat Technol 132 (2–3):256–261. doi:http://dx.doi.org/10.1016/S0257-8972(00)00907-5

Zhou D, McCauley TG, Qin LC, Krauss AR, Gruen DM (1998) Synthesis of nanocrystalline diamond thin films from an Ar-Ch_4 microwave plasma. J Appl Phys 83(1):540

Chapter 7
Growth Mechanism of CVD Silicon

In Chap. 6, the charged cluster model (CCM) was suggested as a new growth mechanism of diamond. Then, a question arises as to whether the CCM applies to other CVD or PVD processes. Since the silicon CVD is one of the most important process in microelectronics, the possibility that the CCM should be applied to the growth of silicon films will be examined in this chapter. As a preliminary examination for this possibility, it was mentioned in Chap. 4 that the cauliflower structure of silicon prepared by CVD is the indirect indication that the structure should be formed by the incorporation of the gas phase nuclei. In establishing the CCM, the seemingly paradoxical phenomenon of diamond deposition with simultaneous etching play a critical role. Interestingly, a similar phenomenon of simultaneous deposition and etching occurs also in the silicon CVD process.

7.1 Simultaneous Deposition and Etching of Silicon During CVD

For deposition of silicon films by CVD, silane (SiH_4) or dichlorosilane (SiH_2Cl_2) is used as a source of silicon and hydrochlorine or hydrogen is used to control the decomposition of the reactant. Silicon is produced typically through the following decomposition reactions.

$$SiH_4 \rightarrow Si + 2H_2 \tag{7.1}$$

$$SiH_2Cl_2 \rightarrow Si + 2HCl \tag{7.2}$$

Addition of H_2 and HCl will make respectively chemical reactions (7.1) and (7.2) proceed in the backward direction and decrease the driving force for precipitation of silicon, thereby decreasing the deposition rate.

© Springer Science+Business Media Dordrecht 2016
N.M. Hwang, *Non-Classical Crystallization of Thin Films and Nanostructures in CVD and PVD Processes*, Springer Series in Surface Sciences 60,
DOI 10.1007/978-94-017-7616-5_7

In chemical reactions (7.1) and (7.2), forward and backward reactions represent respectively deposition and etching of silicon. In this sense, SiH_4 and SiH_2Cl_2 contribute to deposition; H_2 and HCl contribute to etching.

Many articles mention that deposition and etching of silicon occur simultaneously (Shi and Seinfeld 1992; Osenbach et al. 1991; Goulding 1991; Carlsson 1990). This statement is misleading and violates the second law of thermodynamics just as diamond deposition with simultaneous graphite etching does in Chaps. 5 and 6. This misleading statement might come from the confusion between reversible and irreversible deposition or etching. According to the second law of thermodynamics, the irreversible deposition and etching cannot occur simultaneously. Therefore, if such a phenomenon is macroscopically observed, an important discovery can be made as to the growth mechanism because of the constraint that the second law of thermodynamics cannot be violated. In other words, the growth process should be interpreted in such a way not to violate the second law.

In the selective epitaxial growth of silicon, silicon grows selectively on a conducting surface patterned on an insulating substrate (Hampden-Smith and Kodas 1994). After preparing slightly conducting SiN_x patterned on an insulating SiO_2 substrate, Kumomi and Yonehara (1990) made the selective deposition of silicon on the SiN_x portions over SiO_2 surface in period of 10 μm. The substrate temperature was 950 °C with the pressure of 150 Torr. The ratio of the supplied gas was $SiH_2Cl_2 : HCl : H_2 = 0.53 : 1.8 : 100$.

Figure 7.1 shows the microstructure evolved with deposition time (Kumomi and Yonehara 1990). After 480 s, many fine silicon particles were selectively deposited on square $4 \times 4\ \mu m^2$ portions of SiN_x with x = 0.56 (Fig. 7.1a). After 720 s, however, one large particle appeared among fine particles, while other fine particles

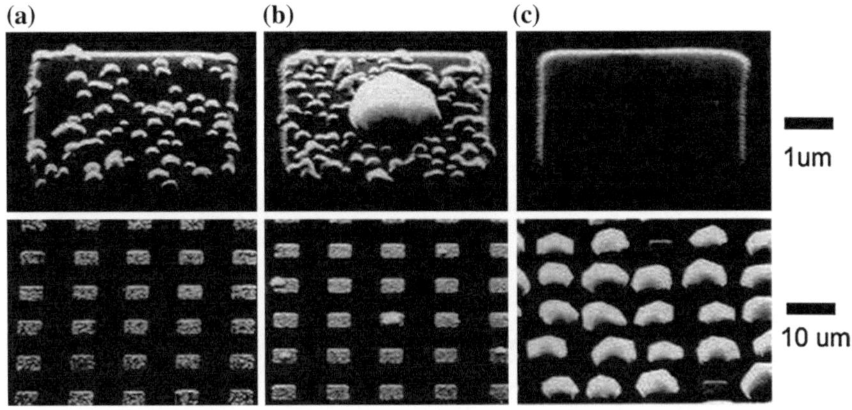

Fig. 7.1 SEM images of the Si-deposited surface at three deposition times. **a** At 480 s, many fine nanoparticles are formed selectively on all SiN_x portions. **b** At 720 s, a large nanoparticle emerges among the fine ones on some SiN_x portions. **c** At 960 s, most SiN_x portions were covered with a single large nanoparticle, but there were a few portions without a nanoparticle (Kumomi and Yonehara 1990)

disappeared due to etching (Fig. 7.1b). After 960 s, only one large particle remains on each SiN_x portion with some exception that all particles had been etched away on a few SiN_x portions, which were vacant (Fig. 7.1c).

Figure 7.1 shows that one particle grows exclusively and at the same time other particles shrink. This coarsening phenomenon looks similar to Ostwald ripening, where large particles grow at the expense of small ones and, as a consequence, the average particle size increases while the total number of particles decreases during coarsening (Lifshitz and Slyozov 1961). However, the detailed analysis indicates that this phenomenon is quite different from Ostwald ripening (Kumomi and Yonehara 1990).

What happened actually is that all the fine silicon particles, which had been deposited in the initial stage, etched away except for one large silicon particle, which continued to grow. Definitely, deposition and etching, which are two irreversible processes in opposite directions, take place simultaneously. The driving force should be either for deposition or for etching; it cannot be for both deposition and etching.

The irreversible deposition of silicon means that the chemical potential of silicon in the gas phase is higher than that in solid, which can be written as

$$\mu_{Si}^{solid} < \mu_{Si}^{gas}. \tag{7.3}$$

The irreversible etching of silicon means that the chemical potential of silicon in the gas phase is lower than that in solid, which can be written as

$$\mu_{Si}^{solid} > \mu_{Si}^{gas}. \tag{7.4}$$

Since inequalities of (7.3) and (7.4) cannot hold simultaneously, simultaneous deposition and etching violate the second law of thermodynamics. The situation is similar to that of the diamond CVD process in that the experimental observation seemingly violates the second law of thermodynamics.

As in the case of diamond CVD, if gas phase generated nanoparticles deposit into a crystal as a building block and etching takes place by the atomic unit, this phenomenon can avoid the violation of the second law of thermodynamics. The necessary condition for this to happen is that silicon should have a retrograde solubility in the gas phase. This possibility can be checked by the CVD phase diagram of the Si–Cl–H under 150 Torr. Figure 7.2 shows the equilibrium solubility of silicon in the gas phase calculated by the Thermo-Calc (Sundman et al. 1985). In Fig. 7.2, the abscissa represents the mole fraction of silicon with the fixed amount of chlorine and hydrogen: 2.86 mol of Cl and 201.06 mol of H, which are based on the composition used by Kumomi and Yonehara (1990).

The equilibrium solubility line which represents the boundary for deposition and corresponds to the supersaturation ratio of 1 is drawn as the solid line in Fig. 7.2. Two additional dashed lines are for the supersaturation ratios of 0.1 and 10. Figure 7.2 shows that silicon has the retrograde solubility, where solubility increases with decreasing temperature. The composition used by Kumomi and

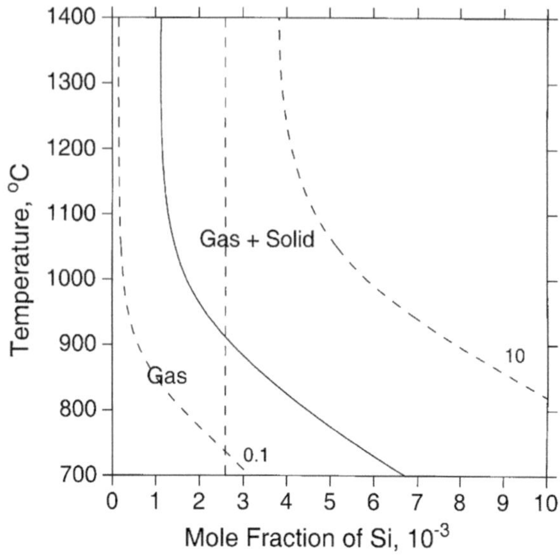

Fig. 7.2 Temperature dependence of solubility of silicon (*solid line*) in the gas phase in the Si–Cl–H system. The supersaturation ratios in the gas phase and the gas-solid regions are, respectively, 0.1 and 10 along the *dashed lines*. The thermodynamic calculation was done under the conditions: pressure of 150 Torr, 2.86 mol of Cl and 201.06 mol of H (Hwang 1999)

Yonehara (1990) is designated as the vertical dashed line, where the fraction of silicon is about 0.006. This composition falls into the two phase region of gas and solid at the substrate temperature of 950 °C. Therefore, the driving force is for deposition if the gas phase nucleation does not occur. For the composition used by Kumomi and Yonehara (1990), the supersaturation ratio at 950 °C is 1.39. For comparison, the supersaturation ratio at 1100 °C is 2.93.

If the gas-phase nucleation takes place, however, the thermodynamic situation changes drastically. The phase separation occurs and the silicon composition in the gas phase becomes depleted, reaching the solubility line drawn as the solid line in Fig. 7.2. If the gas depleted of silicon approaches the substrate at 950 °C, the silicon on the substrate will be etched. Therefore, the driving force is changed for etching.

This aspect becomes clearer if we examine how the equilibrium amount of solid silicon changes with temperature as shown in Fig. 7.3, which is also calculated by Thermo-Calc (Sundman et al. 1985). The conditions for calculation are pressure of 150 Torr, 0.53 mol of SiH_2Cl_2, 1.8 mol of HCl and 100 mol of H_2.

The equilibrium amount of precipitation at 1100 °C is 0.267 mol. It decreases with decreasing temperature and becomes 0.096 mol at 950 °C. The difference between 1100 and 950 °C is 0.171 mol. Therefore, if the gas-phase nucleation occurs at 1100 °C, the driving force is for etching this difference at the substrate temperature of 950 °C.

Since the overall composition falls in the two-phase region, all the silicon precipitated in the gas phase would not etch away. At least, the equilibrium amount of precipitation at 950 °C, which is 0.096 mol, would remain unetched at the substrate temperature. This unetched portion of gas phase nuclei would contribute to deposition with a building block of nanoparticles. In other words, deposition occurs by

Fig. 7.3 Temperature dependence of the equilibrium amount of precipitation of silicon in the gas phase. The thermodynamic calculation was done under the conditions: pressure of 150 Torr, 0.53 mol of SiH_2Cl_2, 1.8 mol of HCl and 100 mol of H_2 (Hwang 1999)

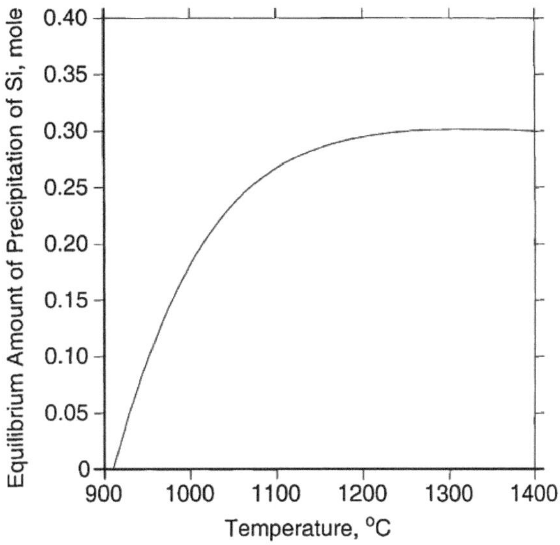

the nanoparticle unit and simultaneously etching occurs by the atomic unit at the substrate. That explains the macroscopic observation of simultaneous irreversible deposition and etching of silicon without violating the second law of thermodynamics.

According to this analysis, the atomic contribution to deposition is negative because etching occurs. Therefore, a large silicon particle evolved after 960 s in Fig. 7.1 should grow 100 % by nanoparticles generated in the gas phase. Since a large silicon particle is a single crystal, the analysis indicates that a single crystal can be grown by the building block of nanoparticles, which indicates again that nanoparticles should be liquid-like. If this analysis is combined with the suggestion made in Chaps. 1 and 6 that if nanoparticles are charged, being liquid-like, and the puzzling phenomenon in Fig. 7.1 can also be explained by the CCM.

The question that remains to be solved in the microstructure evolution of Fig. 7.1 is why only one particle grows while other fine particles etch away or why some SiN_x portions are vacant with all particles being etched away. This question is important because it can be explained by assuming that nanoparticles are electrically charged. Besides, this question is related with the important deposition behavior of CNPs formed in the gas phase.

The exclusive growth of one particle with simultaneous etching of other particles is a kind of instability phenomena in that large particles continue to grow while small particles continue to etch. To understand the deposition behavior of CNPs, the electrostatic interaction should be considered. In the electrostatic interaction between ions or electrons, the charge center is fixed and a Coulomb equation is used. In the electrostatic interaction between CNPs, however, the charge center changes, which depends on the geometrical parameters and on the relative magnitudes of the charges. Therefore, the equation for CNPs is more complicated than a

Coulomb equation and expressed for two conducting spherical particles by a series (Dove 1964):

$$force = \frac{q_1 q_2}{4\pi\varepsilon_o d^2} - \frac{q_1^2 r_2 d}{4\pi\varepsilon_o \left(d^2 - r_2^2\right)^2} - \frac{q_2^2 r_1 d}{4\pi\varepsilon_o \left(d^2 - r_1^2\right)^2} + \cdots, \tag{7.5}$$

where one sphere has a net charge q_1 and radius r_1, and the other has charge q_2 and radius r_2; d is the distance between centers and $1/4\pi\varepsilon_o$ is the permittivity.

In order to evaluate (7.5) by putting numeric values, let's rewrite (7.5) with $q_1 = n_1 e$, $q_2 = n_2 e$, where n_1 and n_2 are the number of charges per sphere, and e is the electronic charge, then

$$force = \frac{n_1 n_2 e^2}{4\pi\varepsilon_o d^2} - \frac{n_1^2 e^2 r_2 d}{4\pi\varepsilon_o \left(d^2 - r_2^2\right)^2} - \frac{n_2^2 e^2 r_1 d}{4\pi\varepsilon_o \left(d^2 - r_1^2\right)^2} + \cdots, \tag{7.6}$$

The first term is the well-known Coulomb equation, which can be attractive or repulsive depending on the signs of the two interacting particles. The second and third terms come from the image force, which are attractive regardless of the sign of the charge. This equation reveals that if a large spherical particle has a large amount of excess charges, it will attract not only charged particles of opposite sign but also small charged particles of the same sign. This aspect can be analyzed more specifically by inserting numerical values as follows.

If $r_1 = 2\,\text{nm}$, $r_2 = 1\,\mu\text{m}$, $d = 1.02\,\mu\text{m}$, $4\pi\varepsilon_o = 1.11 \times 10^{-10}$, and $e = 1.60 \times 10^{-19}$,

$$force\ (newtons) = 3.5 \times 10^{-16} n_1 n_2 - 2.2 \times 10^{-11} n_1^2 - 6.7 \times 10^{-19} n_2^2 + \cdots, \tag{7.7}$$

When CNPs of opposite sign approach a large charged silicon particle, the interaction is only attractive. Let's consider when CNPs of like sign approach a large charged silicon particle. Even in this case, (7.7) shows that the second term is dominant and the net force is an attraction.

Let us assume that Si particles are conducting at 950 °C and thereby (7.6) can be applied to the electrostatic interaction between incoming particles and particles on the substrate in Fig. 7.1. Let us further assume that the particle on the substrate has the radius of $r_2 = 1\ \mu$m with the charge of $n_2 = 1$ and the incoming nanoparticle has the radius of $r_1 = 2$ nm with the charge of $n_1 = 1$. Then the integration of (7.6) from the inter-particle distance d = 1.2 to 1.002 μm leads to the value of $\sim 2.9 \times 10^{-20}$ J, which is the energy of attraction. This value is comparable to the thermal energy of the nanoparticle, which is $\sim 1.7 \times 10^{-20}$ J at 950 °C. If r_2 decreases from 1 to 0.05 μm, the attraction decreases from 2.9×10^{-20} J to 2.2×10^{-20} J. If r_1 decreases to 1 nm, the attraction increases from 2.9×10^{-20} J to 5.7×10^{-20} J. The attraction increases with increasing r_2 and decreasing r_1.

Therefore, once one of the silicon particles survives, it will have progressively increasing attraction with the incoming charged silicon nanoparticles and will undergo the exclusive growth, while the other small particles, which will have the smaller attraction or the repulsion with the incoming CNPs, will etch away. This would explain exclusive coarsening of one silicon particle in Fig. 7.1.

7.2 Deposition Behaviour Between Conducting, Semi-conducting and Insulating Substrates

Phenomenologically, there is a tendency that deposition occurs on the conducting area but does not occur on the non-conducting area in the CVD process. This phenomenon is used in the selective CVD process, which is extensively utilized for the formation of patterned films in microelectronics (Hampden-Smith and Kodas 1994). In this process, a given material is made to deposit selectively on a specific pattern of the growth surface in the presence of a non-growth surface. The growth surface is typically a metal or a semiconductor and the non-growth surface is a dielectric material such as SiO_2.

Frequently, the selective growth is epitaxial on the pattern and is called selective epitaxial growth (SEG) (Fitch et al. 1992) or selective nucleation based epitaxy (SENTAXY) (Kumomi et al. 1990; Kumomi and Yonehara 1990, 1997). It is known that selective growth can be changed to non-selective growth by introducing plasma CVD (Hampden-Smith and Kodas 1994; Tsubouchi and Masu 1992). The reason for the loss of selectivity is not also clearly understood. On the other hand, the analysis in Sect. 7.1 implies the possibility that the main deposition flux in silicon CVD should be CNPs. Based on this possibility, the selective deposition may be newly understood. If charged silicon nanoparticles approach the substrate, the electrostatic interaction would depend on the conductivity or the dielectric property of the substrate. The image force will be relatively weak for the dielectric substrate and relatively strong for the conducting substrate. When nanoparticles are in the gas flow, gas molecules will collide with nanoparticles and exert a drag force. When this drag force acts against the deposition of CNPs on the substrate, the deposition would not occur if the attractive force to the substrate is not strong enough.

With this possibility in mind, Hwang et al. (1999) compared the deposition behavior during silicon CVD between five different substrates; two insulating ones of SiO_2 and Si_3N_4, one semi-conducting one of Si and two conducting ones of Mo and Pt. The time evolution of the microstructure for each substrate was examined. A gas mixture of 1 % SiH_4–2 % HCl–97 % H_2 was used, with a substrate temperature of 950 °C and a reactor pressure of 100 Torr.

On the conducting Mo and Pt substrates, silicon deposited without an incubation time. This was confirmed by both the mass increase of the substrates and SEM observation. After 3, 6, 12 and 24 min, the weight increases were 0.15, 0.19, 0.19

Fig. 7.4 Deposition behavior of silicon on the Mo substrate after deposition of **a** 3 min and **b** 24 min (Hwang et al. 1999)

and 0.24 mg/cm^2 for the Mo substrate and 0.19, 0.20, 0.25 and 0.34 mg/cm^2 for the Pt substrate, respectively. Figure 7.4a, b shows the deposition of silicon on the Mo substrate respectively after 3 and 24 min.

The deposition behavior of Si on the semi-conducting silicon substrate was found to be midway between the insulating and the conducting substrates. The deposition continued on the Si substrate but the growth rate was much lower than that on Mo and Pt substrates.

The deposition behavior changes drastically when the insulating substrate of SiO_2 is used. Figure 7.5a–d shows SEM images of SiO_2 substrates after deposition of 3, 6, 12 and 24 min, respectively. After 3 min, silicon particles were hardly seen on the entire substrate as shown in Fig. 7.5a. This time period for the first 3 min, where deposition hardly occurs, will be referred to as the incubation time for deposition. After 6 min, an appreciable number density of silicon particles of sub-micron size was observed (Fig. 7.5b). The number density of silicon particles was $\sim 10^9/\text{cm}^2$. However, after 12 min, most of the silicon particles had disappeared. The number density decreased drastically and the size of the remaining particles increased as shown in Fig. 7.5c. After 24 min, the silicon particles were hardly seen on the substrate. Figure 7.5d shows one of the rare sites on the substrate where a remaining silicon particle was observed.

The deposition behavior on the Si_3N_4 substrate is quite similar to that on SiO_2 substrate: no deposition after 3 min, an appreciable number density of silicon particles after 6 min and their being etched away almost completely after 24 min. The weight change was negligible for both SiO_2 and Si_3N_4 substrates during deposition.

The time evolution of Fig. 7.5, which is difficult to explain by the conventional growth mechanism based on the growth unit of individual silicon atom, is similar to that of Fig. 7.1. Numerous silicon particles shown in Fig. 7.5b disappear in Fig. 7.5c, which indicates that etching is taking place. At the same time, some

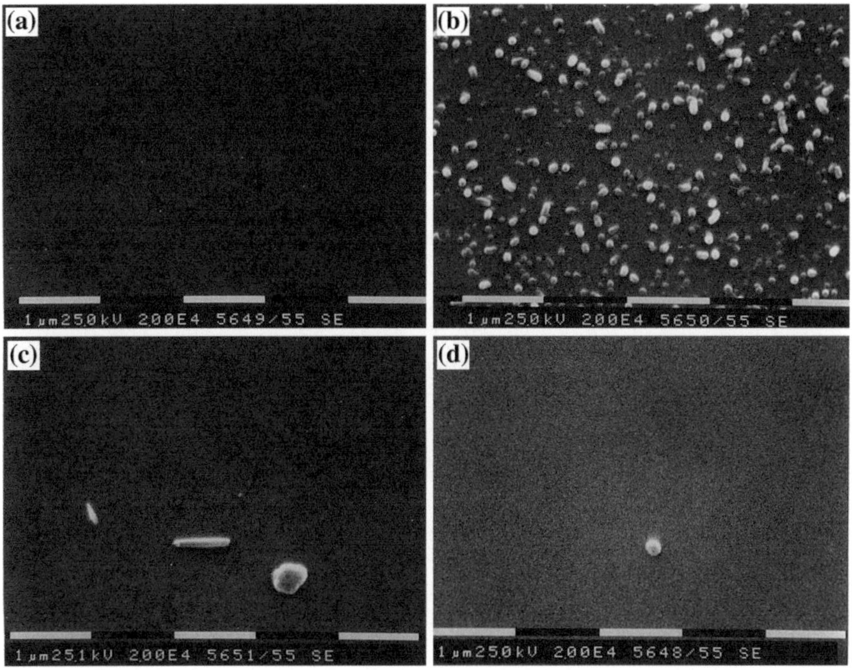

Fig. 7.5 Deposition behavior of silicon on the SiO_2 substrate with deposition times of **a** 3 min, **b** 6 min, **c** 12 min and **d** 24 min. The gas mixture of $SiH_4 : HCl : H_2 = 1 : 2 : 97$ was used under the pressure of 100 Torr at the substrate temperature of 950 °C and with the flow rate of 100 sccm. The *scale bar* in the photograph is 1 μm (Hwang et al. 1999)

particles grow larger, indicating that growth is taking place. Therefore, Fig. 7.5b, c definitely shows that two irreversible processes of deposition and etching occur simultaneously.

Again, in order to avoid violating the second law of thermodynamics, the gas phase nucleation should occur, which changes the driving force for etching according to Figs. 7.2 and 7.3. And the deposition flux should come from the gas phase nuclei. Under the situation where two irreversible fluxes of deposition and etching are involved simultaneously, the macroscopic deposition or etching will be determined by the relative magnitude of deposition and etching fluxes. If the deposition flux from nanoparticles is larger than the etching flux, the particle will grow and vice versa.

The fact that only a few of particles in Fig. 7.5b grow with other particles being etched implies that the silicon particle should be electrically charged. The time evolution of microstructure from Fig. 7.5b–c can only be understood by considering the electrostatic interaction of charged silicon particles as treated in Sect. 7.1.

Since the SiO_2 substrate is insulating, electric charges would build up as the deposition of charged silicon nanoparticles goes on. Then incoming CNPs tend to be repelled. Then, etching will dominate over deposition, which explains why most particles disappear or are etched away (Fig. 7.5c, d). According to (7.6), however, some large particles can attract incoming CNPs and can continue to grow, which explains why some particles continue to grow (Fig. 7.5b, c).

Figure 7.5d shows that in the time interval between 12 and 24 min, most of the silicon particles etched away. In this stage, it appears that the charge becomes saturated and most of the incoming CNPs are repelled. The etching flux by the atomic unit is dominant over the deposition flux by the particle unit. This aspect is different from that of Fig. 7.1, where one large particle remains. This difference would be attributed to the fact that the deposition in Fig. 7.1 occurs on SiN_x, which is somewhat conducting, whereas the deposition in Fig. 7.5 occurs on SiO_2, which is insulating.

Considering all these, it is clear why the selective deposition occurs. Deposition on the conducting substrate is favored over the insulating substrate because the CNPs have much higher attracting image force on the conducting surface than on the insulating surface. Then, it can be understood also why the selective growth can be changed to non-selective growth by introducing plasma CVD. Plasma is a way of transferring electricity through the gas phase. In other words, plasma makes the gas phase conducting. Besides, all the chamber walls are negatively charged because the mobility of electrons is much higher than that of ions. This means that even the surface of the insulating substrate is also negatively charged, attracting the positively charged ions or nanoparticles.

As a rough measure of the conductivity in the gas phase, the amount of charges generated in the gas phase can be measured by the electric current between the conducting substrate or electrode and the ground. Typically, $\sim$mA/cm^2, $\sim\mu$A/cm^2 and $\sim$nA/cm^2 are measured respectively in plasma, hot filament and thermal CVD processes. Considering that the amount of charges generated in the hot filament CVD process is three orders of magnitude larger than that in the thermal CVD process, the gas phase of hot filament CVD is much more conducting than that of thermal CVD and there is a possibility that non-selective growth may occur in the hot wire CVD process.

In order to check this possibility that the hot filament CVD should change the deposition behavior on the insulating substrate, the deposition was done at the hot tungsten filament of 1600 °C. The gas ratio was $SiH_4 : HCl : H_2 = 1 : 1 : 98$ and the other deposition conditions were the same as before.

Figure 7.6a, b shows SEM images after deposition on SiO_2 and Si_3N_4 substrates, respectively, for 6 min. The deposition behavior changed completely. The entire surface of the substrate was covered by the deposited silicon. In this case, the deposition started without any incubation time. Therefore, when the hot filament is used at least under a certain condition, there seems to be no selective nature in deposition between the conducting and insulating substrates.

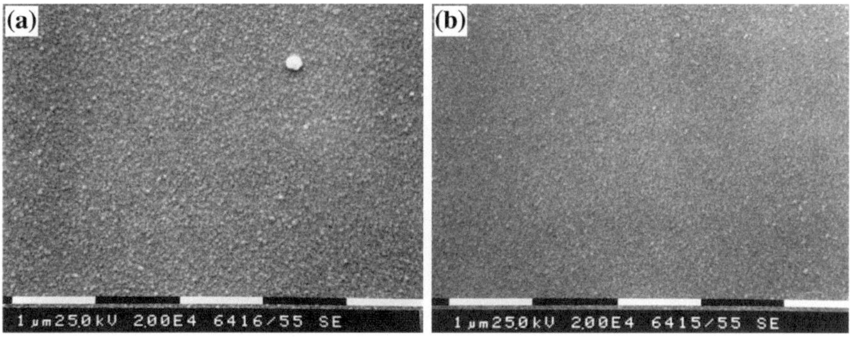

Fig. 7.6 Deposition behavior of silicon on SiO_2 (**a**) and Si_3N_4 (**b**) substrates after 6 min in the hot filament reactor with the filament temperature of 1600 °C for the gas mixture of SiH_4 : HCl : H_2 = 1 : 1 : 98. The other processing conditions were the same as in Fig. 7.5 (Hwang et al. 1999)

7.3 Deposition Behaviour on Substrates with Different Charge Transfer Rates

In Chap. 6, it was shown that charged diamond nanoparticles deposit as a dense film on the silicon substrate with a low charge transfer rate whereas they deposit as porous skeletal soot on the iron substrate with a high charge transfer rate. A similar deposition behavior is expected for silicon although silicon has no allotropic transition like carbon.

Since drastically different deposition behavior was observed between silicon and iron substrates, on which diamond films and porous skeletal soot particles deposit respectively in Fig. 6.1, Cheong et al. (2000) compared the deposition behavior of silicon between the silicon and iron substrates. The result is shown in Fig. 7.7.

The deposition time was 10 min. A highly porous structure is evolved on the Fe substrate while a dense film is evolved on the Si substrate. It should be noted that Fig. 7.7a is a plan view and Fig. 7.7b is a cross section view. In Fig. 7.7b, the bottom half is the substrate. The film thickness on the Fe substrate in Fig. 7.7a was ~5 mm while that on the Si substrate in Fig. 7.7b was only ~3 μm. This difference in film growth rate between the two substrates is related to the fact that CNPs have difficulty in landing on a semi-conducting or insulating substrate, leading to preferential deposition on a conducting substrate over a semi-conducting or insulating one (Hwang et al. 1999). It should be noted that the silicon substrate, unless specially treated, usually has a thin insulating layer of silicon dioxide.

As mentioned in Chap. 6, the CTR is high in the order of Pd, Rh, Ir, Cu, Cr and Ti. The deposition behaviors on these six substrates were compared as shown in Fig. 7.8. The deposition conditions were the same as those for Fig. 7.7. Silicon films on all substrates were approximately 5 mm in thickness with some variations among substrates.

Fig. 7.7 SEM images of silicon deposits on **a** Fe and **b** Si substrates with the SiH_4 : HCl : H_2 gas ratio of 1 : 1 : 98 under a reactor pressure of 10 Torr at a substrate temperature of 850 °C (Cheong et al. 2000)

Fig. 7.8 SEM images of silicon deposited on **a** Pd, **b** Rh, **c** Ir, **d** Cu, **e** Cr, and **f** Ti substrates with the same processing conditions as those for Fig. 7.7 (Cheong et al. 2000)

On the Pd substrate (Fig. 7.8a), which has the highest CTR, the microstructure is very porous with cocoon-shaped grains. On the Cu substrate with a medium CTR (Fig. 7.8d), the microstructure is also porous but with ball-shaped grains. Each ball consists of fine grains, resembling a cauliflower-like structure. On the Ti substrate with the lowest CTR (Fig. 7.8f), the film is dense although the morphology is typical of a cauliflower-like structure. The overall trend is that films become denser with decreasing CTR.

Once the substrate surface is covered with a deposited material, further landing of CNPs would not be affected by CTR of the substrate. Instead, the CTR of the deposited material will affect the deposition behavior. For example, in the CVD diamond process, the porous soot was deposited on Fe in the initial stage but in a

Fig. 7.9 SEM images of silicon deposited after **a** 3 and **b** 30 min on the Ni substrate with other conditions being the same as those for Fig. 7.7 (Cheong et al. 2000)

later stage diamond crystal grew on the porous soot coating (Hwang et al. 1996; Hahn et al. 1996). A similar behavior is expected in the deposition of charged Si nanoparticles.

To confirm this, Cheong et al. (2000) compared the morphology of silicon films deposited on a Ni substrate between the initial and later stages of deposition, as shown in Fig. 7.9a, b. The deposition times were 3 and 30 min for Fig. 7.9a, b, respectively. Figure 7.9a shows a porous structure with relatively fine grains while Fig. 7.9b shows large grains with well-developed facets on the previously formed fine grains. Figure 7.9b is very similar to the morphological evolution of large diamond crystals on the initially formed soot structure on Fe in Fig. 6.7b (Hwang et al. 1996).

By comparing the microstructure evolution between Fig. 7.9a, b, the role of electric charge carried by the nanoparticles can be deduced. First, the charge induces the self-assembly, which is attributed to the electrostatic interaction. Second, the charge induce a dense structure without any void. Third, the charge induce an epitaxial deposition of nanoparticles. From the second and third roles, it can be inferred that the charge should enhance atomic diffusion. This analysis again supports the suggestion in Chaps. 1 and 6 that CNPs are liquid-like.

7.4 Growth of Silicon Nanowires

Hwang et al. (2000) could grow silicon nanowires without any catalytic metals and in a reducing atmosphere, using the CVD reactor shown in Fig. 7.10. In this CVD process, the silicon nanowires without any metal impurity could be grown not only at a high rate but in a large scale.

The substrate was heated radiantly by a halogen lamp, which was placed below the substrate holder. Another halogen lamp heater was placed above the substrate. It should be noted that in a conventional epitaxial reactor, the halogen lamp is placed above a quartz plate, which separated a reactor from the lamp. In the new set-up, the

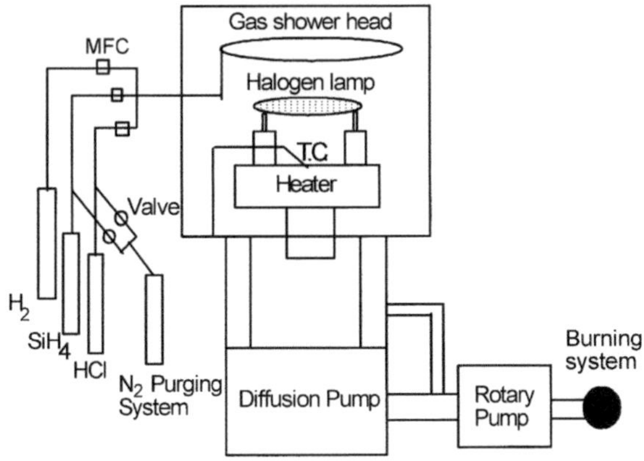

Fig. 7.10 Schematic of the CVD reactor for the growth of silicon nanowires (Hwang et al. 2000)

halogen lamp is placed directly above the substrate without a quartz plate in-between as shown in Fig. 7.10. This set-up appears to have enhanced the growth of silicon nanowires.

Since it was believed that the growth of silicon nanowires might be related with the electrostatic interaction of CNPs as in the case of the selective deposition in Sect. 7.2, Hwang et al. (2000) compared the growth behavior of silicon nanowires between Mo, Si, SiO_2 and SiN_3 substrates. The processing condition for the growth of nanowires was a gas ratio of $SiH_4 : HCl : H_2 = 3 : 1 : 96$, a reactor pressure of 10 Torr and the substrate temperature of 950 °C. It should be noted that the gas ratio for the growth of silicon nanowires has a little bit higher percentage of SiH_4 than a gas ratio of $SiH_4 : HCl : H_2 = 1 : 2 : 97$ which was used for the study of simultaneous deposition and etching in Sect. 7.2. The higher percentage of SiH_4 would mean the larger size of CNPs, which would be a bad condition for selective deposition because the magnitude of electrostatic interaction decreases.

Figure 7.11a–d shows the microstructures evolved after 3 min respectively on Mo, Si, SiO_2, and Si_3N_4 substrates. A silicon film was deposited on the Mo substrate in Fig. 7.11a. On the Si substrate, however, nanowires started to form as shown in Fig. 7.11b. Moreover, an appreciable number of nanowires formed on SiO_2 and Si_3N_4 substrates as shown respectively in Fig. 7.11c and d. The diameter of nanowires was 20–70 nm. Considering the length of nanowires grown after 3 min, the growth rate is estimated to be higher than at least $\sim 50\ \text{nms}^{-1}$.

Considering that charged silicon nanoparticles are generated in the gas phase, Fig. 7.11a shows that CNPs are deposited on the conducting Mo substrate, producing a dense film. Figure 7.11b shows that CNPs are reluctant to landing on the Si substrate, which has a native oxide on the surface and thereby is somewhat insulating. It should be noted that CNPs did not land on a considerable portion of the surface area and tended to land on the pre-deposited silicon particles. This selective landing of CNPs appears to induce the growth of silicon nanowires. These results imply that the growth of silicon nanowires should be related to the selective deposition, where CNPs deposit on the conducting substrate but do not on the

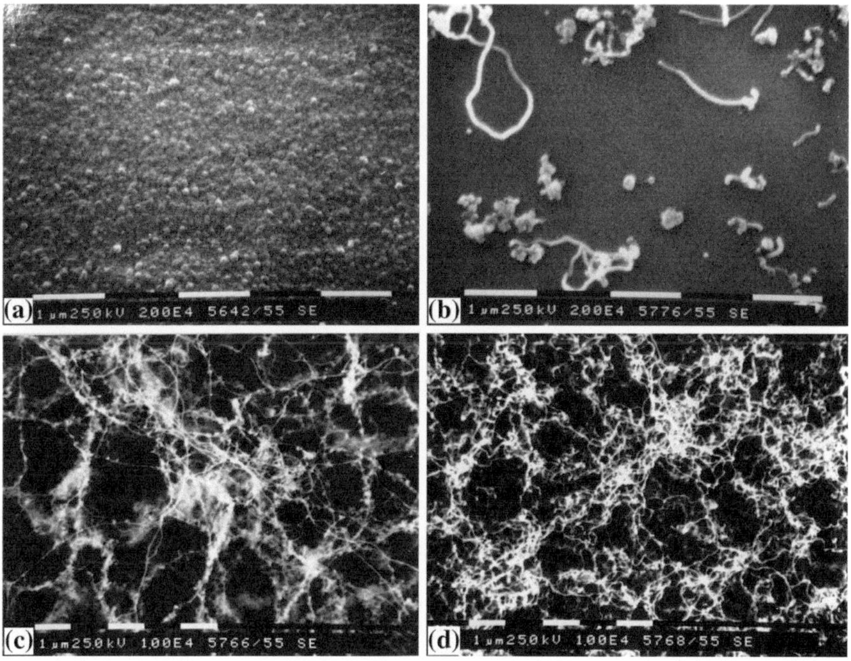

Fig. 7.11 SEM images of microstructures evolved on **a** Mo, **b** Si, **c** SiO_2 and **d** Si_3N_4 substrates after deposition for 3 min under 10 Torr at a substrate temperature of 950 °C with a gas ratio of $SiH_4 : HCl : H_2 = 3 : 1 : 96$ (Hwang et al. 2000)

insulating substrate. This aspect might also be related with the extensive growth of nanowires on the two insulating substrates SiO_2 and SiN_4 respectively shown in Fig. 7.11c, d.

Consider the electrostatic interaction between incoming CNPs and the pre-deposited short nanowires on the Si substrate in Fig. 7.11b. According to (7.6), when a small charged nanoparticle approaches the large one, the interaction can be attractive whereas when it approaches the small one, the interaction can be repulsive.

Consider the case where the charged particle is the short nanowire as shown in Fig. 7.11b. When a positively-charged small nanoparticle approaches a positively-charged short nanowire in the radial direction, the positive charge will move backward and the negative charge will move forward as illustrated in Fig. 7.12a. The electrostatic interaction would be repulsive. When a positively-charged small charged nanoparticle approaches a positively-charged short nanowire in the axial direction, the positive charge will move backward and the negative charge will move forward as illustrated in Fig. 7.12b. The electrostatic interaction would be attractive. In this situation, the small CNPs will be selectively deposited in the axial direction and this selective deposition will be accelerated as the length of nanowires increases, leading to extensive growth of nanowires.

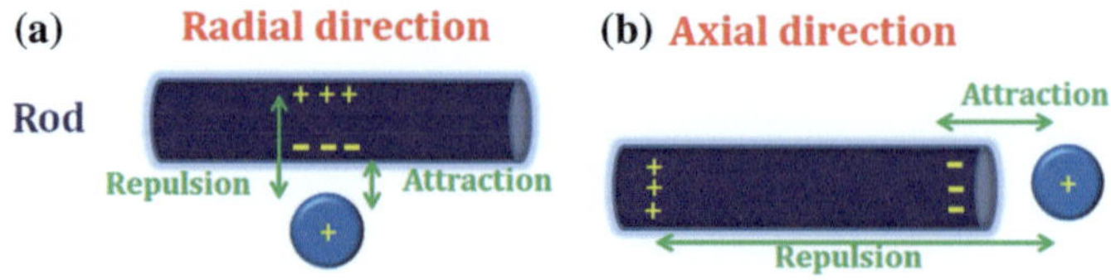

Fig. 7.12 Coulomb interaction between short nanowire and approaching small nanoparticle (both of which are charged positively) is **a** repulsive when the nanoparticle approaches in the radial direction of rod, but **b** attractive when the nanoparticle approaches in the axial direction (Youn et al. 2014)

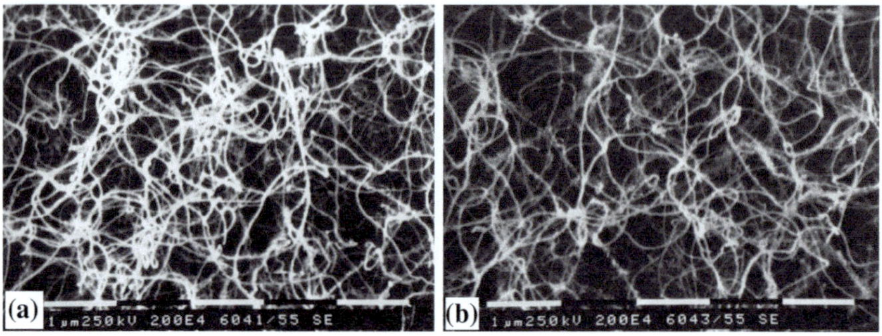

Fig. 7.13 SEM images on **a** Si and **b** SiO_2 substrates after 6 min of deposition with the other conditions being the same as those for Fig. 7.11 (Hwang et al. 2000)

As expected, after 6 min of deposition, the short nanowires in Fig. 7.11b grew extensively as shown in Fig. 7.13a. For comparion, the nanowires on the SiO_2 substrate in Fig. 7.11c continues to grow after 6 min of deposition, as shown in Fig. 7.13b.

Silicon nanowires or whisker are usually grown by CVD using gold nanoparticles upon a silicon substrate. The one-dimensional growth is achieved by preferential growth of silicon on Au–Si eutectic liquid droplets. This phenomenon is explained by the vapour-liquid-solid (VLS) mechanism, which was proposed in 1964 (Wagner and Ellis 1964). However, Figs. 7.11 and 7.13 show that nanowires could be grown even by the electrostatic interaction of CNPs without catalytic metal particles. The growth of nanowires without catalytic metal particles was also reported in many other studies (Lee et al. 1999a, b, 2000; Zhang et al. 1999; Shi et al. 2000; Jin et al. 2001; Wang et al. 1998a, b, 1999; Feng et al. 2000; Hwang et al. 2000). Especially, Cheng and Cheung (2003, 2004) proposed a charge-assisted mechanism to explain the growth of Si nanowires without catalytic metal particles. They reported that the electric charge accumulated at the tip of the nanowires produces a strong electric field and attracts most of the SiO vapor, forcing it to land on the tip. However, the origin of the charge and the electric field was not clear. Without thinking of the possibility for the generation of CNPs, it would be difficult to find out the origin of the electrostatic interaction for the nanowire growth without catalytic metal particles. The growth of nanostructures by the electrostatic interaction of CNPs appears to be very general and the related experimental results will be shown in Chap. 9.

References

Carlsson J-O (1990) Selective vapor-phase deposition on patterned substrates. Crit Rev Solid State Mater Sci 16(3):161–212. doi:10.1080/10408439008244628

Cheng S-W, Cheung H-F (2003) Role of electric field on formation of silicon nanowires. J Appl Phys 94(2):1190

Cheng S-W, Cheung H-F (2004) Temperature profile and pressure effect on the growth of silicon nanowires. Appl Phys Lett 85(23):5709–5711

Cheong WS, Yoon DY, Kim D-Y, Hwang NM (2000) Effect of substrates on morphological evolution of a film in the silicon CVD process: Approach by charged cluster model. J Cryst Growth 218(1):27–32. doi:http://dx.doi.org/10.1016/S0022-0248(00)00542-X

Dove D (1964) Possible influence of electric charge effects on the initial growth processes occurring during the vapor deposition of metal films onto substrates inside the electron microscope. J Appl Phys 35(9):2785–2786

Feng SQ, Yu DP, Zhang HZ, Bai ZG, Ding Y (2000) The growth mechanism of silicon nanowires and their quantum confinement effect. J Cryst Growth 209(2–3):513–517. doi:http://dx.doi.org/10.1016/S0022-0248(99)00608-9

Fitch J, Denning D, Beard D (1992) The pattern dependence of selectivity in low pressure selective epitaxial silicon growth. J Electron Mater 21(4):455–462

Goulding M, R (1991) The selective epitaxial growth of silicon. J Phys IV 02 (C2):745–778

Hahn JH, Hwang NM, Yoon DY (1996) Formation of soot or diamond on the iron substrate in the chemical vapour deposition process of diamond. J Mater Sci Lett 15(14):1240–1242

Hampden-Smith MJ, Kodas TT (1994) The chemistry of metal CVD. VCH

Hwang NM (1999) Deposition and simultaneous etching of Si in the chemical vapor deposition (CVD) process: Approach by the charged cluster model. J Cryst Growth 205(1–2):59–63. doi:http://dx.doi.org/10.1016/S0022-0248(99)00247-X

Hwang NM, Hahn JH, Yoon DY (1996) Charged cluster model in the low pressure synthesis of diamond. J Cryst Growth 162(1):55–68

Hwang NM, Cheong WS, Yoon DY (1999) Deposition behavior of Si on insulating and conducting substrates in the CVD process: Approach by charged cluster model. J Cryst Growth 206(3):177–186

Hwang NM, Cheong WS, Yoon DY, Kim D-Y (2000) Growth of silicon nanowires by chemical vapor deposition: Approach by charged cluster model. J Cryst Growth 218(1):33–39. doi:http://dx.doi.org/10.1016/S0022-0248(00)00543-1

Jin S, Li Q, Lee CS (2001) Direct growth of amorphous silicon oxide nanowires and crystalline silicon nanowires from silicon wafer. Phys Status Solidi (a)188(2):R1–R2. doi:10.1002/1521-396X(200112)188:2<R1::AID-PSSA99991>3.0.CO;2-W

Kumomi H, Yonehara T (1990) Coarsening phenomenon of Si clusters. In: MRS Proceedings, 1990. Cambridge Univ Press, p 83

Kumomi H, Yonehara T (1997) Selective nucleation-based epitaxy (SENTAXY): A novel approach for thin film formation. Jpn J Appl Phys 36(3S):1383

Kumomi H, Yonehara T, Nishigaki Y, Sato N (1990) Selective nucleation based epitaxy (sentaxy): Investigation of initial nucleation stages. Appl Surf Sci 41:638–642

Lee ST, Wang N, Zhang YF, Tang YH (1999a) Oxide-assisted semiconductor nanowire growth. MRS Bull 24(08):36–42. doi:10.1557/S088376940005288X

Lee ST, Zhang YF, Wang N, Tang YH, Bello I, Lee CS, Chung YW (1999b) Semiconductor nanowires from oxides. J Mater Res 14(12):4503–4507. doi:10.1557/JMR.1999.0611

Lee ST, Wang N, Lee CS (2000) Semiconductor nanowires: synthesis, structure and properties. Mater Sci Eng: A 286(1):16–23. doi:http://dx.doi.org/10.1016/S0921-5093(00)00658-4

Lifshitz IM, Slyozov VV (1961) The kinetics of precipitation from supersaturated solid solutions. J Phys Chem Solids 19(1):35–50

Osenbach JW, Schimmel DG, Feygenson A, Bastek JJ, Tsai JCC, Praefcke HC, Bonato EW (1991) Selective epitaxial growth of silicon. J Mater Res 6(11):2318–2323. doi:10.1557/JMR.1991.2318

Shi FG, Seinfeld JH (1992) Selective nucleation of silicon clusters in CVD. J Mater Res 7(07):1809–1815. doi:10.1557/JMR.1992.1809

Shi WS, Peng HY, Zheng YF, Wang N, Shang NG, Pan ZW, Lee CS, Lee ST (2000) Synthesis of large areas of highly oriented, very long silicon nanowires. Adv Mater 12(18):1343–1345. doi:10.1002/1521-4095(200009)12:18<1343::AID-ADMA1343>3.0.CO;2-Q

Sundman B, Jansson B, Andersson J-O (1985) The Thermo-Calc databank system. Calphad 9(2):153–190. doi:http://dx.doi.org/10.1016/0364-5916(85)90021-5

Tsubouchi K, Masu K (1992) Selective aluminum chemical vapor deposition. J Vac Sci Technol A 10(4):856–862

Wagner RS, Ellis WC (1964) Vapor-liquid-solid mechanism of single crystal growth. Appl Phys Lett 4(5):89. doi:http://dx.doi.org/10.1063/1.1753975

Wang N, Tang YH, Zhang YF, Lee CS, Lee ST (1998a) Nucleation and growth of Si nanowires from silicon oxide. Phys Rev B 58(24):R16024–R16026

Wang N, Zhang YF, Tang YH, Lee CS, Lee ST (1998b) SiO_2-enhanced synthesis of Si nanowires by laser ablation. Appl Phys Lett 73(26):3902. doi:http://dx.doi.org/10.1063/1.122930

Wang N, Tang YH, Zhang YF, Lee CS, Bello I, Lee ST (1999) Si nanowires grown from silicon oxide. Chem Phys Lett 299(2):237–242. doi:http://dx.doi.org/10.1016/S0009-2614(98)01228-7

Youn W-K, Lee S-S, Lee J-Y, Kim C-S, Hwang N-M, Iijima S (2014) Comparison of the deposition behavior of charged silicon nanoparticles between floating and grounded substrates. J Phys Chem C 118(22):11946–11953. doi:10.1021/jp5001144

Zhang YF, Tang YH, Wang N, Lee CS, Bello I, Lee ST (1999) One-dimensional growth mechanism of crystalline silicon nanowires. J Cryst Growth 197(1–2):136–140. doi:http://dx.doi.org/10.1016/S0022-0248(98)00953-1

Chapter 8
Other Works Related to Non-classical Crystallization of Thin Films and Nanostructures

Growth of films and nanostructures based on the concept similar to non-classical crystallization was suggested by many scientists including Yoshida, Cabarrocas and Ostrikov. All of them used plasma for the growth of their films or nanostructures. In this chapter, their works are briefly introduced.

8.1 Hot Cluster Epitaxy by Thermal Plasma Flash Evaporation

A concept of the film growth by nanoparticles was suggested by Yoshida and his colleagues (Terashima et al. 1998; Takamura et al. 1997, 1998; Hayasaki et al. 1997a; Yamaguchi et al. 1999) who made extensive studies on the epitaxial growth of films with a building block of nanoparticles by the method called 'thermal plasma flash evaporation' (TPFE). The schematic of TPFE is illustrated in Fig. 8.1. In this process, the starting material was powder with particles of ~ 1 μm, which was continuously injected into a thermal plasma for complete evaporation and precipitation into nanoparticles above the substrate. These clusters or nanoparticles are the building block for the growth of films.

The advantages of this process are the high deposition rate and the deposition of multi-component films with the same composition as the fed powder. Actually they demonstrated the deposition rate of ~ 2.2 μm/min of high quality epitaxial high T_c superconducting oxide $YBa_2Cu_3O_{7-x}$ (YBCO) films with $T_c = 90$ K on $SrTiO_3$ (100) from clusters of 0.8 nm diameter.

In order to estimate the size of clusters, they used sub-micron trenches fabricated on a Si wafer put on the substrate holder and could estimate the size of the depositing nanoparticles to be about 0.3–10 nm (Terashima et al. 1998). Figure 8.2 shows scanning tunneling microscopy (STM) images at various substrate

© Springer Science+Business Media Dordrecht 2016

N.M. Hwang, *Non-Classical Crystallization of Thin Films and Nanostructures in CVD and PVD Processes*, Springer Series in Surface Sciences 60,
DOI 10.1007/978-94-017-7616-5_8

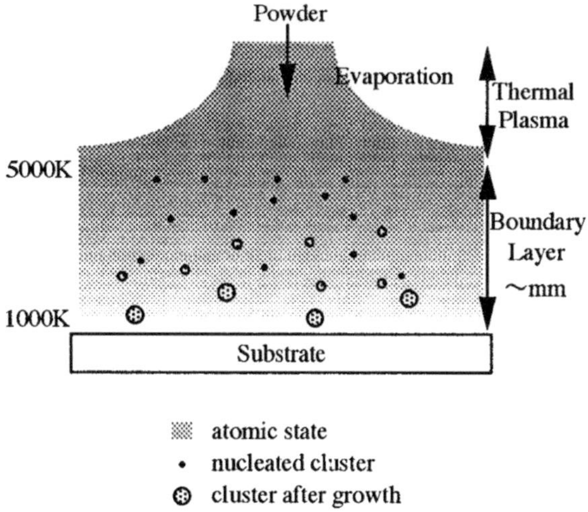

Fig. 8.1 The concept of cluster deposition in the plasma flash evaporation process (Hayasaki and Takamura 1997)

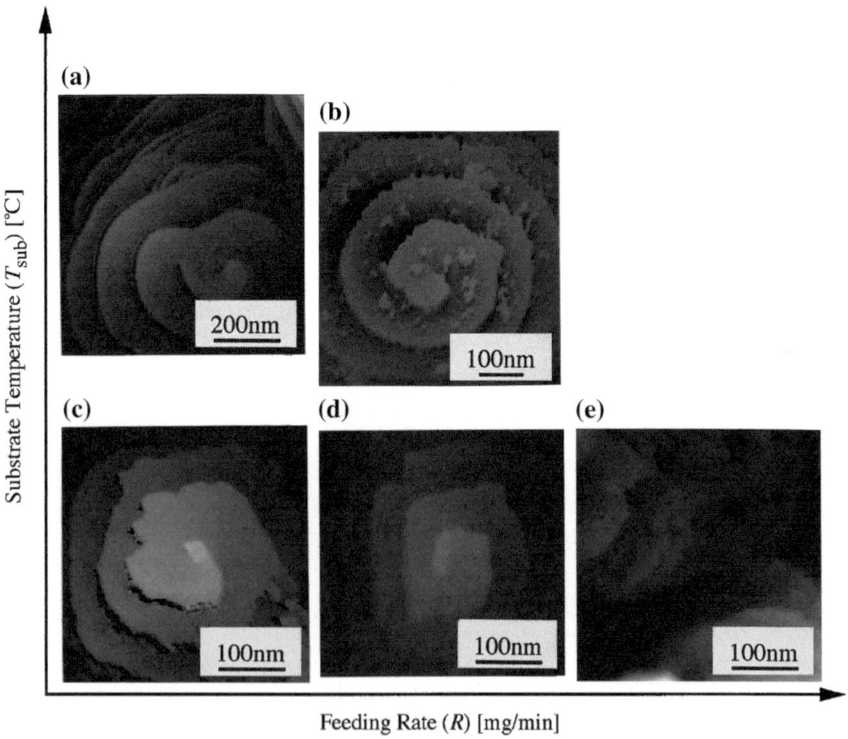

Fig. 8.2 STM images of YBCO films grown on $SrTiO_3$ (100). **a** $R = 60$ mg/min, $T_{sub} = 700$ °C, **b** $R = 150$ mg/min, $T_{sub} = 695$ °C, **c** $R = 60$ mg/min, $T_{sub} = 670$ °C, **d** $R = 150$ mg/min, $T_{sub} = 670$ °C, and **e** $R = 200$ mg/min, $T_{sub} = 670$ °C (Hayasaki et al. 1997b)

temperatures (T_{sub}) and feeding rates (R) with the distance between the torch exit and the substrate of $L = 301$ mm.

Figure 8.2a–d clearly shows spiral growth. According to Hayasaki et al. (1997b), the films with $R = 60$ mg/min ($T_{sub} = 700$–660 °C) and $R = 150$ mg/min ($T_{sub} = 695$–670 °C) show spiral growth originated from the edges of the screw dislocations at the substrate surfaces and the vertical step height is equal to one unit cell (1.2 nm) as shown in Fig. 8.2a–d.

On the other hand, in the case of $R = 200$ mg/min (Fig. 8.2e), the spiral growth feature is not observed and is irregular in shape, but still regular 1.2 nm step-height terraces are observed. These facts indicate that this film grew with a non-spiral epitaxial growth like normal two-dimensional nucleus growth (2DNG).

By combining STM and the size determination by the microtrench method, Takamura et al. (1998) observed that small 1–2 nm nanoparticles underwent epitaxial spiral growth, medium size 3 nm nanoparticles became epitaxial 2-dimensional nuclei, and large nanoparticles over 3 nm produced non-epitaxial island grains.

Their observation that even the spiral growth can be grown by small 1–2 nm nanoparticles is important because the spiral growth is believed as the evidence for the classical crystallization with an atom or molecule as a growth unit. For example, the morphology of spiral growth very similar to Fig. 8.2a–d is reported in films prepared by sputtering and CVD. The spiral growth was regarded as a classical spiral growth mechanism (Frank 1949; Burton et al. 1951). However, the morphology of spiral growth does not guarantee the classical spiral growth which is valid under conditions where gas phase nucleation does not occur. Especially, the epitaxial YBCO thin films with spiral growth reported by Luo et al. (1993) were grown under a reactor pressure of 4 Torr by CVD and had a growth rate of 4 μm/h, which is so high that the gas phase nucleation is expected. Such spiral growth morphology is frequently reported in other methods such as sputtering (Hawley et al. 1991; Gerber et al. 1991) and CVD (Luo et al. 1993).

Takamura and Yamaguchi (1998) called the related phenomenon ‘hot cluster epitaxy’, where the terminology ‘hot cluster’ should be understood in comparison with the clusters used by Tagaki in their ionized cluster beam deposition (ICBD) (Takagi 1988). In the ICBD process, the clusters should be made charged by electron impact and the electric field should be used to accelerate charged clusters toward the substrate for epitaxial deposition. However, clusters used by Takamura and Yamaguchi (1998) need not be accelerated for epitaxy because clusters are hot whereas cold clusters in ICBD should be accelerated for epitaxial growth.

According to understanding based on the theory of charged nanoparticles (TCN), however, the electric charge plays an important role in epitaxial deposition of clusters in addition to temperature and acceleration. The charge is believed to play a critical role also in the high rate of epitaxial deposition by TPFE. If depositing clusters are charged, the deposition rate would be retarded on the insulating substrate.

In agreement with this possibility, Hattori et al. (1999) reported various experimental results implying that clusters in the TPFE process are electrically charged. The growth rate of $YBa_2Cu_3O_{7-x}$ epitaxial films was decreased markedly on the

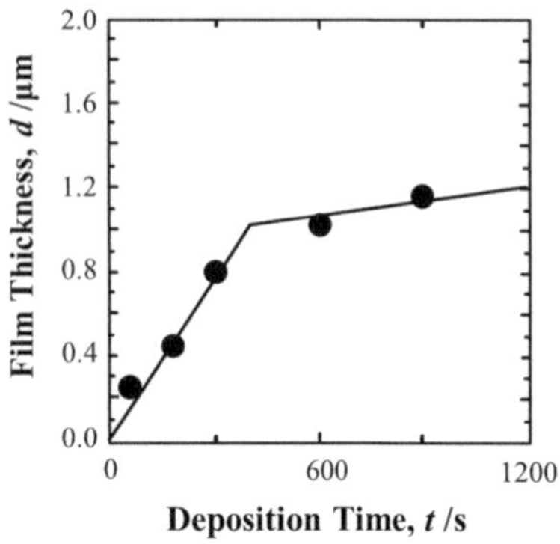

Fig. 8.3 Deposition time dependence of thickness of film deposited on $SrTiO_3$ (Hattori et al. 1999)

insulating $SrTiO_3$ substrate when the film thickness became thicker than 1 μm as shown in Fig. 8.3. Besides, the films thicker than 2 μm could not be grown on the $SrTiO_3$ substrate although the films could be deposited at a rate as high as 16 nm/s initially. When the conducting Nb-doped $SrTiO_3$ substrate was used, however, films thicker than 3 μm could be deposited at a constant rate. When the DC bias of −40 or +40 V was applied to the conducting Nb-doped $SrTiO_3$ substrate, the film growth rate for the positive bias was 2–3 times higher than that for the negative bias.

By using a probe made of Au foil on a tungsten tip placed on the substrate holder, they measured the current between the probe and the ground during substrate heating and deposition. The probe current was ~nA during substrate heating but increased to 5–6 μA during deposition. Based on these results, Hattori et al. (1999) suggested that clusters during TPFE are mainly negatively charged.

8.2 Polymorphous Si Films by Plasma Enhanced CVD

As to the possibility for the incorporation of particles generated in the gas phase during PECVD, Cabarrocas et al. (2000, 2002, 2004a, b) made extensive studies and suggested that these nanoparticles can be incorporated into films. They also suggested that incorporation of nanocrystallites produced in the gas phase increased the film growth rate and the crystalline fraction and thereby improved transport properties and stability compared with hydrogenated amorphous Si (a–Si:H) films. They called such films polymorphous silicon films, which refer to a–Si:H matrix embedded with silicon nanocrystallites produced in the gas phase.

According to Cabarrocas et al. (2008), the incorporation of gas phase crystalline nuclei increased the film growth rate, which is correlated with high electron mobility. This aspect is shown in Fig. 8.4, where the x-axis is the product of the

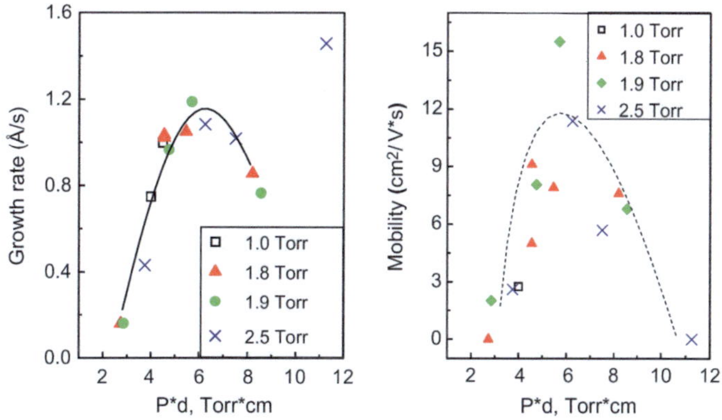

Fig. 8.4 Deposition rate and electron mobility deduced from time resolved microwave conductivity measurements as functions of the product of the total pressure and the inter-electrode distance. Note that the highest mobility is achieved in the films deposited at the highest rate (Roca i Cabarrocas et al. 2008)

total pressure and the inter-electrode distance. Both film growth rate and electron mobility are maximum when the product is ~6 Torr*cm.

They could deposit fully crystallized films, consisting of 68 % small grain materials and 28 % monocrystalline silicon with a 4 % fraction of voids at the substrate temperature at 200 °C, the RF power at 20 W and the gas flow rates at 1, 5 and 36 sccm respectively for SiF_4, H_2 and Ar. Considering the substrate temperature of 200 °C, the microcrystalline film deposition cannot be explained by the classical crystallization because of the low diffusivity of silicon atom at 200 °C. Therefore, the high crystallinity of their films is attributed to the incorporation of crystalline particles formed in the gas phase.

Another question can be raised as to how particles in the gas phase can have a crystalline structure under the condition the temperature of the plasma is also so low. This question is generally applicable to the plasma CVD, by which not only the decomposition of precursors but also the deposition of crystalline films occurs at such low temperatures. All these enhanced kinetics can be explained if the charge weakens the bond strength of molecules and nanoparticles as described in Chap. 1.

In order to check whether nanoparticles formed in the gas phase should be charged or not, Cabarrocas et al. (2008) compared two identical depositions on $5 \times 5\ cm^2$ glass substrates partly coated with Cr. They also compared the deposition behavior with and without a hydrogen plasma pretreatment as shown in Fig. 8.5.

In the case where the deposition was performed without any plasma treatment there is no film deposited on the uncoated part of the glass substrate as shown in the right of Fig. 8.5. In both cases of left and right in Fig. 8.5, deposition only took place on the Cr-coated portion, which is a kind of selective deposition. On the contrary, for the same type of substrate exposed to a hydrogen plasma prior to microcrystalline-Si:H process, the entire surface of the substrate was deposited.

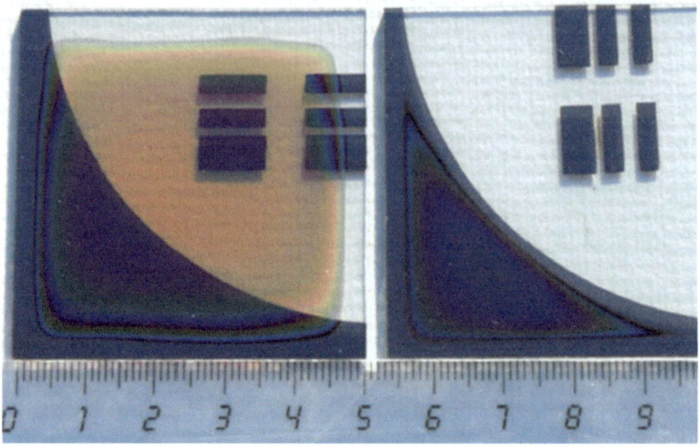

Fig. 8.5 Pictures of two depositions carried out on partly Cr-coated glass substrates with a hydrogen plasma pretreatment (*left*) and without plasma pretreatment (*right*). The strong effect of the plasma treatment is attributed to the contribution of charged nanocrystals to the growth (Roca i Cabarrocas et al. 2008)

These effects strongly support their hypothesis that deposition mainly takes place from charged species. While deposition from positively charged nanocrystals has no difficulty in the deposition on conductive substrates, it can have difficulty on insulating substrates such as glass or a-SiN:H as required for thin film transisters (TFTs).

Using radio-frequency PECVD at 13.56 MHz under conditions where nanocrystals are the primary contributors to film growth, Johnson et al. (2008) demonstrated that Ge nanocrystals generated in the gas phase land on a GaAs (100) wafer at 175 °C, adopting the orientation of the underlying lattice. This result is important because it provides the evidence of the epitaxial recrystallization by CNPs at low temperature, indicating that CNPs are liquid-like.

Figure 8.6 shows the Ge nanocrystals captured on the carbon membrane of the TEM grid. Ge nanocrystals on the surface in the form the two-dimensional chains like a pearl-necklace indicate that they have arrived individually at the surface and have undergone self-assembly rather than agglomerating in the plasma, which would result in a three-dimensional assembly of particles. This one-dimensional self-assembly is the initial process for the growth of nanowires. If the Ge crystal maintains the liquid-like property by maintaining its charge, the nanowires with a smooth surface would have been evolved instead of the chain-like structure. The chain-like growth would be the result of the electrostatic interaction between charged Ge nanoparticles. The particles are quite uniform in size and circular with a diameters of ~5 nm as shown in the inset of Fig. 8.6.

Figure 8.7 shows the cross-sectional HRTEM images of the RF PECVD-deposited Ge on a {100} GaAs wafer. The image in Fig. 8.7 shows the

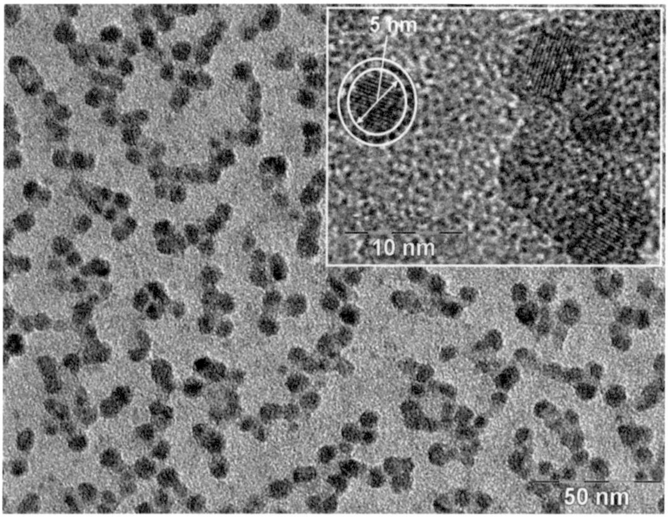

Fig. 8.6 Ge nanocrystals on C-membrane Cu grid. Image shows that the particles arrive at the surface as independent and mobile objects. *Inset* shows an image taken at greater magnification of completely crystallized Ge particles (Johnson et al. 2008)

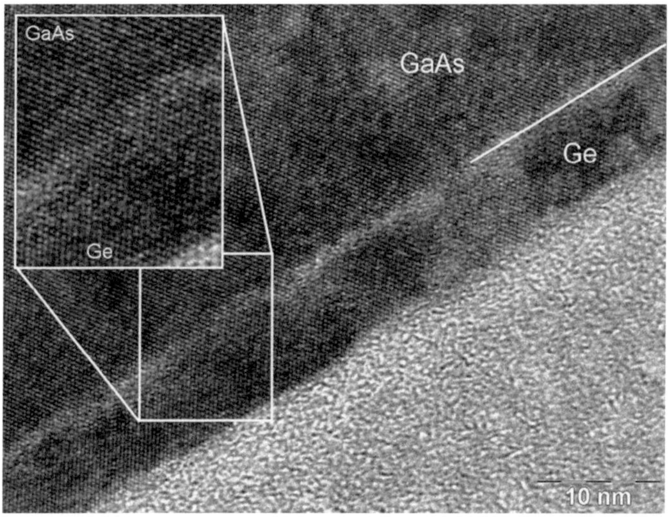

Fig. 8.7 Cross-sectional image of sample N9 deposited on GaAs. The *inset* is a digital zoom on the interface region. Despite an amorphous region at the interface, directional growth of Ge still occurs (Johnson et al. 2008)

epitaxial lattice match between the Ge film and the underlying GaAs substrate. The epitaxial growth of charged Ge nanocrystals at 175 °C would be difficult to explain unless CNPs are liquid-like.

Dusty plasma has traditionally unwanted side-effect but turns out to be very useful because nanometer-sized particles increase the deposition rate of films and produce crystalline films with epitaxial recrystallization at low temperature. It should be reminded that the electric charges generated from plasma enhance kinetics considerably but do not affect the thermodynamics. In other words, plasma enhances gas phase precipitation when the driving force is for precipitation and enhances etching when the driving force is for etching. It does not enhance etching when the driving force is for deposition. When irreversible etching and precipitation occurs simultaneously, however, it should be checked if the system has a retrograde solubility in the temperature range of processing as in the case of C–H and Si–Cl–H systems.

8.3 Nano-Fabrications Using the Building Blocks Generated in Plasma

As explained in Chap. 1, the enhanced atomic diffusivity of CNPs is attributed to the excess electrostatic energy described by (1.2), the neutral but electronically-excited molecules would have the high reactivity because they also have extra energy compared with their non-excited counterparts. Since PECVD is abundant with charged or excited species, the possible way to fabricate films and nanostructures would be much more diverse than that of thermal CVD. In relation with this point, Vladimirov and Ostrikov (2004) suggested that by controlling systematically fine clusters and particulates generated during PECVD, various nanostructures can be tailored, opening a new possibility for numerous challenging applications in fundamental science as well as in nanotechnology and other leading high-tech industries.

Ostrikov (2005), Ostrikov et al. (2013) made an extensive review on the incorporation of gas-phase nuclei into nanostructures for a reactive-plasma assisted nanoassembly process and indicated the importance of detection and control for a building block in plasma-aided nanofabrication. Figure 8.8 shows the schematics indicating how plasma CVD has a variety of building units in comparison with thermal CVD, which has a limited number of building units, such as thermally activated atoms or molecules and nanoclusters.

The electric potential of the surface affects the flux of charged nanoparticles, which is more dominant in the case of an insulating surface. Depending on the surface potential, the charged nanoparticles either deposit on the convex surface or fill up the concave surface. The sharp convex surface effectively attracted charged nanoparticle fluxes, whose phenomenon is closely related to the growth of nanotubes or nanowires. Studies in this field of plasma-aided nanofabrication are rapidly growing, which has been reviewed by Ostrikov and Murphy (2007).

As an example of the nanofabrication using plasma, using the thermophoretic force caused by varying the near-substrate temperature gradient, Rutkevych et al.

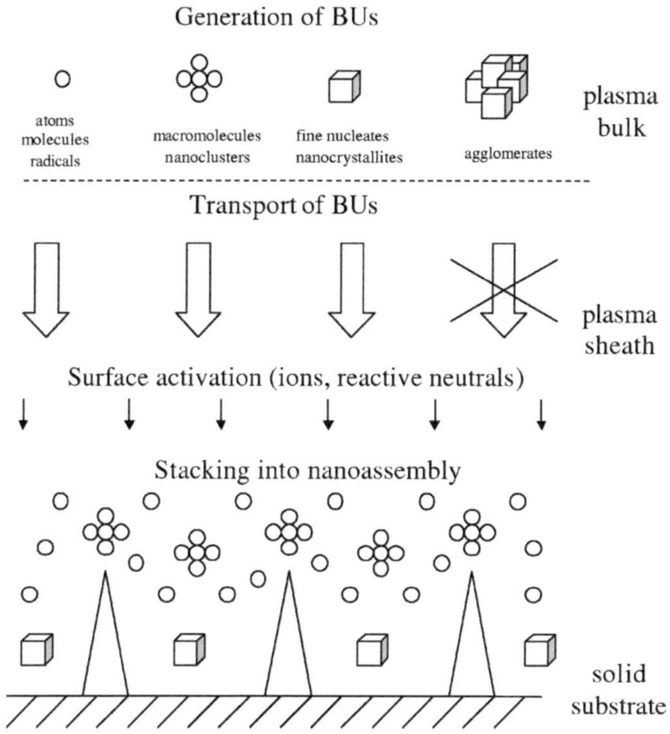

Fig. 8.8 Schematics of the reactive-plasma-based nanofabrication environment. BU = building unit (Ostrikov 2005)

(2004) could selectively deposit or levitate carbon-based nanoparticles grown in a low-temperature reactive plasma of Ar–H_2–CH_4 gas mixture. When the nanoparticles were levitated in the plasma presheath, arrays of vertically aligned carbon nanotips were assembled, whereas the enhanced deposition of the building units from the gas phase favoured the formation of polymorphous nanostructured films. Cheng et al. (2007a, b, 2010) also reported that this concept could be applied to the low-temperature growth of a nanocrystalline SiC film. The effect of surface morphology on the deposition behaviour of charged nanoparticles was numerically simulated in PECVD (Rutkevych et al. 2007).

References

Burton WK, Cabrera N, Frank FC (1951) The growth of crystals and the equilibrium structure of their surfaces, Phil Trans R Soc A 243:299–358 doi:10.1098/rsta.1951.0006

Cheng Q, Xu S, Long J, Ostrikov K (2007a) Low-temperature PECVD of nanodevice-grade nc-3C-SiC. Chem Vapor Depos 13(10):561–566

Cheng Q, Xu S, Long J, Ostrikov KK (2007b) Deterministic plasma-aided synthesis of high-quality nanoislanded nc-SiC films. Appl Phys Lett 90(17):173112

Cheng Q, Tam E, Xu S, Ostrikov KK (2010) Si quantum dots embedded in an amorphous SiC matrix: nanophase control by non-equilibrium plasma hydrogenation. Nanoscale 2(4):594–600

Frank FC (1949) The influence of dislocations on crystal growth. Discuss Faraday Soc 5(0):48–54. doi:10.1039/DF9490500048

Gerber C, Anselmetti D, Bednorz JG, Mannhart J, Schlom DG (1991) Screw dislocations in high-Tc films. Nature 350:279–280. doi:10.1038/350279a0

Hattori T, Yamaguchi N, Majima T, Terashima K, Yoshida T (1999) Deposition of high-critical-temperature superconductor $YBa_2Cu_3O_{7-x}$ epitaxial thick film by hot cluster epitaxy. J Jpn I Met 63:68–73

Hawley M, Raistrick ID, Beery JG, Houlton RJ (1991) Growth mechanism of sputtered films of $YBa_2Cu_3O_7$ studied by scanning tunneling microscopy. Science 251:1587–1589. doi:10.1126/science.251.5001.1587

Hayasaki K, Takamura Y (1997) Scanning tunneling microscopy of epitaxial $YBa_2Cu_3O_{7-x}$ films prepared by thermal plasma flash. J Appl Phys 81(3):1222

Hayasaki K, Takamura Y, Yamaguchi N, Terashima K, Yoshida T (1997a) Scanning tunneling microscopy of epitaxial $YBa_2Cu_3O_{7-x}$ films prepared by thermal plasma flash evaporation method. J Appl Phys 81(3):1222–1226

Hayasaki K, Takamura Y, Yamaguchi N, Terashima K, Yoshida T (1997b) Scanning tunneling microscopy of epitaxial $YBa_2Cu_3O_{7-x}$ films prepared by thermal plasma flash evaporation method. J Appl Phys 81(3):1222–1226. doi:http://dx.doi.org/10.1063/1.364143

Johnson EV, Patriarche G, Roca i Cabarrocas P (2008) Directional growth of Ge on GaAs at 175 °C using plasma-generated nanocrystals. Appl Phys Lett 92(10):103108. doi:10.1063/1.2895636

Luo L, Hawley ME, Maggiore CJ, Dye RC, Muenchausen RE, Chen L, Schmidt B, Kaloyeros AE (1993) Spiral growth in epitaxial $YBa_2Cu_3O_{7-x}$ thin films produced by high deposition rate chemical vapor deposition. Appl Phys Lett 62(5):485

Ostrikov K (2005) Colloquium: Reactive plasmas as a versatile nanofabrication tool. Rev Mod Phys 77(2):489

Ostrikov K, Murphy A (2007) Plasma-aided nanofabrication: where is the cutting edge? J Phys D Appl Phys 40(8):2223

Ostrikov K, Neyts E, Meyyappan M (2013) Plasma nanoscience: from nano-solids in plasmas to nano-plasmas in solids. Adv Phys 62(2):113–224

Roca i Cabarrocas P (2000) Plasma enhanced chemical vapor deposition of amorphous, polymorphous and microcrystalline silicon films. J Non-Cryst Solids 266–269, Part 1:31–37. doi:http://dx.doi.org/10.1016/S0022-3093(99)00714-0

Roca i Cabarrocas P (2002) Plasma enhanced chemical vapor deposition of silicon thin films for large area electronics. Curr Opin Solid State Mate Sci 6(5):439–444. doi:http://dx.doi.org/10.1016/S1359-0286(02)00112-2

Roca i Cabarrocas P (2004a) New approaches for the production of nano-, micro-, and polycrystalline silicon thin films. Phys Status Solidi (c)1(5):1115–1130. doi:10.1002/pssc.200304328

Roca i Cabarrocas P, Chaâbane N, Kharchenko AV, Tchakarov S (2004b) Polymorphous silicon thin films produced in dusty plasmas: application to solar cells. Plasma Phys Contr F 46(12B):B235

Roca i Cabarrocas P, Djeridane Y, Bui VD, Bonnassieux Y, Abramov A (2008) Critical issues in plasma deposition of microcrystalline silicon for thin film transistors. Solid State Electron 52(3):422–426. doi:http://dx.doi.org/10.1016/j.sse.2007.10.028

Rutkevych P, Ostrikov K, Xu S, Vladimirov S (2004) Thermophoretic control of building units in the plasma-assisted deposition of nanostructured carbon films. J Appl Phys 96(8):4421–4428

Rutkevych P, Ostrikov K, Xu S (2007) Two-dimensional simulation of nanoparticle deposition from high-density plasmas on microstructured surfaces. Phys Plasmas (1994-present) 14(4):043502

Takagi T (1988) Ionized-cluster beam deposition and epitaxy. William Andrew Publishing, Norwich, NY

Takamura Y, Yamaguchi N (1998) High-rate deposition of $YBa_2Cu_3O_{7-x}$ films by hot cluster epitaxy. J Appl Phys 84(9):5084–5088

Takamura Y, Hayasaki K, Terashima K, Yoshida T (1997) Cluster size measurement using microtrench in a thermal plasma flash evaporation process. J Vac Sci Technol B 15(3):558–565

Takamura Y, Yamaguchi N, Terashima K, Yoshida T (1998) High-rate deposition of $YBa_2Cu_3O_{7-x}$ films by hot cluster epitaxy. J Appl Phys 84(9):5084–5088

Terashima K, Yamaguchi N, Hattori T, Takamura Y, Yoshida T (1998) High rate deposition of thick epitaxial films by thermal plasma flash evaporation. Pure Appl Chem 70(6):1193–1197. doi:http://dx.doi.org/10.1351/pac199870061193

Vladimirov SV, Ostrikov K (2004) Dynamic self-organization phenomena in complex ionized gas systems: new paradigms and technological aspects. Phys Rep 393(3–6):175–380. doi:http://dx.doi.org/10.1016/j.physrep.2003.12.003

Yamaguchi N, Sasajima Y, Terashima K, Yoshida T (1999) Molecular dynamics study of cluster deposition in thermal plasma flash evaporation. Thin Solid Films 345(1):34–37

Chapter 9
Experimental Confirmation of Charged Nanoparticles During Atmospheric CVD Using Differential Mobility Analyser

In Chaps. 6 and 7, it was suggested that diamond and silicon films grow by CNPs formed in the gas phase. With the assumption that CNPs are liquid-like, the film growth by CNPs explains many puzzling phenomena. Then, a question arise, "How about other CVD systems?" Would the non-classical crystallization by the building block of CNPs be applicable only to exceptional cases such as diamond and silicon or a general growth mechanism of CVD? In generalizing this new paradigm of the thin film growth, the most important work to do would be to confirm the generation of CNPs in other CVD processes. In this chapter, it will be checked experimentally whether gas phase generated nanoparticles exist or not in some important CVD processes. And if they exist, it will be checked whether they are charged or not.

9.1 Differential Mobility Analyzer and Faraday Cup Electrometer

Using a differential mobility analyzer (DMA) combined with Faraday cup electrometer (FCE), not only the generation of CNPs can be confirmed but also their size distribution can be measured in the atmospheric CVD process. Nano-DMA (nano DMA; TSI model 3085) is specially designed to measure the size distribution and concentration of particles in the size range of 1–100 nm whereas other DMA (long DMA; TSI model 3081) is normally designed for measurements in the size range of 10–1000 nm. Since the size of CNPs generated in the CVD process is normally less than ~ 100 nm with some exception, nano DMA is appropriate. The size-classified CNPs are detected as a current on a FCE. At first, the principle of the DMA-FCE system will be briefly explained.

A DMA separates charged particles according to their electrical mobility and can be used to produce aerosols of known size and number concentration. Figure 9.1 shows the schematic of DMA.

© Springer Science+Business Media Dordrecht 2016

N.M. Hwang, *Non-Classical Crystallization of Thin Films and Nanostructures in CVD and PVD Processes*, Springer Series in Surface Sciences 60,
DOI 10.1007/978-94-017-7616-5_9

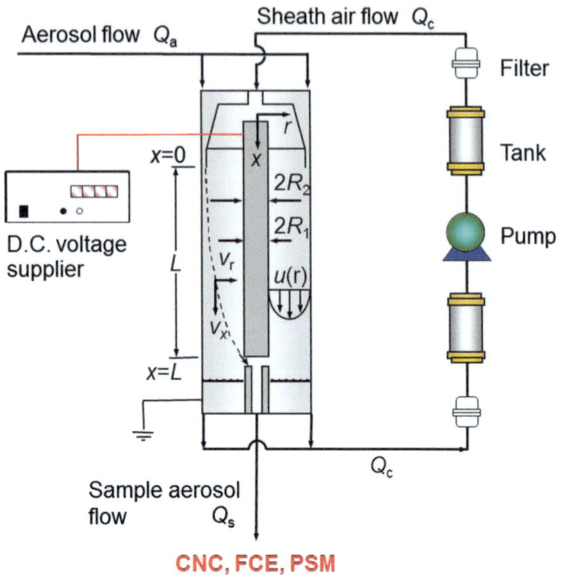

Fig. 9.1 Schematic of a differential mobility analyzer

A laminar particle-free sheath air flow Q_c is introduced through a cylindrical capacitor (x-direction) consisting of an inner electrode (high voltage rod) and an outer electrode. The aerosol flow Q_a (x-direction) is fed into the cylindrical capacitor close to the outer electrode. An electric field is applied between the two electrodes (r-direction) and charged particles are driven toward the inner one (Dhaniyala et al. 2011).

Electrically charged particles move in an electric field according to their electrical mobility. The horizontal particle velocity in the r direction is given by

$$v_z = dr/dt = Z_p E \tag{9.1}$$

where Z_p is the electrical mobility and E is the field strength. The electrical mobility of a particle is defined as the ratio of the constant limiting velocity a charged particle will reach in a uniform electric field to the magnitude of this field (Dhaniyala et al. 2011).

$$Z_p = v_z/E \tag{9.2}$$

The electrical mobility depends mainly on the particle size and electrical charge. The smaller the particle and/or the higher the electrical charge the higher is the electrical mobility. The electrical mobility is given by

$$Z_p = \frac{Q_c \ln(R_1/R_2)}{2\pi LV} \tag{9.3}$$

where Q_c is the flow rate of sheath gas, R_1 and R_2 are respectively radii of inner and outer electrodes, L is the DMA length and V is the voltage between inner and outer rod. The particle diameter D_p is related with the electrical mobility Z_p by

$$D_p = \frac{peC_c}{3\pi\eta Z_p} \tag{9.4}$$

where p is the number of elementary charge units, e is an elementary unit of charge (1.6×10^{-19} C), C_c is a Cunningham correction factor and η is a gas viscosity.

Under the applied voltage of V to the inner electrode and the given sheath flow rate, charged particles satisfying (9.3) and (9.4) would be withdrawn through the central electrode at a downstream location, while most of the flow would be exhausted. Those size-selected particles according to (9.3) and (9.4) are transported from the DMA for subsequent use such as FCE, condensation nuclei counter (CNC) and particle size magnifier (PSM).

The number of size-selected charged particles is then measured as electrical current by the FCE shown in Fig. 9.2. In the metal housing of a Faraday cup, a filter made of stainless mesh is mounted on an insulator of glass fiber. Charged particles collected by the filter generate an electric current which is measured by the electrometer, which has a detection limits of 10^{-16} A.

Figure 9.3 shows the schematic of the DMA-FCE system installed to the CVD reactor with a horizontal tube for in situ measurements of CNPs. In some CVD processes, reactant gases as well as nitrogen or hydrogen gas are supplied. In other

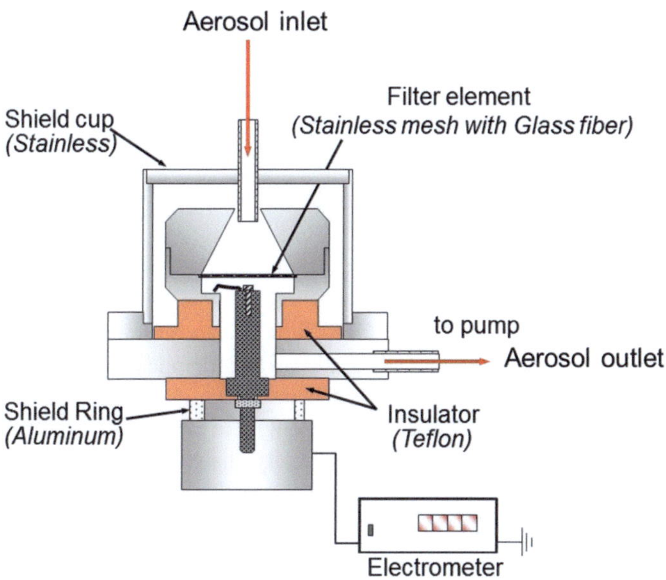

Fig. 9.2 Schematic of a Faraday cup electrometer

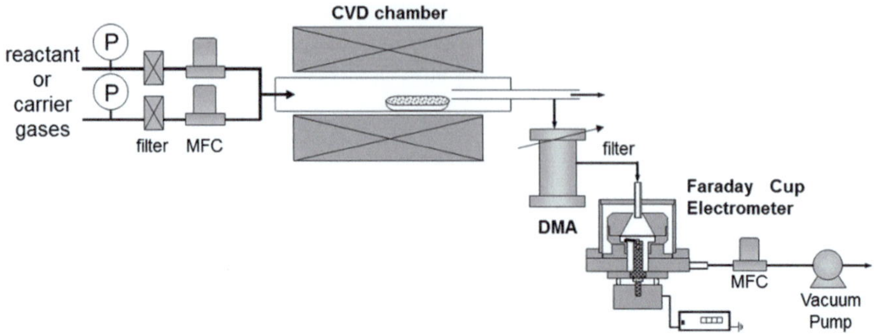

Fig. 9.3 Schematic of experimental set-up for the CVD reactor with the DMA-FCE system for measurements of CNPs generated during atmospheric CVD

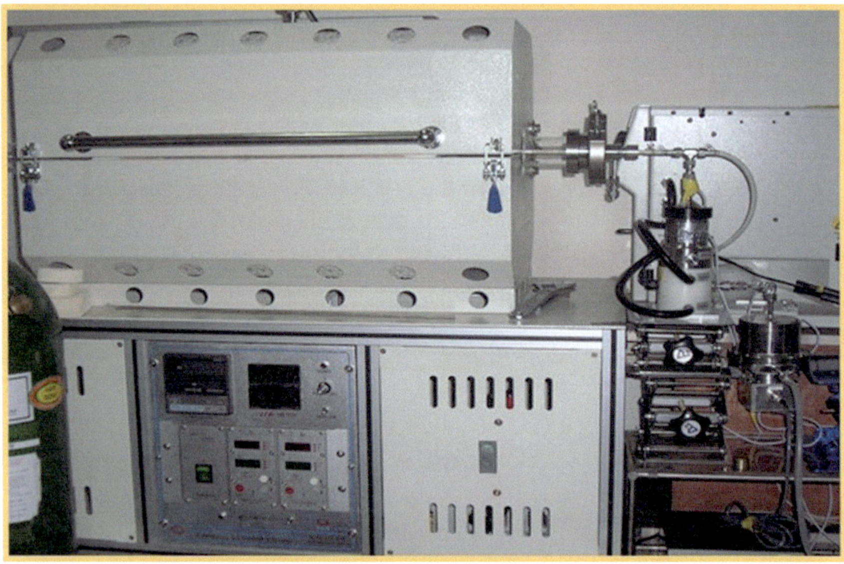

Fig. 9.4 Photograph of the CVD reactor installed with the nano-DMA-FCE system

CVD processes synthesizing ZnO or GaN films or nanostructures by carbothermal reduction of oxides, the alumina boat loaded with graphite and oxide is placed in the hot zone of the furnace. The reduced metals such as Zn and Ga react respectively with oxygen and nitrogen to produce ZnO and GaN films or nanostructures.

Normally, particles are made to be electrically charged by the artificial charging system before DMA measurements because a DMA can measure only charged particles. However, an artificial charger are not used in this experiment in order to check whether the gas phase generated nanoparticles are charged or not in the CVD reactor.

Figure 9.4 shows the photograph of the horizontal tube CVD reactor on the power supply and the controller. On the right of the reactor, the nano DMA is installed. On the right of and below the nano DMA, the FCE is installed. At the back of the nano-DMA, the flow system for the sheath air is installed.

9.2 Generation of Charged Nanoparticles During the Synthesis of ZnO Nanostructures

Since nanostructure shapes have a great effect on electrical and optical properties, ZnO nanostructures of various shapes such as nanowires (Yao et al. 2002), nanobelts (Ago et al. 2001; Xiang Yang Kong and Zhong Lin 2004; Kong et al. 2004), nanocoils (Korgel 2004), nanorings (Xiang Yang Kong and Zhong Lin 2004; Kong et al. 2004), and nanocombs (Wang et al. 2003) have been synthesized by CVD processes.

As a growth mechanism of ZnO nanostructures, various models such as vapor-liquid-solid (VLS) growth, vapor-solid (VS) growth and redox growth were proposed to understand their growth (Fu et al. 2007). However, many microstructural features such as spontaneous coiling could not be understood from the conventional concept of crystal growth. To explain the growth of the ZnO nanobelts and nanorings, Kong et al. (2004), and Zhong Lin (2004) suggested the charging model, where the electrostatic energy plays a critical role in the growth of ZnO nanorings by a spontaneous self-coiling process of polar nanobelts. Their polar charge model is based on the polar nature of ZnO. However, there is a possibility that the origin of charge comes from the CNPs, which are expected to be generated during the synthesis of ZnO nanostructures. This possibility is examined in the following section.

9.2.1 In-situ Measurements of CNPs by DMA During the Synthesis of ZnO Nanostructures

ZnO nanostructures were synthesized using a typical carbothermal reduction process (Kim et al. 2009a). 2 g of ZnO (99.9 %) powder and 2 g of graphite powder (99.99 %) were mixed. 4 g of the resultant mixture in an alumina boat was placed at the center of a quartz-tube furnace. Above the alumina boat, a silicon substrate coated with 50 Å-thick gold was placed. High purity nitrogen gas (99.9999 %) was supplied as a carrier gas together with four different oxygen flow rates of 0, 2, 10, and 90 sccm. The total flow rate was fixed as 1000 sccm.

Figure 9.5 shows the FESEM images of ZnO nanostructures formed at different furnace temperatures.

Figure 9.5a shows that at a furnace temperature of 800 °C, nothing was deposited on the silicon substrate and only isolated gold nanoparticles were formed

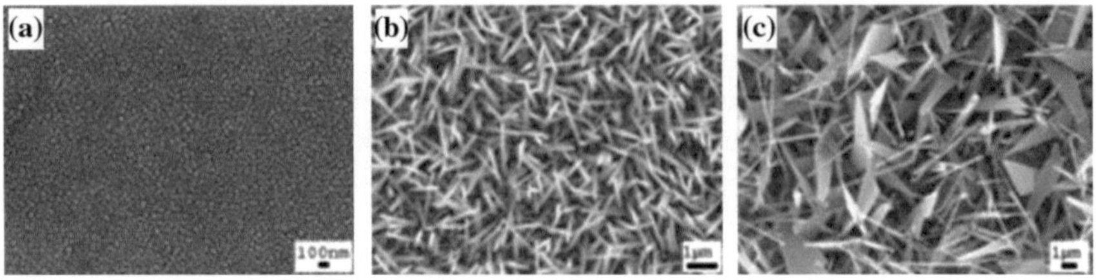

Fig. 9.5 FESEM images of ZnO nanowires at furnace temperatures of **a** 800 °C, **b** 900 °C, and **c** 1000 °C and at an oxygen flow rate of 10 sccm (Kim et al. 2009a)

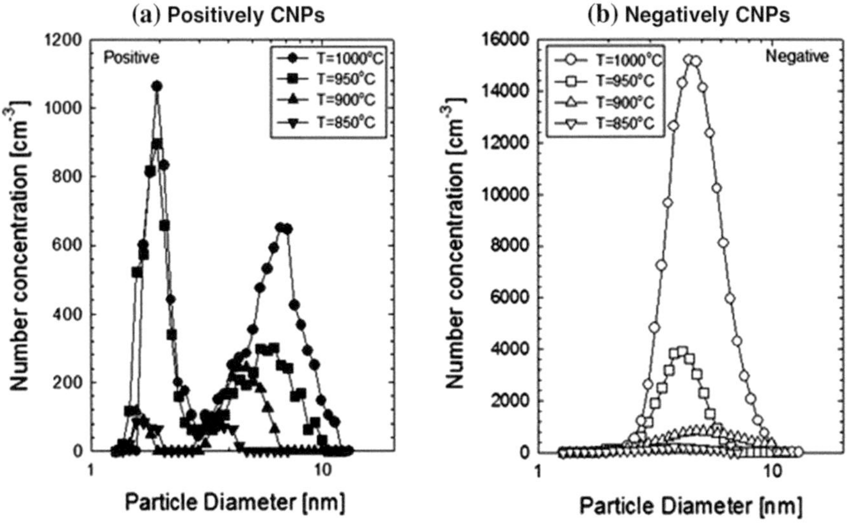

Fig. 9.6 The size distribution of **a** positively and **b** negatively charged ZnO nanoparticles at various furnace temperatures and at an oxygen flow rate of 10 sccm (Kim et al. 2009a)

from the film deposited by sputtering. When the furnace temperature increased to 900 °C, nanowires were produced with a diameter of ~100 nm as shown in Fig. 9.5b. As the furnace temperature increased to 1000 °C, nanosheets began to form along the basal nanowires but the diameter of nanowires was almost the same as that formed at 900 °C as shown in Fig. 9.5c.

Figure 9.6 shows the DMA data of the size distribution for ZnO CNPs generated in the gas phase with various furnace temperatures with other processing conditions being the same as those of Fig. 9.5.

Both positive and negative CNPs are generated under the typical processing conditions of ZnO nanostructures. At the furnace temperature below 850 °C, CNPs were not detected. This result agrees with no deposition at 800 °C in Fig. 9.5a.

Figure 9.6 shows that both positive and negative CNPs were generated at the furnace temperature of 850 °C and had a single peak at 4–5 nm. With increasing furnace temperature above 900 °C, both positive and negative CNPs tend to have a bimodal size distribution.

Most CNPs are smaller than 10 nm. At all temperatures, the number concentration of negative CNPs was higher than that of the positive CNPs. The particle number concentration of the negative CNPs increased with increasing furnace temperature, while that of the positive CNPs tended to decrease, passing through a minimum as the furnace temperature increased. However, the particle diameter at the peak did not change sensitively with the furnace temperature.

9.2.2 Formation Mechanism of Charged Nanoparticles

There are two possible mechanisms for the formation of CNPs measured in Fig. 9.6. One is that ions are formed first by surface ionization in the reactor and nanoparticles are formed in the gas phase by ion-induced nucleation, which is the case of formation of CNPs in HF CVD (Hwang et al. 1996; Hwang and Kim 2004) or plasma CVD (Girshick et al. 1996; Haaland et al. 1996; De Bleecker et al. 2004) However, in the synthesis condition of ZnO nanostructures, the temperature of the reactor is so low that the number density of ions would not be so high. The other possible mechanism is that nucleation takes place first and then the nuclei undergo surface ionization on any surface such as the quartz tube of the reactor (Hwang and Kim 2004). Magnusson et al. (1999) reported similar unusual charging of gold nanoparticles during reshaping at high temperature.

The positive and negative surface ionizations are described by the Saha–Langmuir equations (Langmuir and Kingdon 1925) as

$$\frac{n^+}{n^0} = \frac{g^+}{g^0} \exp\left(-\frac{IP - WF}{kT}\right) \tag{9.5}$$

$$\frac{n^-}{n^0} = \frac{g^-}{g^0} \exp\left(-\frac{WF - EA}{kT}\right) \tag{9.6}$$

where IP, EA, WF, and k are ionization potential, electron affinity of nanoparticles, work function of the quartz wall, and Boltzmann constant, respectively. n^0, n^+, and n^- are the numbers of neutral, positive, and negative species, respectively. g^0, g^+, and g^- indicate the statistical weight of neutral, positive, and negative species, respectively. Therefore, WF of the reactor wall and IP and EA of nanoparticles generated in the gas phase are the key factor in determining the polarity of charged nanoparticles. Since the nanoparticles generated in this experiment are sufficiently large, both IP and EA of nanoparticles approach the work function value of their bulk (Seidl et al. 1991). The difference between (IP-WF) and (WF-EA) in the above equations would make the difference between the number concentrations of positively and negatively CNPs.

This analysis shows that charging mechanism of nanoparticles in the CVD reactor is similar to contact charging or triboelectrification which we experience in a daily life. When we rub plastics again our hair, plastics are electrically charged, which means that contact charging occurs at room temperature. Therefore, it is not surprising that when the gas phase nucleation takes place in the CVD reactor, the nuclei can be charged by surface ionization through contact with the reactor wall at low temperature.

It is not known how much the surface polar charge of ZnO, which was suggested by Kong and Wang (Xiang Yang Kong and Zhong Lin 2004), contributes to the electrostatic energy for the evolution of various ZnO nanostructures. However, it is expected that the contribution from the surface polar charge would be much less than that from CNPs measured in Fig. 9.6. Therefore, it can be concluded safely that the origin of the electrostatic energy responsible for the evolution of various ZnO nanostructures should come from the CNPs.

9.3 Generation of Charged Nanoparticles During the Synthesis of Carbon Nanotubes

In Chap. 6, charged diamond nanoparticles were shown to be generated in the gas phase in the concentration range of 1 % CH_4–99 % H_2 to 5 % CH_4–95 % H_2. Considering that carbon nanotubes (CNTs) are synthesized using much higher concentrations of CH_4, it is rather evident that charged carbon nanoparticles should be generated during the synthesis of CNTs. Besides, Chiang and Sankaran (2007) reported the generation of carbon nanoparticles in the gas-phase synthesis of CNTs using C_2H_2 at reactor temperatures above 700 °C without a catalyst.

It has been reported that the application of electric fields affects the growth behavior in various CNT syntheses, such as CVD (Avigal and Kalish 2001; Ural et al. 2002; Lee et al. 2003), flame (Merchan-Merchan et al. 2004; Xu et al. 2006; Qiaoliang et al. 2006), HFCVD (Takahito et al. 2002; Lee and Hwang 2008) and plasma CVD (Bower et al. 2000; Merkulov et al. 2000; Teo et al. 2001b; Delzeit et al. 2002; Meyyappan et al. 2003; AuBuchon et al. 2006; Lim et al. 2006). Judging from the fact that the electric field has an effect on the growth behavior of CNTs, there is a possibility that the moving charged carriers might be involved in CNT growth. Therefore, CNPs appear to be closely related to the growth mechanism of CNTs. In this section, the generation of CNPs and its relation with the growth of CNTs will be investigated.

9.3.1 Hot Filament CVD Process

To examine the growth mechanism of CNTs, some important thermodynamic aspects of the C–H system, which was described in detail in Chaps. 5 and 6, will be mentioned again. According to the phase diagram of the carbon–hydrogen system, the solubility of carbon in the gas phase increases with decreasing temperatures below ~900 °C (Hwang and Yoon 1996). In other words, the C–H system has a retrograde carbon solubility in the gas phase. Because of this retrograde solubility, the formation of nanoparticles in the gas phase of the high temperature region near the hot filament depletes carbon in the low-temperature region below 900 °C, so that carbon becomes undersaturated. As a result, the driving force at the substrate temperature is not for deposition but for etching of solid carbon.

The synthesis of CNTs by HFCVD has much in common with that of diamonds by HFCVD, except that in the former, catalytic metal nanoparticles are used together with a much higher concentration of hydrocarbons. For example, a 1–5 % CH_4 and 95–99 % H_2 gas mixture is used in the HFCVD for diamond whereas a 20 % CH_4 and 80 % H_2 gas mixture or an even higher methane concentration is used in the HFCVD for CNTs. Considering the experimental confirmation of the generation of CNPs in the HFCVD for diamond in Chap. 6, it is expected that CNPs should also be generated during the synthesis of CNTs by HFCVD. In addition, the CNPs that are generated must be the main growth flux for CNTs, because carbon is undersaturated in the temperature region around the substrate.

A typical HFCVD process was employed for CNT growth: a tungsten-filament temperature of 1900 °C, a reactor pressure of 170 Torr, and a 20 % CH_4 and 80 % H_2 gas mixture. A stainless steel substrate was placed 2 cm below the tungsten filament. The substrate temperature was maintained at 600 °C. The electric current produced during the synthesis was measured on a 1×1 cm^2 stainless steel substrate, using a picoammeter (Keithley 486). The effect of bias on the deposition behavior was examined by the application of an electric bias to the substrate during the growth of both CNTs and carbon nanoparticles.

The electric currents on the substrate measured during the CNT growth are shown in Fig. 9.7. At the filament temperature of 1900 °C, the current was negative and maintained at roughly –2 μA cm^{-2} (Fig. 9.7a), which increased negatively with an increase in the filament temperature (Fig. 9.7b).

The negative charge, regardless of whether its carrier is an electron or an ion such as C^{-2} has a high electrostatic energy, which can be reduced by further attachment of carbon in the gas phase. Thus, the negative charge induces a clustering or nucleation in the gas phase, which is known as ion-induced nucleation as described in detail in Chap. 3. Because the mean free path at a reactor pressure of 170 Torr is much less than 1 μm, the negative-charge carriers undergo numerous collisions with atoms or molecules in the gas phase before they reach the substrate. This implies the possibility that the negative-charge carriers, which are estimated by measuring the electric current on the substrate (Fig. 9.7), might be mainly carbon clusters or nanoparticles.

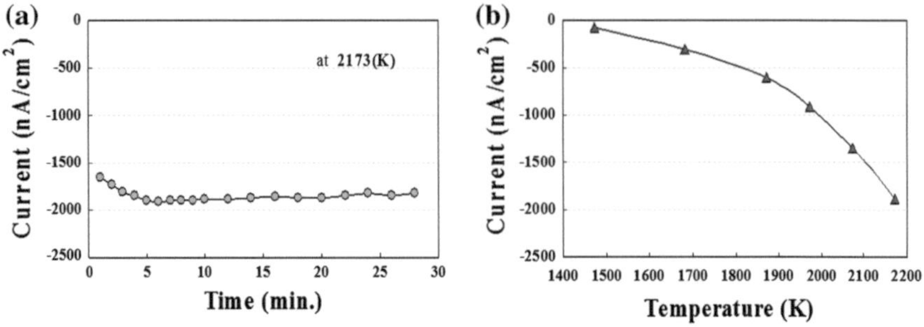

Fig. 9.7 Negative current measured **a** at the filament temperature of 2173 K with time and **b** with varying filament temperatures during CNT growth (Lee and Hwang 2008)

Fig. 9.8 The mass of CNTs or carbon nanoparticles grown at the applied biases of −200, 0, and +25 V for 30 min (Lee and Hwang 2008)

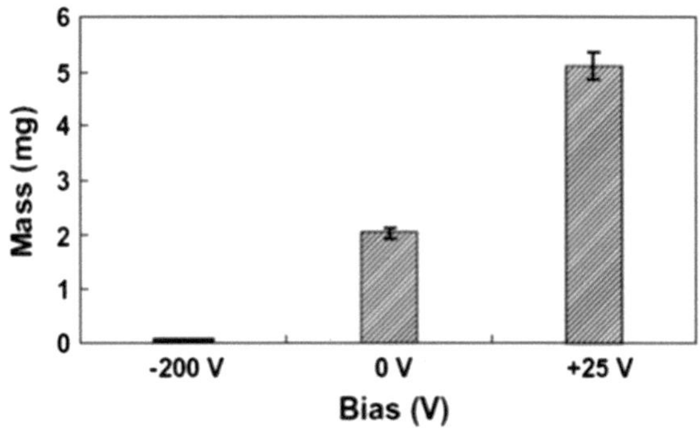

If CNPs contribute to the growth of CNTs, the deposition behavior or the growth rate would be affected by the applied bias. Therefore, the biases of −200, 0, and +25 V were applied between the substrate and the filament. Then the substrate are weighed to check the weight change, which would indicate the deposited flux or the growth rate. The results are compared for different biases in Fig. 9.8.

The mass of CNTs deposited on the substrate biased at +25 V is found to be twofold and tenfold larger than the mass of CNTs or carbon nanoparticles deposited on the substrate biased at 0 and −200 V, respectively. This result suggests that the deposition flux is mainly negatively charged and might be closely related to the negative-charge carriers measured as electric current shown in Fig. 9.7.

FESEM images of the samples grown on the substrate at the applied biases of −200, 0, and +25 V are shown in Fig. 9.9a–c, respectively. CNTs are observed to grow appreciably at an applied bias of +25 V, whereas at the applied biases of 0 and −200 V, they are not formed, but carbon nanoparticles grow instead. This result (a positive bias promoting the CNT growth) implies that the growth flux of CNTs is negatively charged. This result agrees with previous reports that an electric field affected the growth behavior of CNTs (Avigal and Kalish 2001; Ural et al. 2002; Lee et al. 2003).

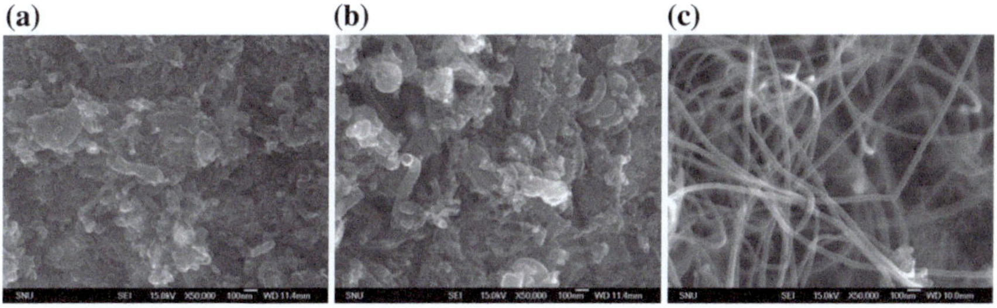

Fig. 9.9 FESEM images of CNTs and carbon nanoparticles grown on the stainless steel substrate at 600 °C at the applied biases of **a** −200, **b** 0 and **c** +25 V (Lee and Hwang 2008)

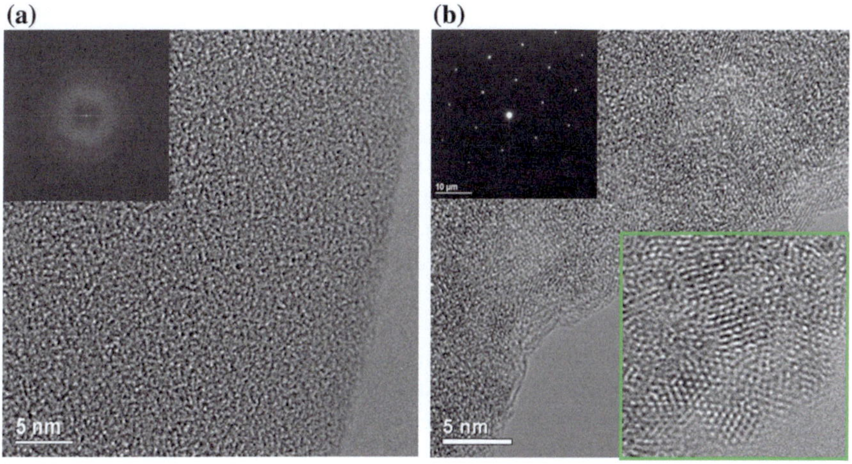

Fig. 9.10 HRTEM microstructures of the TEM grid placed in the second chamber at applied biases of **a** −200 and **b** +25 V (Lee and Hwang 2008)

The proposal that the negative CNPs exist in the gas phase of the reactor has been confirmed as follows: the gas phase is drawn through a 1-mm-diameter orifice into the second, higher vacuum (1 Torr) chamber. In the second chamber, any possible growth flux in the extracted gas is captured on a grid at applied biases of −200 and +25 V for 5 s and is subsequently observed by HRTEM, as shown in Fig. 9.10a, b, respectively. No nanoparticle is detected on the grid at a bias of −200 V (Fig. 9.10a), whereas nanoparticles of 5 nm are observed on the grid at the bias of +25 V (Fig. 9.10b). The structure of the nanoparticles, as examined by electron-diffraction, is shown in the inset of Fig. 9.10b. The lattice indices are consistent with that of graphite, with a 2 % lattice mismatch.

It should be noted that the TEM-grid temperature in the second chamber does not exceed 100 °C. Considering the negligible diffusivity of carbon at such a low temperature, the crystalline graphite nanoparticles in Fig. 9.10b could not have formed on the grid during this process. Instead, they must have landed on the grid from the gas phase. In addition, the absence of nanoparticles on the grid at the bias of −200 V and their presence at the bias of +25 V indicate that the graphite nanoparticles shown in Fig. 9.10b are negatively charged in the gas phase.

Combining all the results and analyses, the negative charge carriers in the gas phase, being the main deposition flux for CNTs, are expected to be mainly nanoparticles rather than individual ions or electrons. Therefore, it can be concluded that CNTs are grown mainly by the negative CNPs formed in the gas phase during HFCVD. These results are consistent with previous reports that negative CNPs were generated abundantly during the synthesis of diamonds by HFCVD using the gas mixture of 1–5 % CH_4 and 95–99 % H_2. Catalytic metals are necessary for CNT growth, and CNPs seem to interact with catalytic metals, in which nanoparticles are dissolved, supplying individual carbon atoms for the growth of CNTs. Electrostatic energy will build up on a floating stainless steel substrate because the growth flux carries an electric charge. Therefore, electrostatic energy might also be involved in CNT growth. It should be noted that the electrostatic energy is much smaller for CNTs, which have a high surface area, than for carbon with its compact structure. Additionally, in most gas-phase syntheses of CNTs, the catalytic metal particles are placed on an insulating substrate such as alumina (Kim et al. 1991).

In other systems such as Si, SiC and GaN nanorod or nanowires rather than nanotubes grow by CNPs. There must be some reason why nanotubes grow in the carbon system. One reason would be that the tube structure is rather stable. Another would be that the surface diffusion of carbon around the catalytic metal particles is higher than the bulk diffusion especially when the catalytic metal remains as a solid. If this is correct, small catalytic metal particles tend to produce carbon rods whereas large particles tend to produce nanotubes and carbon rods tend to be produced when the temperature is high enough to melt the catalytic metal particles.

9.3.2 Measurements of CNPs by DMA During Atmospheric CVD

CNTs were synthesized using a typical atmospheric pressure CVD process at reactor temperatures of 950, 990, 1000 and 1010 °C at CH_4 flow rates of 30, 50 and 100 sccm for 120 min (Kim et al. 2009b). Nitrogen gas was supplied as a carrier gas. Hydrogen gas was additionally supplied at flow rates of 0, 50 and 100 sccm. Although the flow rates of CH_4 and hydrogen varied, the total gas flow rate was fixed at 1000 sccm. A Ni-coated silicon substrate was placed at the center of the quartz-tube furnace.

Figure 9.11 shows the FESEM images of materials formed at different reactor temperatures. At a furnace temperature of 950 °C (Fig. 9.11a), no CNTs were produced; however, solid carbon was formed on the Ni particles. When the reactor temperature increased to 990 °C (Fig. 9.11b), CNTs with a diameter of ~20 nm began to form. As the reactor temperature was further increased to 1000 °C

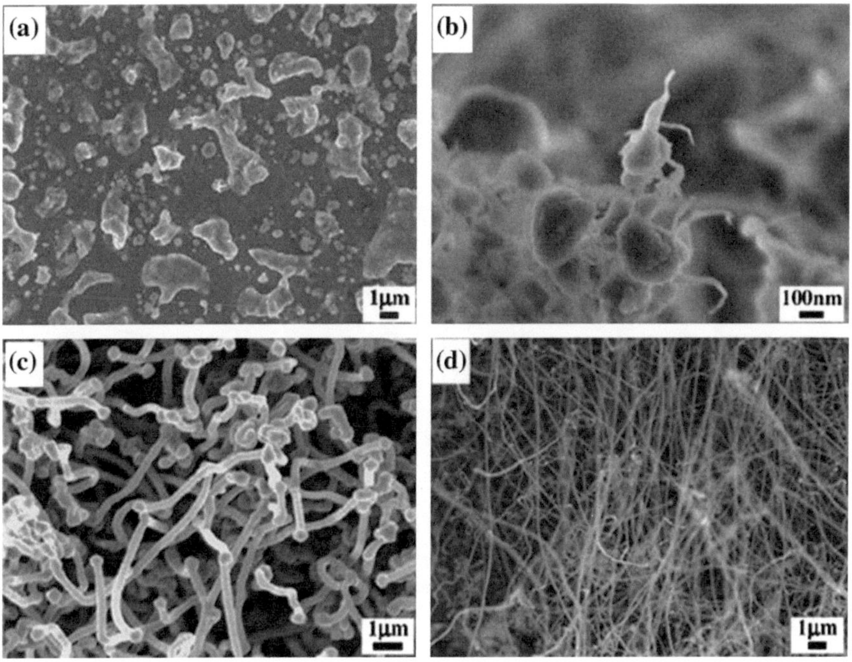

Fig. 9.11 FESEM images of CNTs. The CNTs synthesized at a reactor temperatures of **a** 950 °C, **b** 990 °C, **c** 1000 °C and **d** 1010 °C with a CH_4 flow rate of 100 sccm (Kim et al. 2009b)

(Fig. 9.11c), CNTs with a diameter of ~300 nm were produced. At a reactor temperature of 1010 °C, CNTs became longer, concurrent with a smaller diameter of ~130 nm. These results indicate that the CNT growth is enhanced with increasing reactor temperatures within the examined temperature range.

To confirm the generation of CNPs in the gas phase during the synthesis of CNTs by atmospheric-pressure CVD, in situ measurements of CNPs were carried out using the DMA-FCE system. Figure 9.12 shows the size distribution of CNPs generated in the gas phase at various reactor temperatures.

Both positive and negative CNPs are generated under the typical processing conditions of CNTs. At reactor temperatures below 990 °C, charged nanoparticles were not detected. This result indicates that CNPs were not formed or that their number was small enough not to be detected by the DMA-FCE system. It appears that the decomposition of CH_4 occurred mainly on the surface of the Ni catalyst at reactor temperatures below 990 °C (Kim et al. 1991). Figure 9.12 shows that both positive and negative CNPs were generated at a reactor temperature of 990 °C and peaked at 2–4 nm. The number concentration and the peak diameter of both positive and negative CNPs increased drastically with increasing reactor temperatures. This indicates that nucleation and charging of nanoparticles in the gas phase are enhanced with increasing reactor temperature due to enhancement of the decomposition of CH_4.

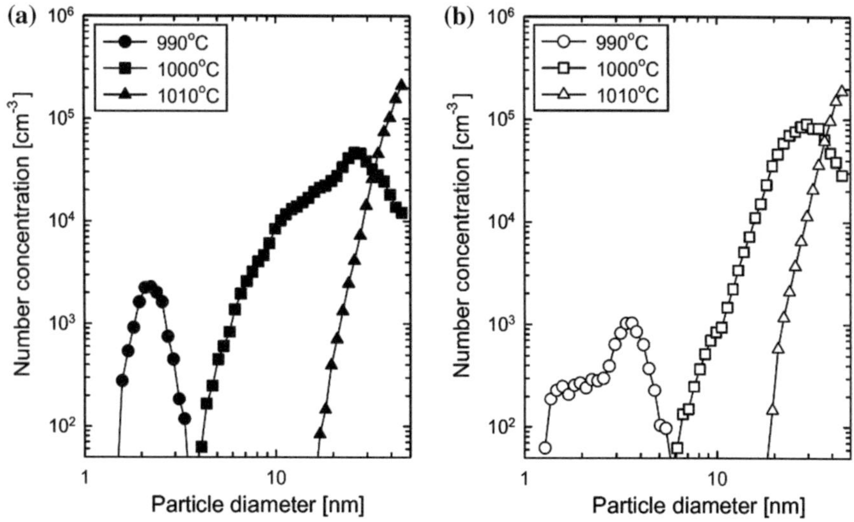

Fig. 9.12 The size distribution of **a** positively and **b** negatively charged carbon nanoparticles at various reactor temperatures with a CH_4 flow rate of 100 sccm (Kim et al. 2009b)

To examine the effect of the CH_4 flow rate on both the CNT growth and size distribution of CNPs, the experiment was repeated under various CH_4 flow rates without a hydrogen supply at a reactor temperature of 1010 °C. Figure 9.13 shows the FESEM images of CNTs formed at different CH_4 flow rates.

As the CH_4 flow rate was increased from 30 to 50 sccm, both the diameter and length of CNTs increased, with the diameter being 210 and 290 nm at CH_4 flow rates of 30 and 50 sccm. As the CH_4 flow rate further increased to 100 sccm, the diameter tended to decrease, though the length drastically increased. The HRTEM image of a sample deposited at a CH_4 flow rate of 100 sccm showed multiwalled CNTs with a diameter of ~60 nm, as shown in Fig. 9.13d. This result indicates that the CH_4 concentration is an important parameter controlling the diameter and length of CNTs. Therefore, the size of the catalyst nanoparticles does not appear to be the only parameter that determines the diameter of the CNTs grown on them. This is in contrast with the previous report by Choi et al. (2000), who reported the dependence of CNT diameter on the size of the catalyst. In agreement with the current results, Garg et al. (2008) reported that the CNT diameters tended to decrease with increasing CH_4 flow rates. They also reported that the quality of the growth product was generally improved with increasing CH_4 flow rate. De Jong and Geus (2000) also found that the diameter of CNTs changed with the operating conditions and the composition of the reactant mixture.

Figure 9.14 shows the particle size distribution of positively (Fig. 9.14a) and negatively (Fig. 9.14b) charged carbon nanoparticles generated at various CH_4 flow rates. In both cases, the particle number concentration and the peak of the size distribution increased with increasing CH_4 concentrations. For the positive CNPs,

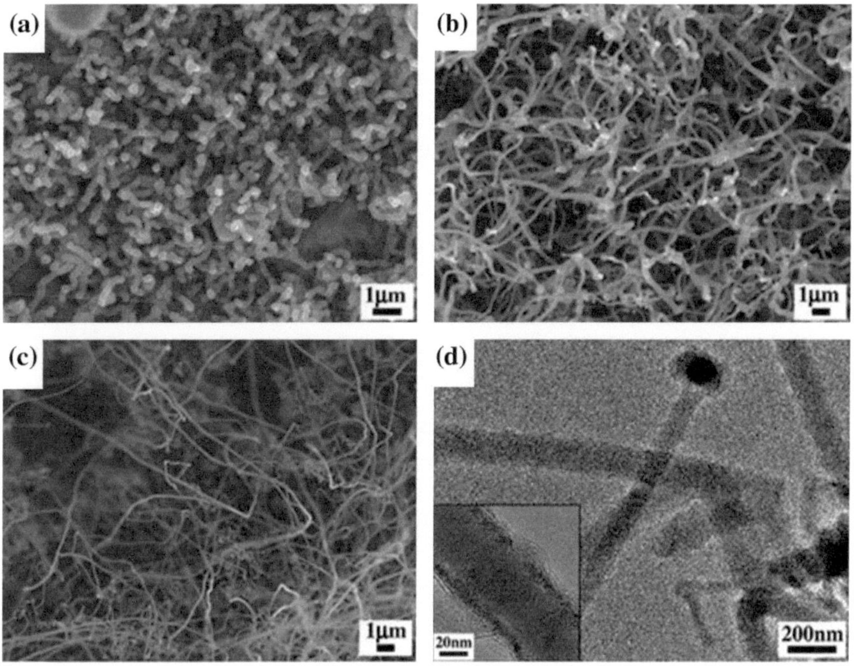

Fig. 9.13 FESEM (**a–c**) and TEM (**d**) images of CNTs at CH_4 flow rates of **a** 30 sccm, **b** 50 sccm, and **c**, **d** 100 sccm at a reactor temperature of 1010 °C (Kim et al. 2009b)

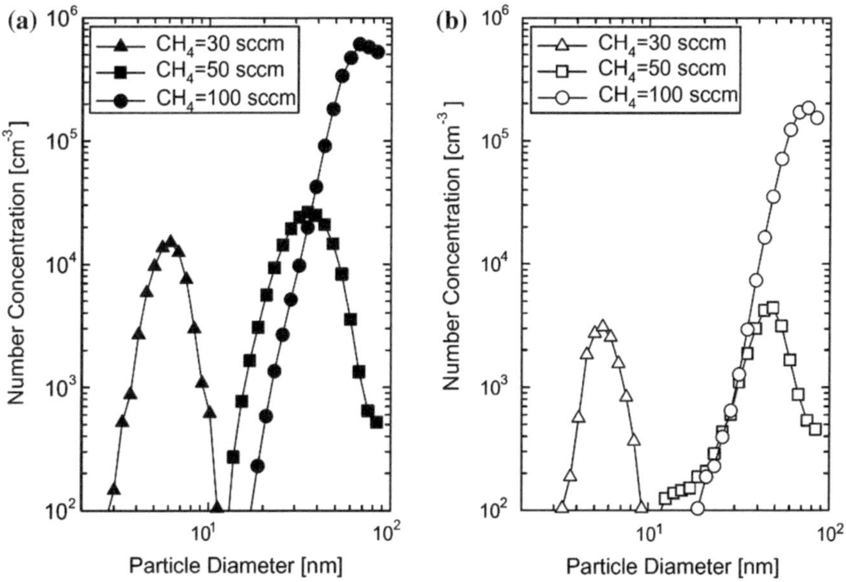

Fig. 9.14 The size distribution of **a** positively and **b** negatively charged carbon nanoparticles as a function of CH_4 flow rates at a reactor temperature of 1010 °C (Kim et al. 2009b)

the peak diameters corresponded to 6.13, 35.04 and 67.13 nm at CH_4 flow rates of 30, 50 and 100 sccm, respectively. For the negative CNPs, the respective peak diameters corresponded to 5.54, 48.34 and 75.03 nm. At a CH_4 flow rate of 100 sccm, the number concentration of CNPs increased abruptly. This behavior shows that the generation of CNPs is sensitive to the amount of CH_4.

9.4 Generation of Charged Nanoparticles During the Synthesis of Si Films

In Chap. 7, for puzzling experimental observation of simultaneous deposition and etching of Si not to violate the second law of thermodynamics, the gas phase nucleation occurs and those nuclei should be the building block of thin films. Therefore, it is necessary to confirm experimentally the generation of CNPs in the Si CVD process.

Silicon films were deposited using a typical atmospheric CVD process at four reactor temperatures of 500, 600, 700, and 800 °C at 10 sccm of helium-diluted silane of 10 % SiH_4–90 % He for 30 min (Kim et al. 2010b). High purity nitrogen gas (99.9999 %) was supplied as a carrier gas at a flow rate of 1000 sccm. The quartz tube reactor was used for deposition. The length of the tube was 100 cm. In the preliminary experiment, the maximum deposition of silicon took place on the quartz tube wall 35 cm away from the entrance of the tube whereas the maximum reactor temperature was at the center zone, which was 50 cm away from the entrance. A quartz substrate of $10 \times 10 \times 1$ mm^3 was placed at 35 cm from the entrance of the tube to obtain the maximum deposition rate. The substrate temperatures were 439, 541, 641, and 750 °C, respectively, for the reactor temperatures of 500, 600, 700, and 800 °C.

Figure 9.15a–d shows the FESEM image of silicon films deposited, respectively, at reactor temperatures of 500, 600, 700, and 800 °C. At a reactor temperature of 500 °C, silicon film was not deposited as shown in Fig. 9.15a. As the reactor temperature was increased, the surface morphology of silicon films was changed as shown in Fig. 9.15.

Figure 9.16 shows the size distribution of CNPs measured at various reactor temperatures. Both positively (Fig. 9.16a) and negatively charged (Fig. 9.16b) nanoparticles were generated. The number concentration of positive CNPs was higher than that of the negative CNPs at the reactor temperature of 500 °C. As the reactor temperature was increased, the number concentration of negative CNPs increased, and became higher than that of positive CNPs at reactor temperatures above 600 °C. The particle diameter at the peak of the size distribution tended to decrease, passing through a minimum as the reactor temperature was increased.

At the reactor temperature at or above 600 °C, the particle number concentration increased drastically from that at 500 °C. For example, the number density for the particle size of 100 nm at 600, 700, and 800 °C was more than an order of magnitude larger than that at 500 °C. This result indicates that SiH_4 decomposed only

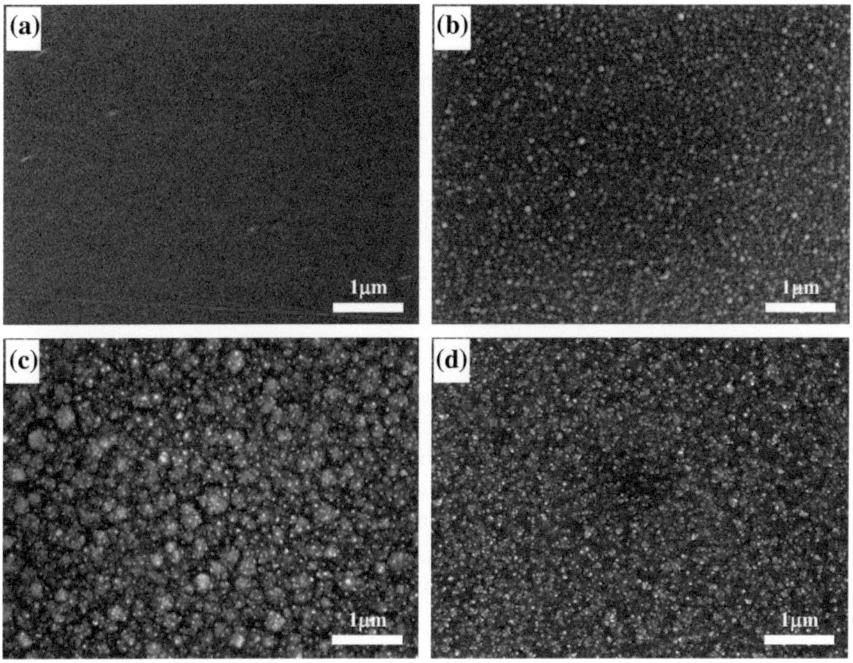

Fig. 9.15 FESEM images of silicon films at reactor temperatures of **a** 500, **b** 600, **c** 700, and **d** 800 °C with a SiH_4 flow rate of 10 sccm (Kim et al. 2010b)

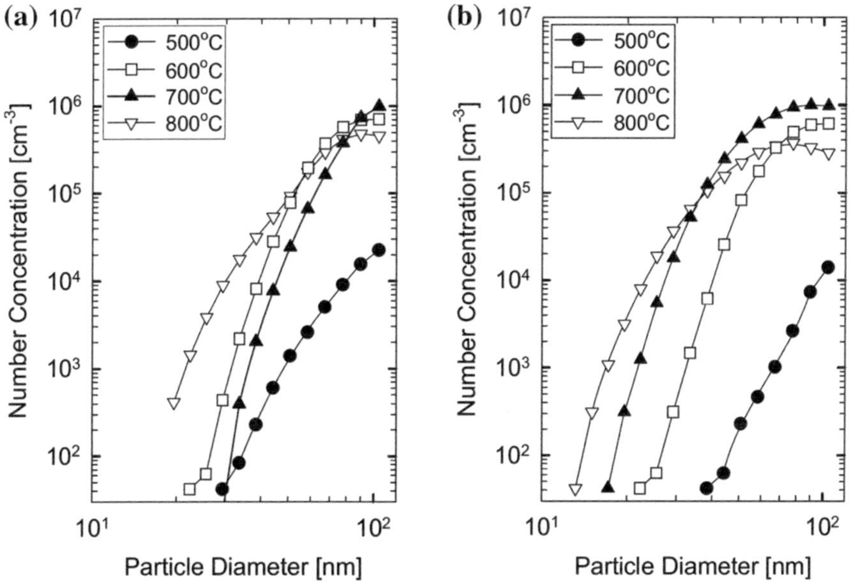

Fig. 9.16 The size distribution of **a** positively and **b** negatively charged nanoparticles at various reactor temperatures with a SiH_4 flow rate of 10 sccm (Kim et al. 2010b)

partially at the reactor temperature of 500 °C but tended to decompose fully at or above 600 °C. The partial decomposition of SiH_4 at 500 °C would be related with the fact that the silicon film was not deposited in Fig. 9.15a. It should be noted that even under the condition that the film was not deposited, the CNPs were generated, which implies that under the condition of film deposition, the generation of CNPs in the gas phase is unavoidable.

9.5 Generation of Charged Nanoparticles During the Synthesis of Si Nanowires

Kim et al. (2010a) tried to confirm the generation of CNPs during the growth silicon nanowires using the DMA-FCE system.

They used $SiCl_4$ liquid as a silicon source. Nitrogen gas was supplied directly to the CVD reactor. Hydrogen gas was additionally supplied at flow rates of 5, 10, and 15 sccm to pass through the $SiCl_4$ liquid to carry the saturated $SiCl_4$ vapor into the reactor. Although the flow rate of hydrogen was varied, the total gas flow rate of nitrogen and hydrogen was fixed at 1000 sccm. The temperature of the bottled $SiCl_4$ liquid, which was put in a chiller bath, was maintained by controlling the bath temperature, thus, controlling molar ratio of $SiCl_4$ to H_2. The silicon substrate coated with a 5 nm-thick Au film was placed at the center hot zone of the quartz-tube furnace. Si nanowires were synthesized for 30 min using a typical atmospheric-pressure CVD process at reactor temperatures of 900, 950, 975, and 1000 °C. Since the size of some CNPs was larger than 100 nm, a DMA (TSI model 3081 Long DMA, U.S.A.), which can measure sizes in the range of 10–1000 nm, was used.

Figure 9.17 shows the FESEM images of Si nanowires deposited at different reactor temperatures. At a reactor temperature of 900 °C, no nanowires were produced as shown in Fig. 9.17a; instead, Si nanoparticles were deposited on the gold particles. When the reactor temperature was increased to 950 °C, nanowires with a diameter of ~50 nm and a length of ~500 nm began to form as shown in Fig. 9.17b. As the reactor temperature was further increased to 975 and 1000 °C, nanowires became longer and denser as shown in Fig. 9.17c, d, respectively. These results indicate that Si nanowire growth is enhanced with increasing reactor temperatures within the examined temperature range, which might be due to the increased decomposition of $SiCl_4$ with increasing reactor temperature.

Figure 9.18 shows the measured size distribution of CNPs generated in the gas phase of the CVD reactor at various reactor temperatures. Both positive and negative CNPs were measured under typical processing conditions of Si nanowires. As the reactor temperature was increased, the size distribution of both positive and negative CNPs shifted to smaller particle sizes and the number concentration of CNPs increased. At reactor temperatures above 900 °C, the number concentration of CNPs smaller than ~30 nm increased drastically. These results indicate that nucleation and charging of nanoparticles in the gas phase are enhanced with

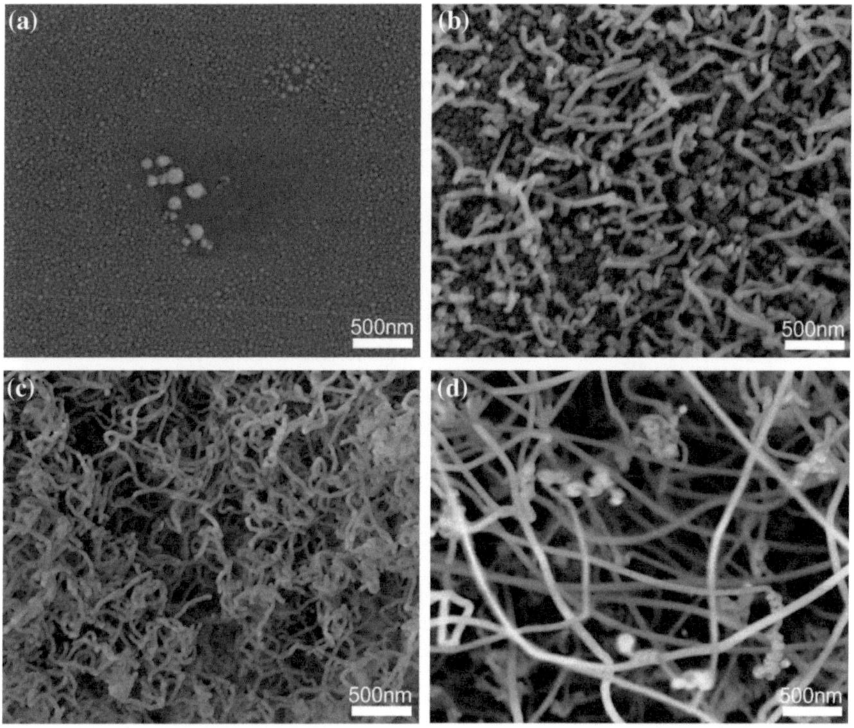

Fig. 9.17 FESEM images of Si nanowires at reactor temperatures of **a** 900, **b** 950, **c** 975, and **d** 1000 °C at a $SiCl_4/H_2$ molar ratio of 0.1 and a hydrogen flow rate of 5 sccm (the scale bar is 500 nm) (Kim et al. 2010a)

increasing reactor temperature due to enhanced thermal decomposition of $SiCl_4$. It should be noted that even at the reactor temperature of 900 °C, where Si nanowires did not grow, CNPs were generated in the gas phase in a huge amount, but CNPs below the size of $\sim$30 nm were absent.

Hydrogen was used as a carrier gas for $SiCl_4$ during Si nanowire growth. The hydrogen flow rate was varied at a constant molar ratio of $SiCl_4/H_2$ (Kim et al. 2010a). Figure 9.19 shows the FESEM images of Si nanowires at various hydrogen flow rates. The hydrogen flow rate changed from 5 to 15 sccm at a fixed $SiCl_4/H_2$ molar ratio of 0.05 and a reactor temperature of 975 °C. As the hydrogen flow rate was increased from 5 to 10 sccm, the diameter of nanowires increased from 24 to 40 nm and the length of nanowires also increased. As the hydrogen flow rate was increased further to 15 sccm, no nanowires were produced, but nanoparticles of $\sim$110 nm were formed, as shown in Fig. 9.19c.

Figure 9.20 shows the size distribution of positive (Fig. 9.20a) and negative (Fig. 9.20b) CNPs generated at three hydrogen flow rates of 5, 10, and 15 sccm. In all cases, the peak of the size distribution increased with increasing hydrogen flow rate. For the positive CNPs, the nanoparticle diameters at the peak were 14.6, 64.0,

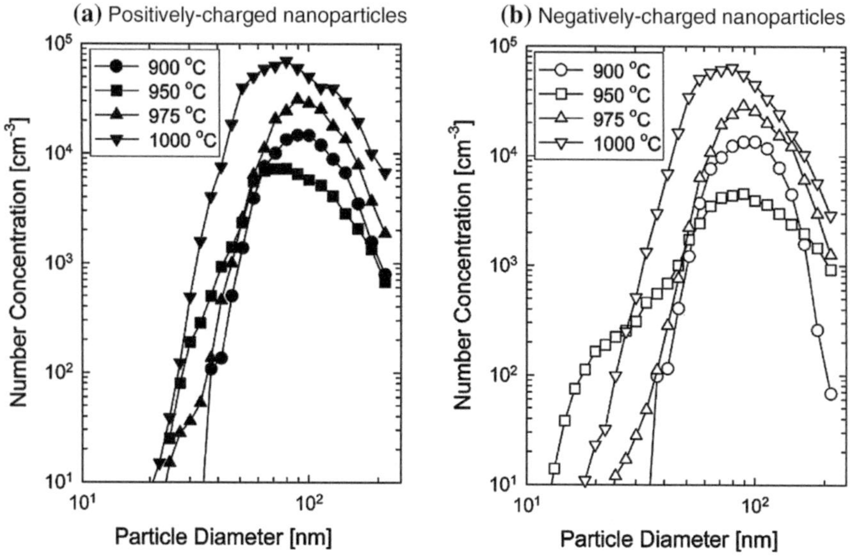

Fig. 9.18 Size distribution of **a** positive and **b** negative CNPs generated at various reactor temperatures at a $SiCl_4/H_2$ molar ratio of 0.1 and a hydrogen flow rate of 5 sccm (Kim et al. 2010a)

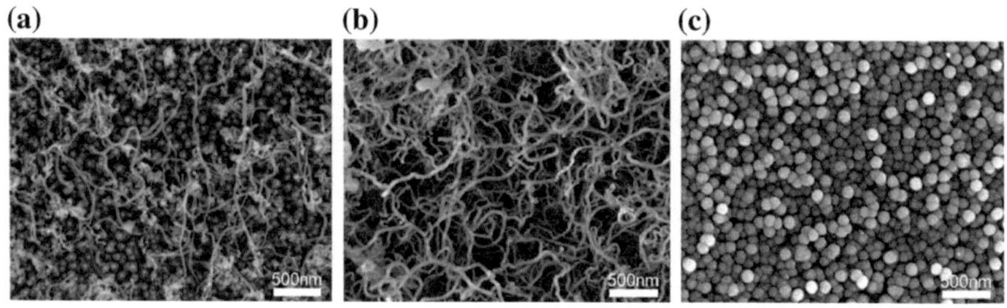

Fig. 9.19 FESEM images of Si nanowires at hydrogen flow rates of **a** 5, **b** 10, and **c** 15 sccm at a $SiCl_4/H_2$ molar ratio of 0.05 and a reactor temperature of 975 °C (the scale bar is 500 nm) (Kim et al. 2010a)

and 113.0 nm at hydrogen flow rates of 5, 10, and 15 sccm, respectively. For the negative CNPs, the respective diameters at the peak were 14.6, 57.3, and 113.0 nm. It should be noted that the nanoparticles size of $\sim$110 nm shown in Fig. 9.19c is close to the size at the peak of the size distribution of CNPs measured in Fig. 9.20a, b.

As the hydrogen flow rate was increased from 5 to 10 sccm, the number concentration of CNPs increased, but it decreased at the hydrogen flow rate of 15 sccm. Especially at a hydrogen flow rate of 15 sccm, where Si nanowires did not grow, CNPs below 40 nm were not detected. These results indicate that the hydrogen flow rate has a sensitive effect on the generation of CNPs as well as the growth behavior of Si nanowires.

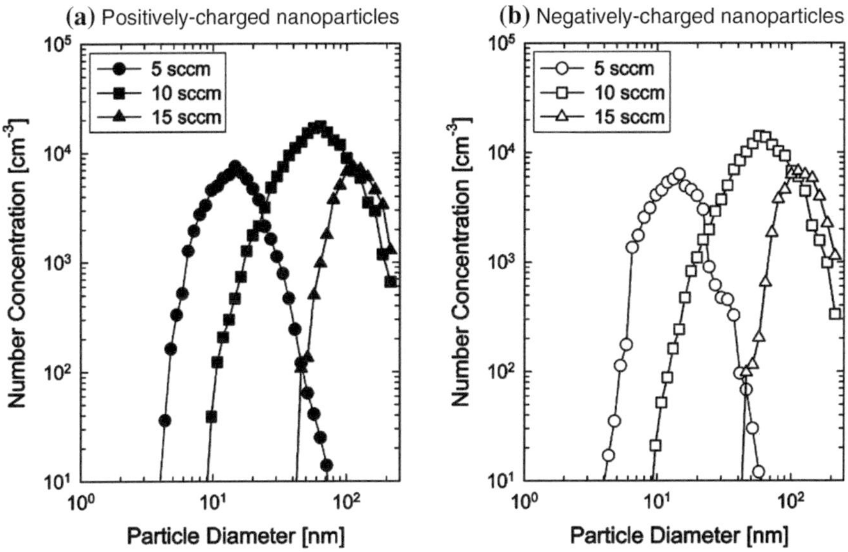

Fig. 9.20 Size distribution of **a** positive and **b** negative CNPs at hydrogen flow rates of 5, 10, and 15 sccm at a $SiCl_4/H_2$ molar ratio of 0.05 and a reactor temperature of 975 °C (Kim et al. 2010a)

Figures 9.18 and 9.20 clearly show that CNPs are generated under typical processing conditions for Si nanowire synthesis. The formation mechanism of charged silicon nanoparticles generated during Si nanowire synthesis by CVD would be similar to those of charged ZnO nanoparticles and of charged carbon nanoparticles in carbon nanotube synthesis explained in previous sections. It should be noted that under the condition where CNPs smaller than 30 nm (Fig. 9.20) were not generated, Si nanowires did not grow.

Although silicon nanowires were grown without catalytic metal nanoparticles in Chap. 7, gold nanoparticles were used in this experiment. Gold nanoparticles enhance the growth of silicon nanowires but not by the VLS mechanism. Here, gold nanoparticles interact electrostatically with CNPs. The nanoparticles carried by the gas flow have difficulty in landing on a surface because of the drag force. For CNPs to land on any surface, the electrostatic attraction between CNPs and the surface should overcome this drag force. It appears that the electrostatic attraction between CNPs and the conducting gold particles is larger than the drag force, whereas the electrostatic attraction between CNPs and the native silicon oxide of the substrate or the silicon surface of nanowires is less than the drag force. This might be why CNPs land preferentially on the conducting gold particles at the tip of nanowires.

After landing on gold nanoparticles, it is not sure how CNPs contribute to the growth of nanowires. One possibility is that they may be dissolved, supplying individual silicon atoms for the growth of nanowires. Another possibility is that they attach themselves as a building block to the tip of nanowires mediated by gold nanoparticles. However, it is clear that the electrostatic energy is involved in the process. This would be why the catalytic metal particles should be placed on an insulating substrate such as quartz or alumina.

Both positive and negative CNPs are believed to participate in the growth of nanowires. Their deposition behavior would depend on the distribution of electric charge in the nanowires. The deposition behavior would also depend on the size of nanoparticles and the gas flow rate. However, further theoretical and experimental studies considering all such related interactions are needed to understand the deposition behavior of CNPs.

9.6 Generation of Charged Nanoparticles During the Synthesis of GaN Nanostructures

GaN, a wide band gap material of $\sim$3.4 eV, belongs to prospective III-nitride semiconductors (Ambacher 1998) and has good electrical and thermal stability (Neudeck et al. 2002; Hudgins et al. 2003) and its light emitting characteristic in a wide range beyond the visible light. Besides, it has good doping characteristics, high carrier mobility, and large peak velocity which are essential for the high frequency devices. Because of these excellent physical properties, GaN nanostructures have potential applications in ultraviolet (UV) or blue photon emitters for light emitting diodes (LEDs) (Orton and Foxon 1998), high-speed field effect transistors of optical telecommunications (Mishra et al. 2002), and high temperature/high power electronic devices for automobile applications (Neudeck et al. 2002).

GaN films or nanostructures have been synthesized by various chemical vapor deposition (CVD) processes, such as atmospheric-pressure CVD (Han et al. 1997), PECVD (Choi et al. 1993), metal-organic CVD (Amano et al. 1986), and HWCVD (Yasui et al. 2003; Peng et al. 2000). In spite of extensive researches, the role of process parameters in the deposition behavior has not been clearly understood. For example, it is not understood why films grow under certain conditions and nanowires grow under other conditions.

As the first step toward the understanding of the GaN CVD process, it is necessary to check whether CNPs are generated in the gas phase or not during the GaN CVD process. Motivated by this background, Lee et al. (2012) tried to confirm the generation of CNPs during the synthesis of the GaN nanostructures by the atmospheric pressure CVD process using a DMA-FCE system.

As a precursor material, 2 g of Ga_2O_3 powder was loaded in an alumina boat, which was placed at the center of a quartz-tube reactor. A Si substrate was placed 17 cm away from the center. The length of the quartz tube was 75 cm. The NH_3 flow rate was varied as 400, 200, 50, and 0 sccm with nitrogen gas used as a carrier gas and the total flow rate (NH_3 + N_2) was fixed at 600 sccm.

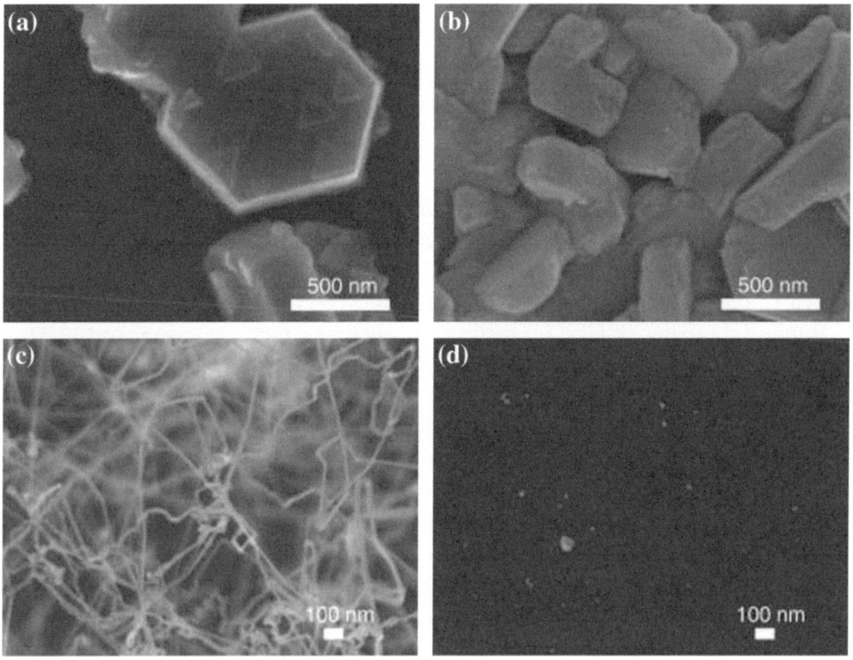

Fig. 9.21 FESEM images of GaN nanostructures deposited at the flow rate of **a** 400 sccm, **b** 200 sccm, **c** 50 sccm, and **d** 0 sccm of the ammonia at a reactor temperature of 1100 °C (Lee et al. 2012)

Since it was reported that GaN nanostructures were affected by the NH_3 flow rate in the CVD process (Cai et al. 2006; Djurišić et al. 2008), the effect of the NH_3 flow rate was examined with respect to the morphology of GaN nanostructure and the size distribution of CNPs. The resultant GaN nanostructures deposited at a reactor temperature of 1100 °C are shown in Fig. 9.21.

At the NH_3 flow rates of 400 and 200 sccm, the hexagonal crystals were synthesized as shown in Fig. 9.21a, b, respectively. This hexagon shape has been already reported as a typical GaN crystal (Grzegory et al. 1995; Li et al. 1996). The deposition was nonuniform on the Si substrate; although the amount of GaN in Fig. 9.21a appeared smaller than that in Fig. 9.21b, the overall amount in the sample in Fig. 9.21a was larger than that in Fig. 9.21b.

Energy dispersive spectroscopy (EDS, Horiba X-Max) of GaN crystals shows the atomic composition ratio of C : N : O : Si : Ga = 6.86 : 19.69 : 40.71 : 1.16 : 31.59. Carbon might be due to contamination during SEM observation and silicon might come from the substrate. However, a very large amount of oxygen was detected reproducibly. The oxygen might be formed by oxidation and absorption in the air when the sample was taken outside as suggested by Wolter et al. (1998) and Tang et al. (2003). Although the EDS is not accurate especially in the analysis of light elements, it poses a possibility that the GaN crystals formed in the reactor might be deficient in nitrogen and the excess Ga is oxidized in the air after the process.

As the NH_3 flow rate decreased from 400 to 200 sccm, the size of hexagonal nanostructures decreased from ~1 μm to ~560 nm. At the NH_3 flow rate of 50 sccm (Fig. 9.21c), however, the one-dimensional structure of nanowires was grown instead of hexagonal crystals. The diameter of these nanowires was in the range of 10–20 nm. The total amount of nanowires deposited at the NH_3 flow rate of 50 sccm was much smaller than that of hexagonal crystals deposited at the flow rates of 400 and 200 sccm. At the NH_3 flow rate of 0 sccm, deposition hardly occurred as shown in Fig. 9.21d, as expected.

The weight losses of Ga_2O_3 precursor powder after the CVD process at 1100 °C were 0.214, 0.176, 0.056, and 0.005 g, respectively, for the NH_3 flow rates of 400, 200, 50, and 0 sccm. After the CVD process, the surface color of the precursor (Ga_2O_3) was changed from white to yellow for the NH_3 flow rates of 400, 200, and 50 sccm, indicating the formation of GaN. For the flow rate of 0 sccm, however, the color remained white.

Figure 9.22 shows the particle size distribution of charged GaN nanoparticles generated in the gas phase of the reactor at various NH_3 flow rates. Both positive and negative CNPs were generated. With decreasing flow rate of NH_3, the number concentration and the size of both positive and negative CNPs tended to decrease.

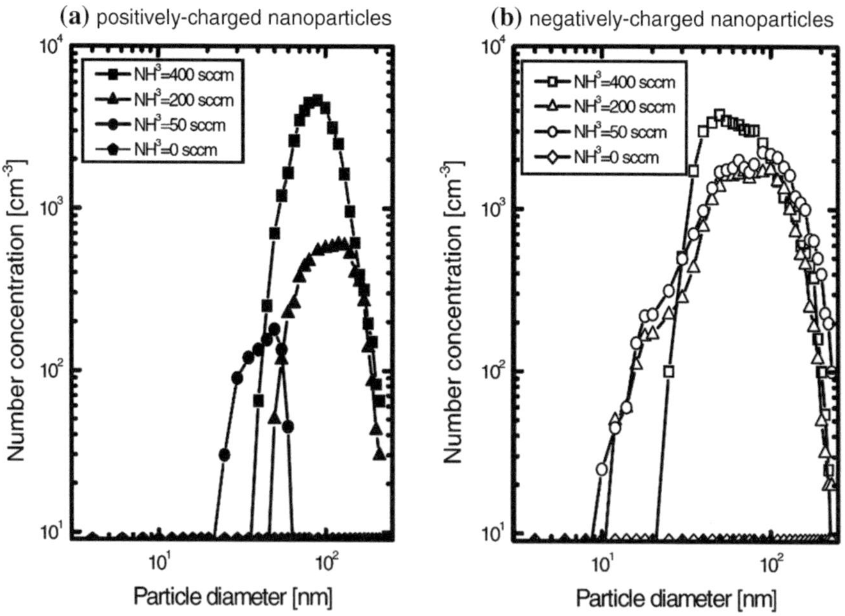

Fig. 9.22 The size distribution of **a** positively and **b** negatively charged GaN nanoparticles with various NH_3 flow rates at a reactor temperature of 1100 °C (Lee et al. 2012)

However, in the absence of NH_3 gas (0 sccm), CNPs were hardly detected as shown in Fig. 9.22, which is consistent with the negligible deposition of Fig. 9.21d. This indicated that nucleation and charging of nanoparticles in the gas phase are enhanced with increasing NH_3 flow rate due to enhancement of the formation of GaN. The NH_3 flow rate affects the deposition behavior of GaN nanostructures, the amount of Ga_2O_3 precursor powder remaining after CVD, and the generation behavior of CNPs. For example, at the flow rate of 400 sccm, the size of GaN hexagonal crystals is largest (Fig. 9.21a); the mass decrease of the precursor powder after CVD and the increase in the amount of generated CNPs (Fig. 9.22) are also largest. At the flow rate of 0 sccm, however, the deposition of GaN was negligible in accordance with the negligible mass decrease of Ga_2O_3 powder and the negligible generation of CNPs.

Figure 9.23 shows the FESEM images of GaN nanostructures synthesized at different reactor temperatures at the NH_3 flow rate of 400 sccm. The size of hexagonal crystals decreased from ~1 μm to ~180 nm as the reactor temperature decreased from 1100 to 900 °C. At the reactor temperature of 500 °C, GaN nanostructures were hardly synthesized as shown in Fig. 9.23d, which might be attributed to the low decomposition rate of NH_3 gas at 500 °C (Di Lello et al. 2002).

Figure 9.24 shows the particle size distributions of positively and negatively charged GaN nanoparticles generated at various reactor temperatures under the

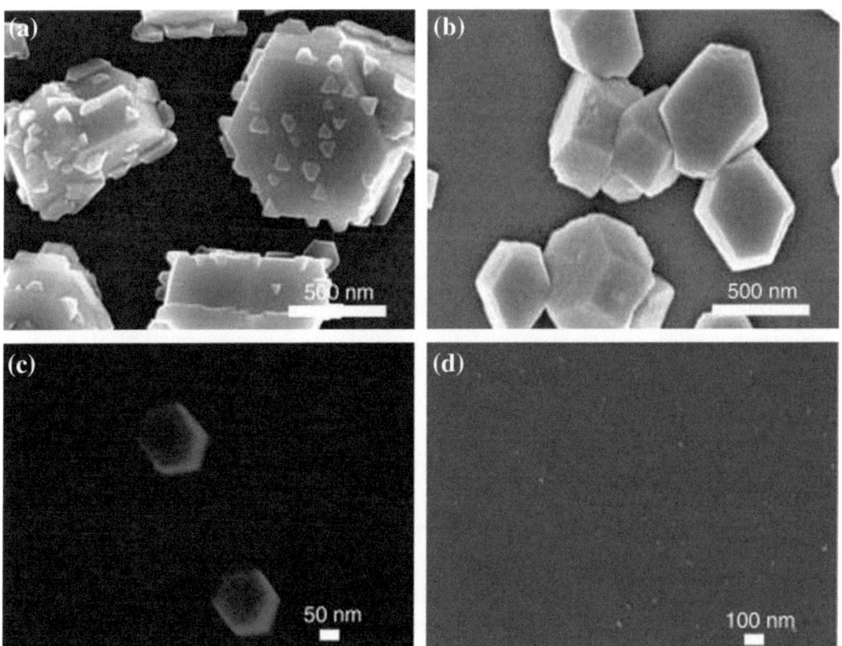

Fig. 9.23 FESEM images of GaN nanostructures deposited at **a** 1100 °C, **b** 1000 °C, **c** 900 °C, and **d** 500 °C at a NH_3 flow rate of 400 sccm (Lee et al. 2012)

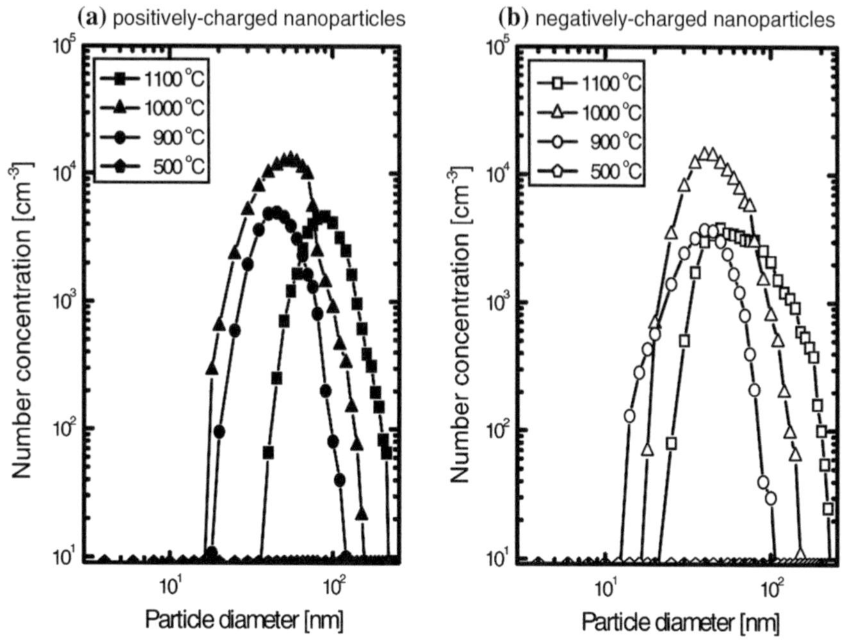

Fig. 9.24 The size distribution of **a** positively and **b** negatively charged GaN nanoparticles with various reactor temperatures at a NH_3 flow rate of 400 sccm (Lee et al. 2012)

processing conditions of the specimens in Fig. 9.23. In both Fig. 9.24a, b, the peak positions of the particle size decreased with decreasing reactor temperature. However, the particle number concentration shows different behavior, being largest at the reactor temperature of 1000 °C. And the number concentrations of both positive and negative CNPs were similar between 900 and 1100 °C. At the reactor temperature of 500 °C, however, the GaN CNPs were hardly detected, which is consistent with the negligible deposition of Fig. 9.23d.

It should be noted that the larger number concentration at 1000 °C than at 1100 °C does not necessarily mean that the total amount of CNPs generated at 1000 °C is larger than that at 1100 °C because the volume of the particle is proportional to the cube of the size. The weight losses were 0.214, 0.183, 0.069, and 0.000 g, respectively, for the reactor temperature of 1100, 1000, 900, and 500 °C. After the CVD process, the surface color of the precursor (Ga_2O_3) was changed from white to yellow for the reactor temperatures of 1100, 1000, and 900 °C with the color becoming lighter with decreasing temperature. For the reaction temperature of 500 °C, however, the color remained white.

It is very difficult to deposit GaN directly on the sapphire or silicon substrate. For this reason, GaN should be made to form initially on the substrate typically by lowering the reactor temperature. By this treatment, some GaN is formed on the substrate. Then, GaN starts to grow. Therefore, the pre-existing GaN enhances the initiation and growth. This phenomenon is not yet understood and difficult to explain by the classical crystallization. Since the generation of CNPs is experimentally confirmed in the gas phase of the GaN CVD process, this phenomenon should be approached by considering the electrostatic interaction of GaN CNPs with a bulk as was done for diamond in Chap. 6.

The interaction of a GaN nanoparticle would be maximum with the bulk GaN among all materials, which explains that the pre-existing GaN enhances the initiation and growth. Besides, it seems that the interaction with a sapphire wafer is stronger than that with a silicon wafer, considering the experimental result that the deposition of GaN is much easier on sapphire than on silicon (Lee 2014). Besides, GaN tends to deposit more easily on the metallic substrates with a high CTR. This means that a CTR is an important parameter in the interaction of CNPs with the substrate or the growing surface.

In conclusion of this chapter, it was without exception that CNPs were generated in all the atmospheric CVD processes that were tested using the DMA-FCE system. The gas phase generation of CNPs appears to be very general at least in the CVD process. The size of CNPs is much larger than expected, especially in contrast with the small size of the clusters confirmed in the diamond HFCVD process. This is why this newly discovered way of thin film and nanostructure growth was called the theory of charged nanoparticles (TCN) instead of the charged cluster model (Hwang and Lee 2010).

References

Ago H, Ohshima S, Uchida K, Yumura M (2001) Gas-phase synthesis of single-wall carbon nanotubes from colloidal solution of metal nanoparticles. J Phys Chem B 105(43):10453–10456. doi:10.1021/jp012084j

Amano H, Sawaki N, Akasaki I, Toyoda Y (1986) Metalorganic vapor phase epitaxial growth of a high quality GaN film using an AlN buffer layer. Appl Phys Lett 48(5):353–355

Ambacher O (1998) Growth and applications of group III-nitrides. J Phys D: Appl Phys 31 (20):2653

AuBuchon JF, Chen LH, Gapin AI, Jin S (2006) Electric-field-guided growth of carbon nanotubes during DC plasma-enhanced CVD. ChemVapor Depos 12(6):370–374. doi:10.1002/cvde.200506444

Avigal Y, Kalish R (2001) Growth of aligned carbon nanotubes by biasing during growth. Appl Phys Lett 78(16):2291–2293

Bower C, Zhu W, Jin S, Zhou O (2000) Plasma-induced alignment of carbon nanotubes. Appl Phys Lett 77(6)

Cai X, Djurišić A, Xie M (2006) GaN nanowires: CVD synthesis and properties. Thin Solid Films 515(3):984–989

Cheong WS, Hwang NM, Yoon DY (1999) Observation of nanometer silicon clusters in the hot-filament CVD process. J Cryst Growth 204(1):52–61

Chiang W-H, Sankaran RM (2007) Microplasma synthesis of metal nanoparticles for gas-phase studies of catalyzed carbon nanotube growth. Appl Phys Lett 91(12):121503. doi:10.1063/1.2786835

Choi S, Bachmann K, Lucovsky G (1993) Growth kinetics and characterizations of gallium nitride thin films by remote PECVD. J Mater Res 8(04):847–854

Choi YC, Shin YM, Lee YH, Lee BS, Park G-S, Choi WB, Lee NS, Kim JM (2000) Controlling the diameter, growth rate, and density of vertically aligned carbon nanotubes synthesized by microwave plasma-enhanced chemical vapor deposition. Appl Phys Lett 76(17):2367–2369

De Bleecker K, Bogaerts A, Gijbels R, Goedheer W (2004) Numerical investigation of particle formation mechanisms in silane discharges. Phys Rev E 69(5):056409

De Jong KP, Geus JW (2000) Carbon nanofibers: catalytic synthesis and applications. Cataly Rev 42(4):481–510

Delzeit L, McAninch I, Cruden BA, Hash D, Chen B, Han J, Meyyappan M (2002) Growth of multiwall carbon nanotubes in an inductively coupled plasma reactor. J Appl Phys 91(9):6027

Dhaniyala S, Fierz M, Keskinen J, Marjamäki M, Kulkarni P, Baron P, Willeke K (2011) Instruments based on electrical detection of aerosols. Aerosol measurement: Principles, techniques, and applications. Wiley, New York, p 393

Di Lello B, Moura F, Solorzano I (2002) Synthesis and characterization of GaN using gas–solid reactions. Mater Sci Eng B 93(1):219–223

Djurišić A, Cai X, Xie M (2008) III-nitride nanomaterials: Growth and properties. III-nitride devices and nanoengineering. World Scientific, London, p399

Fu H, Li H, Jie W, Zhang C (2007) The O_2-dependent growth of ZnO nanowires and their photoluminescence properties. Ceram Int 33(6):1119–1123. doi:http://dx.doi.org/10.1016/j.ceramint.2006.03.024

Garg R, Kim S, Hash D, Gore JP, Fisher T (2008) Effects of feed gas composition and catalyst thickness on carbon nanotube and nanofiber synthesis by plasma enhanced chemical vapor deposition. J Nanosci Nanotechno 8(6):3068–3076

Girshick S, Rao N, Kelkar M (1996) Model for ion-induced nucleation based on properties of small ionic clusters. J Vac Sci Technol A 14(2):529–534

Grzegory I, Jun J, Boćkowski M, Wroblewski M, Łucznik B, Porowski S (1995) III–V Nitrides—thermodynamics and crystal growth at high N 2 pressure. J Phys Chem Solids 56(3):639–647

Haaland P, Garscadden A, Ganguly B (1996) Ionic and neutral growth of dust in plasmas. App Phys Lett 69(7):904–906

Han W, Fan S, Li Q, Hu Y (1997) Synthesis of gallium nitride nanorods through a carbon nanotube-confined reaction. Science 277(5330):1287–1289

Hudgins JL, Simin GS, Santi E, Khan MA (2003) An assessment of wide bandgap semiconductors for power devices. IEEE T Power Electr on 18(3):907–914

Hwang N-M, Kim D-Y (2004) Charged clusters in thin film growth. Int Mater Rev 49(3–4):171–190

Hwang N-M, Lee D-K (2010) Charged nanoparticles in thin film and nanostructure growth by chemical vapour deposition. J Phys DAppli Phy 43(48):483001

Hwang NM, Yoon DY (1996) Thermodynamic approach to the paradox of diamond formation with simultaneous graphite etching in the low pressure synthesis of diamond. J Cryst Growth 160(1–2):98–103. doi:http://dx.doi.org/10.1016/0022-0248(95)00549-8

Hwang NM, Hahn JH, Yoon DY (1996) Charged cluster model in the low pressure synthesis of diamond. J Cryst Growth 162(1):55–68

Kim C-S, Chung Y-B, Youn W-K, Hwang N-M (2009a) Generation of Charged Nanoparticles during Synthesis of ZnO Nanowires by Carbothermal Reduction. Aerosol Sci Tech 43(2):120–125. doi:10.1080/02786820802499068

Kim C-S, Chung Y-B, Youn W-K, Hwang N-M (2009b) Generation of charged nanoparticles during the synthesis of carbon nanotubes by chemical vapor deposition. Carbon 47(10):2511–2518. doi:http://dx.doi.org/10.1016/j.carbon.2009.05.005

Kim C-S, Kwak I-J, Choi K-J, Park J-G, Hwang N-M (2010a) Generation of charged nanoparticles during the synthesis of silicon nanowires by chemical vapor deposition. J Phys Chem C 114 (8):3390–3395. doi:10.1021/jp910242a

Kim C-S, Youn W-K, Hwang N-M (2010b) Generation of charged nanoparticles and their deposition during the synthesis of silicon thin films by chemical vapor deposition. J Appl Phys 108(1):014313. doi:10.1063/1.3452352

Kim M, Rodriguez N, Baker R (1991) The interaction of hydrocarbons with copper nickel and nickel in the formation of carbon filaments. J Catal 131(1):60–73

Kong XY, Ding Y, Yang R, Wang ZL (2004) Single-crystal nanorings formed by epitaxial self-coilingof polar nanobelts. Science 303(5662):1348–1351

Korgel BA (2004) Self-assembled nanocoils. Science 303(5662):1308–1309

Langmuir I, Kingdon KH (1925) Thermionic effects caused by vapours of alkali metals, vol 107. doi:10.1098/rspa.1925.0005

Lee J-I, Hwang N-M (2008) Generation of negative-charge carriers in the gas phase and their contribution to the growth of carbon nanotubes during hot-filament chemical vapor deposition. Carbon 46(12):1588–1592. doi:http://dx.doi.org/10.1016/j.carbon.2008.07.006

Lee K-H, Cho J-M, Sigmund W (2003) Control of growth orientation for carbon nanotubes. Appl Phys Lett 82(3):448–450

Lee S-S (2014) Generation of charged nanoparticles and their contribution to growth of gallium nitride in the thermal chemical vapor deposition process. Ph. D Thesis, Seoul National University, Seoul

Lee S-S, Kim C-S, Hwang N-M (2012) Generation of charged nanoparticles during the synthesis of GaN nanostructures by atmospheric-pressure chemical vapor deposition. Aerosol Sci Tech 46(10):1100–1108. doi:10.1080/02786826.2012.693977

Li H, Yang H, Yu S, Zou G, Li Y, Liu S, Yang S (1996) Synthesis of ultrafine gallium nitride powder by the direct current arc plasma method. Appl Phys Lett 69(9):1285–1287

Lim SH, Park KC, Moon JH, Yoon HS, Pribat D, Bonnassieux Y, Jang J (2006) Controlled density of vertically aligned carbon nanotubes in a triode plasma chemical vapor deposition system. Thin Solid Films 515(4):1380–1384. doi:http://dx.doi.org/10.1016/j.tsf.2006.03.056

Magnusson MH, Deppert K, Malm J-O, Bovin J-O, Samuelson L (1999) Gold nanoparticles: production, reshaping, and thermal charging. J Nanopart Res 1(2):243–251

Merchan-Merchan W, Saveliev AV, Kennedy LA (2004) High-rate flame synthesis of vertically aligned carbon nanotubes using electric field control. Carbon 42(3):599–608. doi:http://dx.doi.org/10.1016/j.carbon.2003.12.086

Merkulov VI, Merkulov VI, Lowndes DH, Lowndes DH, Wei YY, Wei YY, Eres G, Voelkl E (2000) Patterned growth of individual and multiple vertically aligned carbon nanofibers. Appl Phys Lett 76 (24)

Meyyappan M, Lance D, Alan C, David H (2003) Carbon nanotube growth by PECVD: a review. Plasma Sources Sci Tech 12(2):205

Mishra UK, Parikh P, Wu Y-F (2002) AlGaN/GaN HEMTs-an overview of device operation and applications. P-IEEE 90(6):1022–1031

Neudeck PG, Okojie RS, Chen L-Y (2002) High-temperature electronics-a role for wide bandgap semiconductors? P-IEEE 90(6):1065–1076

Orton J, Foxon C (1998) Group III nitride semiconductors for short wavelength light-emitting devices. Rep Prog Phys 61(1):1

Peng H, Zhou X, Wang N, Zheng Y, Liao L, Shi W, Lee C, Lee S (2000) Bulk-quantity GaN nanowires synthesized from hot filament chemical vapor deposition. Chem Phys Lett 327 (5):263–270

Qiaoliang B, Han Z, Chunxu P (2006) Electric-field-induced microstructural transformation of carbon nanotubes. Appl Phys Lett 89(6):063124. doi:10.1063/1.2227620

Seidl M, Meiwes-Broer KH, Brack M (1991) Finite-size effects in ionization potentials and electron affinities of metal clusters. J Chem Phys 95(2):1295

Takahito O, Hidetoshi M, Masayoshi E (2002) Electric-field-enhanced growth of carbon nanotubes for scanning probe microscopy. Nanotechnology 13(1):62

Tang C, Bando Y, Liu Z (2003) Thermal oxidation of gallium nitride nanowires. Appl Phys Lett 83(15):3177–3179

Teo K, Chhowalla M, Amaratunga G, Milne W, Hasko D, Pirio G, Legagneux P, Wyczisk F, Pribat D (2001a) Uniform patterned growth of carbon nanotubes without surface carbon. Appl Phys Lett 79(10):1534–1536

Teo KBK, Chhowalla M, Amaratunga GAJ, Milne WI, Hasko DG, Pirio G, Legagneux P, Wyczisk F, Pribat D (2001b) Uniform patterned growth of carbon nanotubes without surface carbon. Appl Phys Lett 79(10)

Ural A, Li Y, Dai H (2002) Electric-field-aligned growth of single-walled carbon nanotubes on surfaces. Appl Phys Lett 81(18):3464–3466

Wang ZL, Kong XY, Zuo JM (2003) Induced growth of asymmetric nanocantilever arrays on polar surfaces. Phys Rev Lett 91(18):185502

Wolter S, Mohney S, Venugopalan H, Wickenden A, Koleske D (1998) Kinetic study of the oxidation of gallium nitride in dry air. J Electrochem Soc 145(2):629–632

Xiang Yang Kong GC, Zhong Lin W (2004) Polar-surface dominated ZnO nanobelts and the electrostatic energy induced nanohelixes, nanosprings, and nanospirals. Appl Phys Lett 84 (6):975–977. doi:10.1063/1.1646453

Xu F, Liu X, Tse SD (2006) Synthesis of carbon nanotubes on metal alloy substrates with voltage bias in methane inverse diffusion flames. Carbon 44(3):570–577. doi:http://dx.doi.org/10.1016/j.carbon.2005.07.043

Yao BD, Chan YF, Wang N (2002) Formation of ZnO nanostructures by a simple way of thermal evaporation. Appl Phys Lett 81(4):757

Yasui K, Morimoto K, Akahane T (2003) Growth of GaN films on nitrided GaAs substrates using hot-wire CVD. Thin solid films 430(1):174–177

Zhang Y, Zhang Q, Wang N, Yan Y, Zhou H, Zhu J (2001) Synthesis of thin Si whiskers (nanowires) using SiCl4. J Cryst Growth 226(2–3):185–191. doi:http://dx.doi.org/10.1016/S0022-0248(01)01360-4

Zhong Lin W (2004) Zinc oxide nanostructures: growth, properties and applications. J Phys-Condens Mat 16(25):R829

Chapter 10
Experimental Confirmation of Charged Nanoparticles at Low Pressure

In Chap. 9, the generation of CNPs was experimentally confirmed without exceptions in all atmospheric CVD processes that were tested. However, many thin films such as used in microelectronics are deposited at low pressure. Gas phase generation of nanoparticles has been a big issue in the semiconductor fabrication process because they can decrease the product yield. The particle size that results in device failure is called the killer particle size, which is defined as one-half of the gate length. Since the killer particle size continues to decrease with decreasing feature size of semiconductors, the technique to detect such small particles has a great demand. The simplest technique is to capture those particles on a grid membrane and to make an ex situ observation by TEM. For in situ measurements of particles, two instruments have been developed during the low pressure CVD (LPCVD) and PVD processes. One is the low pressure DMA (Kim et al. 2004; Seto et al. 1997) and the other is the particle beam mass spectrometer (Kim et al. 2002; Ziemann et al. 1995). In this chapter, the results of measuring nanoparticles during LPCVD or PVD processes will be shown with a focus on the possibility as to whether they are charged or not and whether they are involved in deposition or not.

10.1 Measurements of Nanoparticles Using a Low Pressure Differential Mobility Analyser

Seto et al. (1997) first reported the measurements of nanometer-sized TiO_2 particles generated in a LPCVD reactor using a low pressure DMA (LPDMA). Figure 10.1 shows the experimental system consisting of a LPCVD reactor, the LPDMA and the FCE. Titanium tetraisopropoxide (TTIP) is used as the reactant. TTIP liquid is evaporated by heating to 30 °C and is introduced into the reactor which consists of a stainless steel tube (inner diameter = 10 mm, length = 10 mm). Titanium dioxide particles are produced by thermal decomposition of TTIP vapor according to the reaction

© Springer Science+Business Media Dordrecht 2016

N.M. Hwang, *Non-Classical Crystallization of Thin Films and Nanostructures in CVD and PVD Processes*, Springer Series in Surface Sciences 60,
DOI 10.1007/978-94-017-7616-5_10

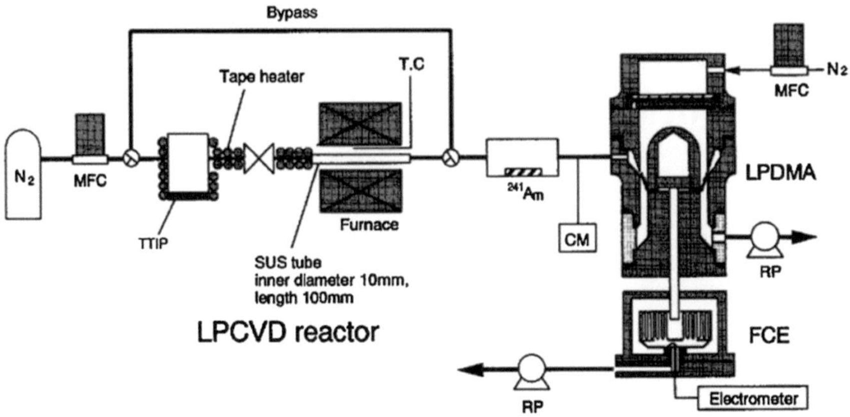

Fig. 10.1 Horizontal LPCVD reactor and LPDMA attached with a FCE used for measurements of TiO_2 particles generated in the reactor (Seto et al. 1997)

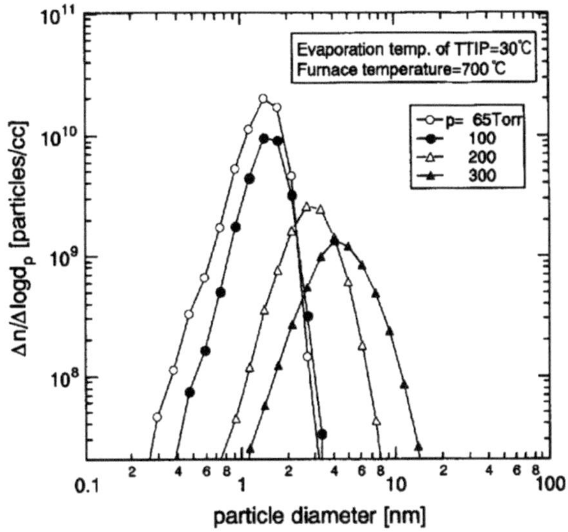

Fig. 10.2 Variation of size distribution of TiO_2 particles at different pressures of the reactor (Seto et al. 1997)

$$TTIP \rightarrow TiO_2 + 4C_3H_8 + 2H_2O. \tag{10.1}$$

Figure 10.2 shows typical size distributions measured using the LPDMA-FCE system. When the reactor pressure is 65 Torr, the peak in the size distribution is at about 1.6 nm. In Fig. 10.2, the peak increases with increasing pressure. The size of TiO_2 nanoparticles tends to increase but the number density tends to decrease with increasing reactor pressure.

Since the thin film deposition was not studied simultaneously in this study, the microstructure evolution of films under the generation condition of TiO_2 particles is not available. When measuring the particle-size distribution using a DMA, the charge distribution on the particles is typically brought to a state of equilibrium (Fuchs 1963). For this purpose, they used a neutralizer containing ^{241}Am of 22.4 MBq as an α-ray source, which generates both positive and negative ions (bipolar ions). Although Seto et al. (1997) did not check whether gas phase generated TiO_2 nanoparticles were electrically charged or not in the CVD reactor, they are expected to be charged, considering the previous results in Chaps. 6–9. Comparing the size distribution in Fig. 10.2 with those in Chap. 9, the size of nanoparticles in LPCVD is much smaller than that in atmospheric CVD.

10.2 Measurements of Particle Size in Plasma Enhanced CVD Using a Low Pressure Differential Mobility Analyzer

Seol et al. (2001) tried to measure the size of SiO_2 nanoparticles generated in the plasma enhanced CVD (PECVD) using a low pressure DMA-FCE system. Since their DMA could operate at pressures as low as 0.7–1.8 Torr, they called it a very low pressure DMA (VLPDMA). By connecting the VLPDMA to a PECVD reactor, where tetraethylorthosilicate (TEOS) and O_2 were used as reactants for amorphous SiO_2 (a-SiO2) film growth, as shown in Fig. 10.3, they could make an in-situ measurement of particles generated during the PECVD process.

Their PECVD reactor was a conventional cylindrical parallel-plate reactor where the gap between electrodes was 3 cm, the diameters of the electrodes were 7.5 cm, and the reactor volume was 10 L. A neutralizer containing ^{241}Am of 22.4 MBq was used as an α-ray source for charging of nanoparticles. By comparing the mobility distribution data of nanoparticles with and without the ^{241}Am α-ray source, however, Seol et al. (2001) could confirm that nanoparticles were charged. The reactor was equipped with an electrostatic collector of charged particles that are used in the TEM analysis of particle size. Charged particles were collected on a carbon film by applying a collector voltage of 400 V to the collector during the PECVD process. A discharge frequency of 13.56 MHz and a rf power of 50–200 W to 0.28–1.12 W/cm^2) were employed in plasma generation. A controlled mass flow rate, 2–8 mg/min, of TEOS was evaporated at 75 °C and was carried by helium gas into the reactor together with oxygen gas. The pressure in the reactor during PECVD was maintained at 0.7–1.8 Torr.

The particles produced during the PECVD process were measured by connecting the VLPDMA to the exhaust gas line of the reactor. The exhaust gas flowing into the VLPDMA includes the particles coming from the reactor.

Figure 10.4 shows the TEM images of particles collected inside the PECVD reactor during the CVD process. The particles formed during PECVD are agglomerates consisting of several primary particles of 10–20 nm in diameter.

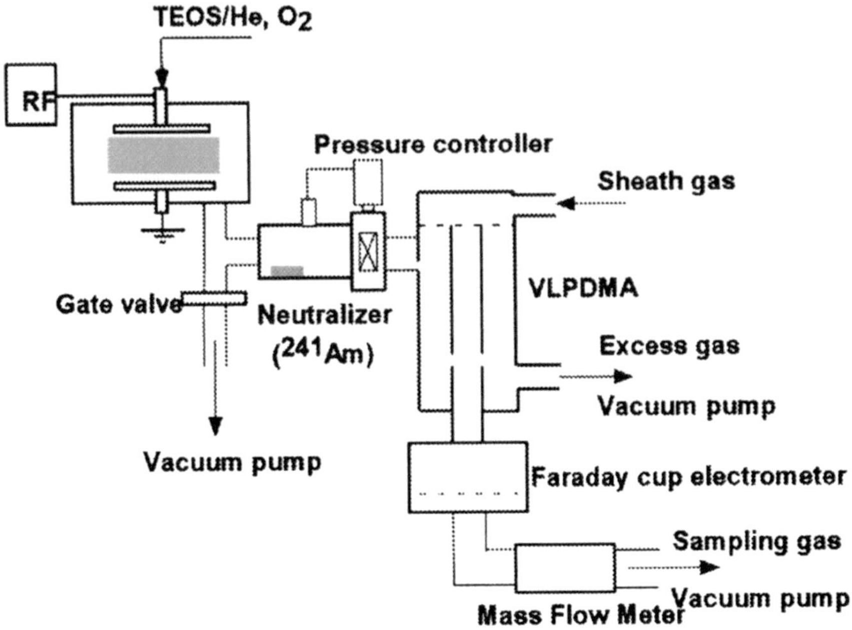

Fig. 10.3 Experimental setup for investigation of particle formation in PECVD of SiO_2 using a VLPDMA (Seol et al. 2001)

Fig. 10.4 TEM image of the particles collected inside the PECVD reactor (Seol et al. 2001)

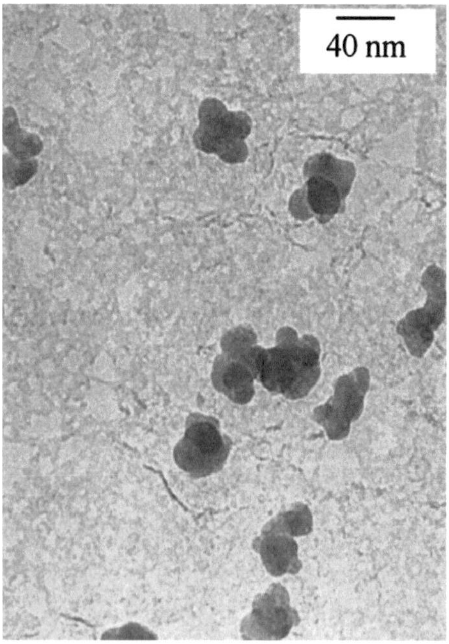

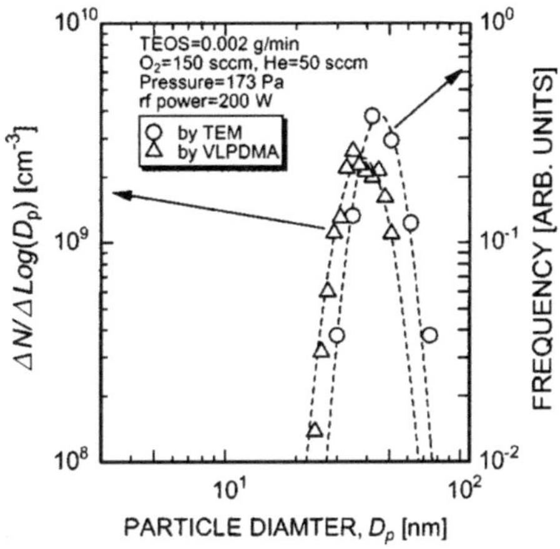

Fig. 10.5 Comparison of the particle size distributions measured by TEM and VLPDMA methods (Seol et al. 2001)

Figure 10.5 shows a comparison of the size distribution (particle concentration or frequency versus particle diameter) obtained from the TEM image and that given by the VLPDMA. The dashed lines shown in Fig. 10.5 are the best fits of a lognormal distribution to the experimental data. The VLPDMA-measured size distribution obeys the lognormal distribution with a geometric median diameter of 40 nm and a geometric standard deviation of 1.3, while the TEM measured size distribution reproduces the lognormal distribution with a geometric mean size of 45 nm and a geometric standard deviation of 1.3. The TEM-measured size distribution is in agreement with the DMA-measured distribution except for a small difference of the mean particle size. Figure 10.5 demonstrates that the VLPDMA system can measure a particle size distribution very close to the actual size of the particles inside a PECVD reactor.

10.3 Measurements of Particle Size During Laser Ablation Using a Low Pressure Differential Mobility Analyzer

Pulsed laser ablation (PLA) , by which a material is evaporated by an intense laser pulse, is a well known technique to produce thin films, nanoparticles and nanomaterials (Lowndes et al. 1996; Chrisey and Hubler 1994). PLA has a unique advantage of the stoichiometric material transport above the threshold fluence, which is utilized for deposition of complex compounds such as YBCO high Tc superconductors. Besides, it is relatively easy to deposit high melting point materials such as C, W and refractory ceramics (Landström et al. 2003). Therefore, it is important to measure the size distribution of nanoparticles generated during PLA.

Kawakami et al. (1999) studied the generation of ultrafine tungsten particles produced during PLA. They used two methods. One is to capture the nanoparticles on the grid membrane for TEM and observe them by TEM. The other is to use the

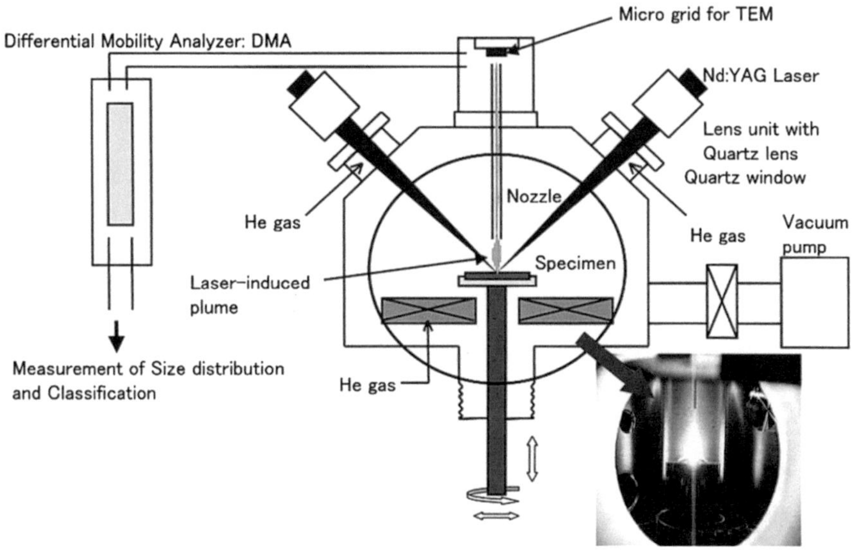

Fig. 10.6 Schematic of the experimental apparatus (Kawakami et al. 1999)

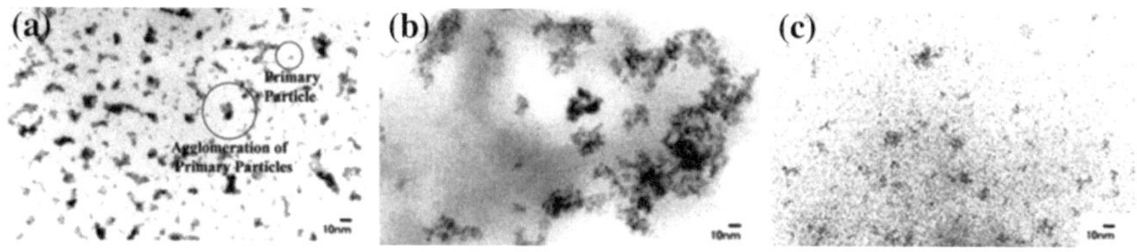

Fig. 10.7 TEM images of ultrafine tungsten particles under different pressures **a** 6.7 kPa; **b** 40 kPa; **c** 67 kPa (Kawakami et al. 1999)

LPDMA at 2.0–67 kPa (15–500 Torr). The schematic of their experimental apparatus is shown in Fig. 10.6.

Two different types of laser, the main laser and an assisting laser were used to synthesize the ultrafine tungsten particles. Irradiation with the main pulsed laser was performed 9 ms after irradiating the sample with the assisting pulsed laser. The laser irradiation conditions were 10 J/pulse for the main laser power, and 46.5–132 J/pulse for the assisting laser power.

Helium gas was introduced into the chamber at a pressure of 1.3–67 kPa after the chamber had been evacuated to 7×10^{-5} Pa. The target was a tungsten disk with a polished surface.

Figure 10.7 shows a TEM micrograph of ultrafine tungsten particles synthesized with E_a = 46.5 J/pulse at pressures from 6.7 to 67 kPa. They could observe an agglomeration of the primary particles generated at 40 kPa. Under higher pressure, the number of synthesized particles decreased due to the cooling effect of the

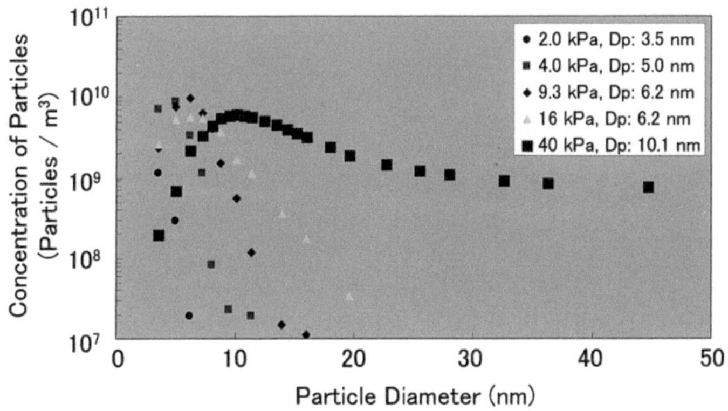

Fig. 10.8 Measurement of size distribution of ultrafine tungsten particles by LP-DMA (Kawakami et al. 1999)

specimen by the high thermal conductivity of the He gas. The size of the primary particles is about several nm in diameter for each pressure.

Figure 10.8 shows the particle size distribution measured by LPDMA. The particles were obtained at 100 J/pulse for the assisting laser power and at a pressure of 2.0–40 kPa. The distribution peak corresponding to the mean diameter of the particles shifts from several nm to tens of nm with increasing ambient pressure. In addition, the distribution profile changes from a narrow distribution to a broad peak distribution. Because the suspended time of the ultrafine particles in the laser induced plume increases with pressure, the time available for particle growth and agglomeration of the ultrafine particles also increases.

10.4 Measurements of CNPs During Si Hot Wire CVD Using a Particle Beam Mass Spectrometer

In many respects, Si hot wire CVD (HWCVD) is analogous to diamond hot filament CVD (HFCVD). Hot wires are believed to have a catalytic effect because microcrystalline silicon can be deposited at low substrate temperature where amorphous silicon is deposited by thermal CVD. This is why sometimes it is called catalytic CVD (cat-CVD) (Matsumura 1991). As in the case of diamond HFCVD, the atomic hydrogen is believed to play an important role and to enhance the kinetics in such a way that crystalline silicon can be deposited at low substrate temperature.

Considering Chaps. 6–9, CNPs are expected to be generated during Si HWCVD. To check where CNPs are generated during Si HWCVD, Hong et al. (2013b) used the particle beam mass spectroscopy (PBMS) (Ziemann et al. 1995). The PBMS can sample particles directly from low-pressure environments (>100 mTorr), detect low

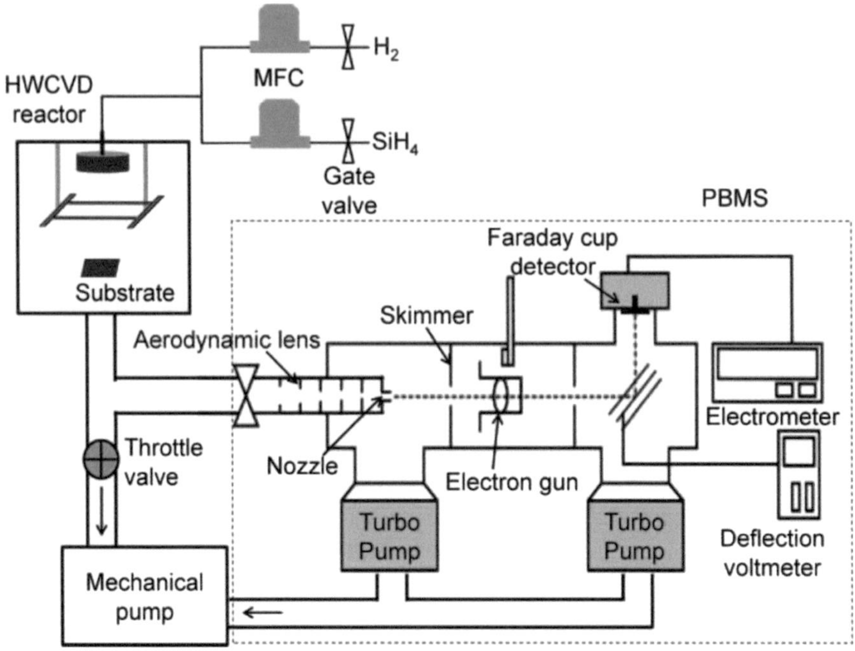

Fig. 10.9 Schematic of experimental set-up for in-situ measurement of CNPs using PBMS during Si HWCVD (Hong et al. 2013a)

particle concentrations (>20 cm^{-3}), and measure size distributions of ultrafine particles (0.005–0.5 μm) in real time (Kim et al. 2002). Figure 10.9 shows the schematic of the PBMS installed to the exhaust line of the HWCVD reactor. The line connected between the exhaust of the HWCVD reactor and the PBMS inlet was short and straight, which turned out to be critical in minimizing the diffusion loss of CNPs. All of the apparatus were electrically grounded to minimize the space charge effect. The reactor pressure was controlled by a throttle valve at the exhaust line.

Normally, the particles are charged by electron gun as shown in Fig. 10.9. In the present study, however, an artificial charger was not used in order to check whether nanoparticles generated in the gas phase are charged or not. The electric charge of nanoparticles was assumed + or −1, considering a soft charging condition in the HWCVD process. The CNPs were classified by an electrostatic deflector which deflects CNPs in proportion to the ratio of the kinetic energy to charge by the induced critical deflection voltage (Kim et al. 2002). In general, Si films are deposited at reactor pressure below 0.5 torr in HWCVD process. However, the PBMS measurements in this study were carried out at 1 and 1.5 torr, which produced reliable data.

Figure 10.10 shows photographs and cross-sectional field emission SEM (FESEM) images of Si thin films deposited for 20 min at wire temperatures of 1800, 1900, and 2000 °C, respectively, with a 5 % SiH_4 concentration under

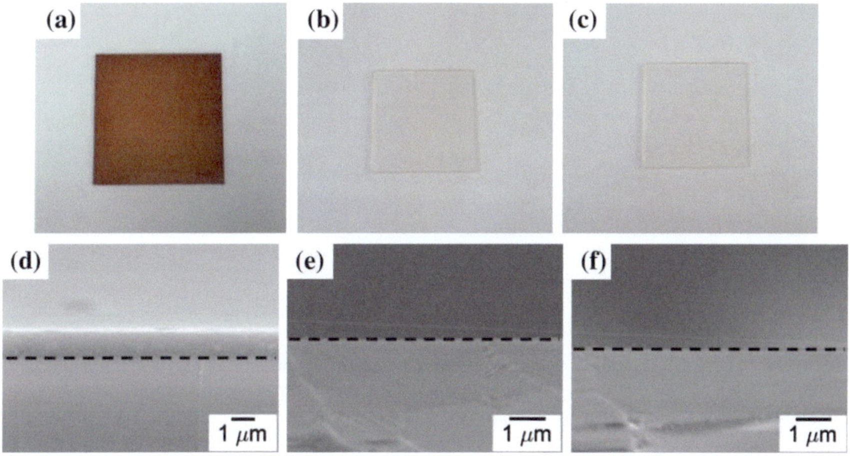

Fig. 10.10 Photographs and corresponding cross-sectional FESEM images of Si films deposited for 20 min on the glass substrate at wire temperatures of **a**, **d** 1800 °C, **b**, **e** 1900 °C, and **c**, **f** 2000 °C under 1.5 torr with a flowing gas mixture of 5 % SiH_4–95 % H_2. The *dashed line* represents the boundary between the substrate and the film (Hong et al. 2013a)

1.5 torr. At a wire temperature of 1800 °C, an appreciable amount of Si was deposited with the color being yellowish as shown in Fig. 10.10a. The film thickness of Fig. 10.10a was about 1.3 μm, as shown in Fig. 10.10d. The black dashed line represents the boundary between the glass substrate and the Si film. As the wire temperature increased, Si thin films were hardly deposited as shown in Fig. 10.10b, c, which did not differ appreciably from a bare glass substrate. As also shown in Fig. 10.10e, f, the cross-sectional FESEM images of Fig. 10.10b, c showed that the Si deposition had hardly occurred. This result shows that the deposition behavior of Si thin films is drastically affected by wire temperatures.

Figure 10.11 shows the size distribution of CNPs measured by the PBMS during the deposition of Si thin films at various wire temperatures under the same deposition conditions of Fig. 10.10. Both positive and negative CNPs were generated. The size distribution of positive CNPs was broader than that of negative CNPs. As the wire temperature increased, the number concentration of both positive and negative CNPs decreased drastically. The particle diameter at the peak of the size distribution is about 10–13 nm. The peak diameters of both positive and negative CNPs slightly decreased with increasing wire temperature. At the wire temperature of 1800 °C, the number concentration of peak diameter for negative CNPs was 2 times higher than that of positive CNPs. As the wire temperature was increased, the number concentration at the peak diameter was nearly the same for both positive and negative CNPs. These results indicate that the generation and growth of CNPs in the gas phase are suppressed with increasing wire temperature. It should be noted that at the wire temperature above 1900 °C, where the Si thin film was hardly deposited, the number concentration of nanoparticles at the peak diameter was

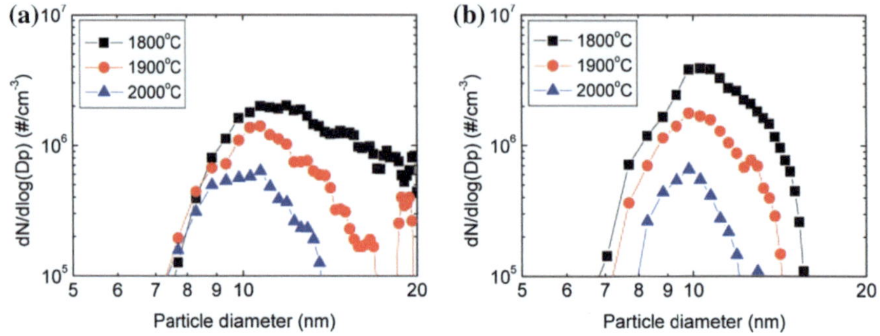

Fig. 10.11 The size distribution of **a** positively and **b** negatively charged Si nanoparticles generated in the gas phase at various wire temperatures with a flowing gas mixture of 5 % SiH_4–95 % H_2 under 1.5 torr (Hong et al. 2013a)

drastically decreased to below 2×10^6 cm^{-3}. This result implies that the generation of CNPs in the gas phase is necessary for film deposition.

As to the effect of the wire temperature on the deposition behavior and on the generation of CNPs, Hong et al. (2013b) reproduced a similar result at 1 Torr. Figure 10.12 shows the FESEM images of the surface morphology of Si thin films deposited for 5 min at wire temperatures of 1800, 1900, 2000, and 2100 °C, respectively, at 1 Torr. At the wire temperature of 1800 °C, an appreciable amount of Si was deposited and showed a typical microstructure of Si films deposited by HWCVD. On the other hand, in Fig. 10.12b–d, continuous Si films were not formed; instead the Si nanoparticles smaller than 20 nm were densely arrayed on the substrate surface. The deposition rates at the wire temperatures above 1900 °C were negligible. This result indicates that the deposition rate of Si films was decreased drastically at the wire temperatures above 1900 °C.

Figure 10.13 shows the size distribution of CNPs measured by the PBMS during the deposition of Si films at the same processing conditions of Fig. 10.12. At the wire temperature of 1800 °C, both positive and negative CNPs were generated. The particle diameter at the peak of size distribution is about 10–14 nm. The size distribution of positive CNPs was slightly broader than that of negative CNPs. The number concentration of negative CNPs was nearly twice higher than that of positive CNPs. However, at the wire temperature above 1900 °C, the number concentration of both positive and negative CNPs decreased drastically.

At wire temperatures of 1900, 2000 and 2100 °C, the decomposition efficiency would be higher than that at 1800 °C. As a result, a larger amount of Si precipitation is expected at the wire temperatures of 1900, 2000 and 2100 °C than that at 1800 °C. However, Figs. 10.11 and 10.13 show that the number concentration of CNPs decreased with increasing wire temperature. This result can be explained if we assume that the Si–H system has a retrograde solubility where the solubility of silicon in the gas phase increases with decreasing temperature. In this case, the amount of precipitation of solid Si in the gas phase decreases with decreasing

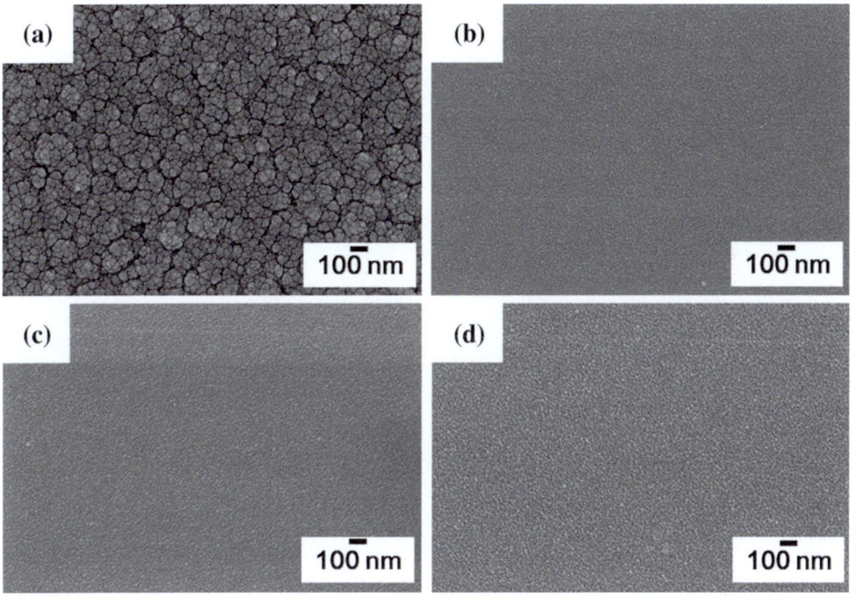

Fig. 10.12 FESEM images of Si films deposited at wire temperatures of **a** 1800 °C, **b** 1900 °C, **c** 2000 °C, and **d** 2100 °C for 5 min at 1 torr (Hong et al. 2013b)

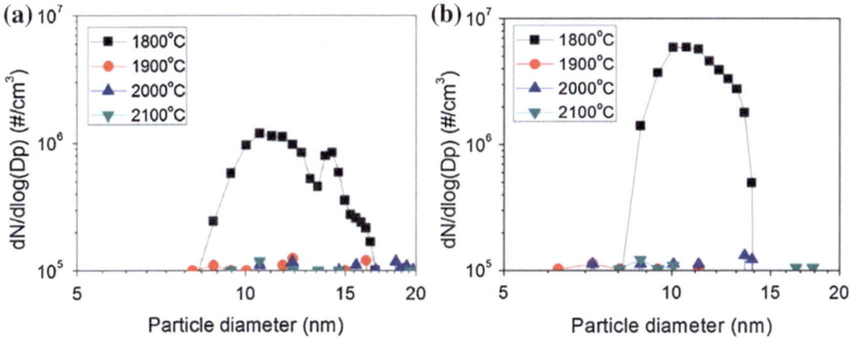

Fig. 10.13 The size distribution of **a** positively and **b** negatively charged Si nanoparticles generated at various wire temperatures at 1 Torr (Hong et al. 2013b)

temperature. In addition, etching of solid Si would depend on the amount of atomic hydrogen. The amount of atomic hydrogen produced by hot wires would increase with increasing wire temperature. Therefore, the size and number concentration of the Si gas phase nuclei would decrease with increasing wire temperature as shown in Figs. 10.11 and 10.13.

This result implies that although a large amount of silicon gas phase nuclei had been generated in the high temperature region near the hot wires, most of them were

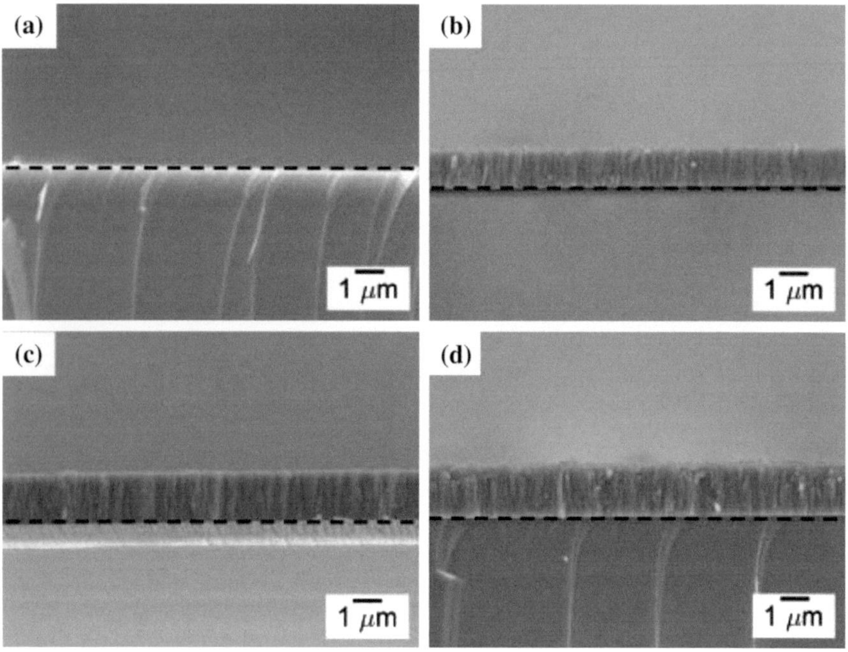

Fig. 10.14 Cross-sectional FESEM images of Si films deposited for 20 min at SiH_4 concentrations of **a** 3 %, **b** 4 %, **c** 5 %, and **d** 6 % with a wire temperature of 1800 °C under 1.5 torr. The *dashed line* represents the boundary between the substrate and the film (Hong et al. 2013a)

etched away around the low temperature substrate. It should be noted that the number concentration of nanoparticles in Fig. 10.13 is closely correlated with the deposition behavior of Si films in Fig. 10.12.

Figure 10.14 shows the FESEM images of the cross-sections of Si thin films at various SiH_4 concentrations. The films were deposited for 20 min with SiH_4 concentration varied from 3 to 6 % at a wire temperature of 1800 °C under 1.5 torr. At 3 % SiH_4, the Si film was hardly deposited as shown in Figs. 10.14a. As the SiH_4 concentration increased, the amount of the deposited Si film increased and the deposition rate was proportional to the SiH_4 concentration. The average film thicknesses at 4, 5, and 6 % SiH_4 were respectively 1.1, 1.3, and 1.6 μm. The increased deposition rate with increasing SiH_4 concentration would be related with the increased amount of Si precipitation.

Figure 10.15 shows the size distribution of positively and negatively charged Si nanoparticles measured by the PBMS under the same conditions as that of Fig. 10.14. The peak diameter and number concentration of both positively and negatively charged Si nanoparticles increased with increasing SiH_4 concentration. The number concentration of negative CNPs was higher than that of positive ones. At 3 % SiH_4, where the deposition rate was negligible, the number concentrations of both positive and negative CNPs were very low. This result implies that the

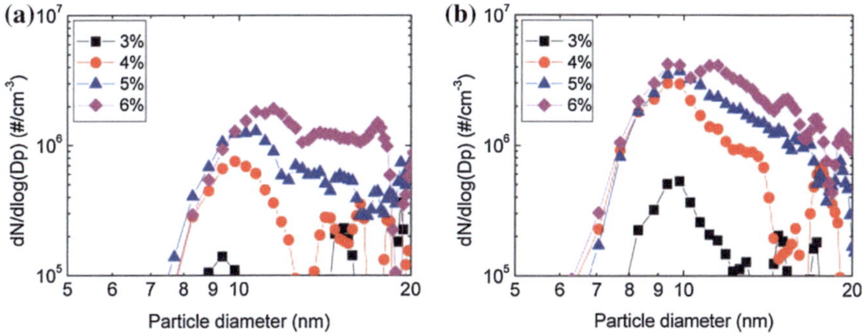

Fig. 10.15 The size distribution of **a** positively and **b** negatively charged Si nanoparticles generated in the gas phase at various SiH_4 concentrations with a wire temperature of 1800 °C under 1.5 torr (Hong et al. 2013a)

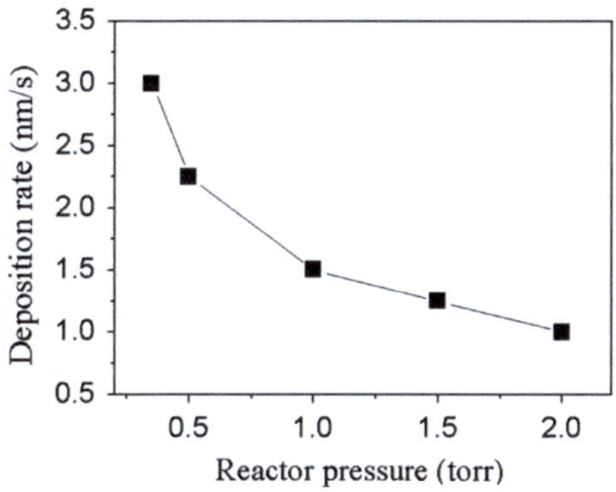

Fig. 10.16 Deposition rate of Si films deposited for 20 min at a wire temperature of 1800 °C with various process pressures from 0.3 to 2 Torr with a flowing gas mixture of 4 % SiH_4–96 % H_2 (Yoo et al. 2014)

nucleation and growth of CNPs in the gas phase are enhanced with increasing SiH_4 concentration.

The film deposition rate decreases with increasing reactor pressure in the range of 0.3–2 Torr (Yoo et al. 2014). Figure 10. 16 shows the dependence of deposition rate of Si films on pressure at a wire temperature of 1800 °C with a flowing gas mixture of 4 % SiH_4–96 % H_2. The deposition rate of Si films were 3, 2.3, 1.5, 1.3, and 1 nm/s respectively at 0.3, 0.5, 1, 1.5, and 2 Torr. This dependence of deposition rate on pressure is different from previous reports, wherein the deposition rate was found to increase with process pressure during the deposition of Si films by HWCVD (Brogueira et al. 1996; Kaneko et al. 1994; Waman et al. 2011). This is due to the difference in the range of the process pressure used in the experiment. In the previous reports, most Si HWCVD processes were done at pressures less than 0.5 Torr. This is supported by the work of Pant et al. (2001) which reported that the

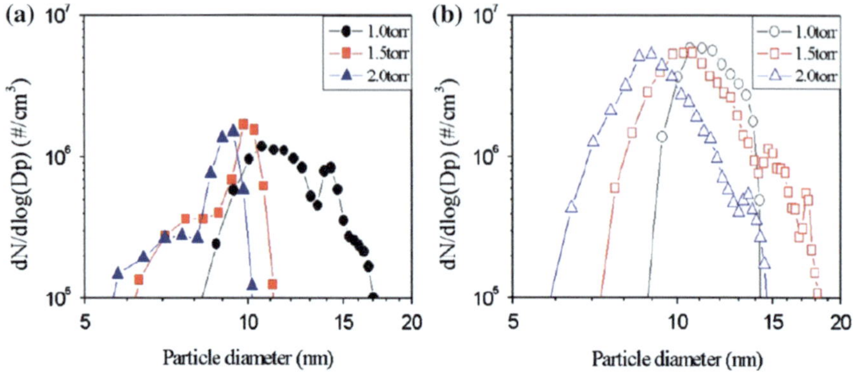

Fig. 10.17 Size distribution of positively (**a**) and negatively (**b**) charged Si nanoparticles at a wire temperature of 1800 °C with various process pressures (Yoo et al. 2014)

deposition rate of Si films increases with pressure from 25 mTorr to 0.25 Torr, but decreases with increasing pressure from 0.25 to 1 Torr. It is generally expected that a higher process pressure would produce a higher deposition rate due to the increase in the equilibrium amount of precipitation; however, Fig. 10.16 shows quite the opposite to be true, with a decrease in deposition rate when increasing pressure between 0.3 and 2 Torr. This implies that there must be a decrease in the equilibrium amount of Si precipitated from the gas phase, indicating that the Si-H system has a retrograde solubility in these pressure and temperature ranges. Fig. 10.17 shows the PBMS data for the size distribution and polarity of charged Si nanoparticles generated in the gas phase of the HWCVD reactor with a 4 % SiH_4–96 % H_2 gas mixture and a wire temperature of 1800 °C at 1.0, 1.5, and 2.0 Torr. The peak of this size distribution is at 9–12 nm at which the concentration values of both positive and negative CNPs are 10^6–10^7/cm^3.

In addition to the PBMS data in Fig. 10.17, the size of the Si gas-phase nuclei was also determined by capturing them on the TEM grid membrane in the HWCVD reactor and observing them ex situ by TEM. Figure 10.18 shows the TEM images of Si nanoparticles captured for 10 s at different pressures under the same deposition conditions as in Fig. 10.16. Numerous Si nanoparticles were observed on the membrane under all conditions, with their size and number decreasing with increasing process pressure. Note that this is consistent with the PBMS data in Fig. 10.17 and the dependence of pressure on the deposition rate demonstrated in Fig. 10.16.

The Si nanoparticles captured at a pressure of 0.3 Torr (Fig. 10.18a) were the greatest in terms of both size and quantity within the pressure range examined, and this corresponds to the highest film growth rate in Fig. 10.16. In contrast, the size and quantity of nanoparticles were the least at 2 Torr (Fig. 10.18e), which corresponds to the lowest film growth rate in Fig. 10.16. It should be noted that the peak diameters of the CNPs at 1.0, 1.5, and 2.0 Torr in Fig. 10.17 are very similar to the most frequently observed nanoparticles captured at the corresponding pressure, as

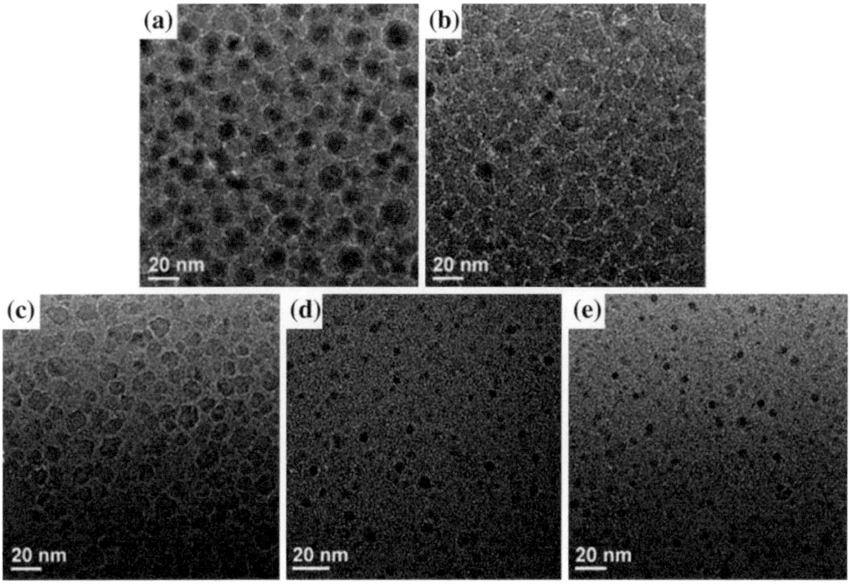

Fig. 10.18 TEM images of Si nanoparticles captured for 10 s on the TEM grid membrane at the process pressures of **a** 0.3, **b** 0.5, **c** 1, **d** 1.5, and **e** 2 Torr, respectively, at the same condition as in Figs. 10.16 and 10.17 (Yoo et al. 2014)

shown in Fig. 10.18c–e, respectively. This dependence of size and quantity on the process pressure is in agreement with the dependence of the growth rate on the pressure demonstrated in Fig. 10.16. It, therefore, appears that the decrease in deposition rate with pressure shown in Fig. 10.16 comes from the decrease in the size and quantity of Si nanoparticles with pressure shown in Figs. 10.17 and 10.18.

The high-resolution TEM images in Fig. 10.19, which are magnified views of those in Fig. 10.18, show that all of the Si nanoparticles have a crystalline lattice, despite the temperature of the TEM grid being as low as ~370 °C. The crystalline lattices shown in Fig. 10.19 have important implications. Since the diffusion of Si atoms would be so low at ~370 °C, it is expected that not crystalline but amorphous Si should be deposited if the nanoparticles in Fig. 10.18 are formed by the atomic diffusion on the surface of the TEM grid membrane. Considering this, it is far more likely that the crystalline Si nanoparticles are instead formed in the high-temperature region near the hot wires, and subsequently land on the TEM grid over a capture time of 10 s. If this capture time is extended beyond the time required to cover the entire surface with Si nanoparticles, then a Si film with a microcrystalline structure will be formed. It should be noted that the low-temperature deposition of crystalline Si films by HWCVD or PECVD has been a quite puzzling phenomenon because of the low diffusivity of Si. However, it can be explained by assuming that these crystalline Si films grow by nonclassical crystallization. That is, they are deposited by crystalline Si nanoparticles formed in the high-temperature region of the gas phase.

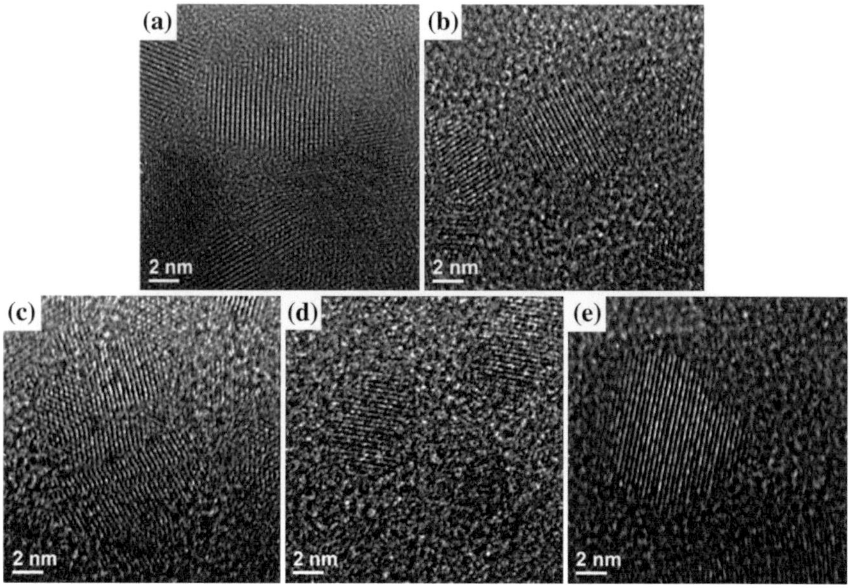

Fig. 10.19 Magnified TEM images of Si nanoparticles in Fig. 10.18 (Yoo et al. 2014)

References

Brogueira P, Conde J, Arekat S, Chu V (1996) Amorphous and microcrystalline silicon films deposited by hot-wire chemical vapor deposition at filament temperatures between 1500 and 1900 C. J Appl Phys 79(11):8748–8760. doi:http://dx.doi.org/10.1063/1.362501

Chrisey D, Hubler G (1994) Pulsed laser deposition of thin films. John Wiley & Sons, New York

Fuchs N (1963) On the stationary charge distribution on aerosol particles in a bipolar ionic atmosphere. Geofisica pura e applicata 56(1):185–193

Hong J-S, Kim C-S, Yoo S-W, Park S-H, Hwang N-M, Choi H-M, Kim D-B, Kim T-S (2013a) In-Situ measurements of charged nanoparticles generated during hot wire chemical vapor deposition of silicon using particle beam mass spectrometer. Aerosol Sci Tech 47(1):46–51. doi:10.1080/02786826.2012.725959

Hong J-S, Kim C-S, Yoo S-W, Park S-H, Lee S-S, Hwang N-M, Choi H-M, Kim D-B, Kim T-S (2013b) Real time measurements of charged gas phase nuclei during the deposition of silicon thin films by hot wire chemical vapor deposition. Curr Appl Phys 13, Supplement 2:S45–S49. doi:http://dx.doi.org/10.1016/j.cap.2012.12.016

Kaneko T, Wakagi M, Onisawa Ki, Minemura T (1994) Change in crystalline morphologies of polycrystalline silicon films prepared by radio-frequency plasma-enhanced chemical vapor deposition using SiF_4 + H_2 gas mixture at 350 °C. Appl Phys Lett 64(14):1865–1867

Kawakami Y, Seto T, Ozawa E (1999) Characteristics of ultrafine tungsten particles produced by Nd:YAG laser irradiation. Appl Phys A-Mater 69(1):S249–S252. doi:10.1007/s003399900286

Kim T, Suh SM, Girshick SL, Zachariah MR, McMurry PH, Rassel RM, Shen Z, Campbell SA (2002) Particle formation during low-pressure chemical vapor deposition from silane and oxygen: Measurement, modeling, and film properties. J Vac Sci Technol A 20(2):413–423. doi:10.1116/1.1448506

Kim J, Seto T, Sakiyama K, Kim D (2004) Characterization of low pressure DMA system for the size selection of magnetic nano-particles. Appl Phys A-Mater 79(4–6):1497–1499. doi:10.1007/s00339-004-2829-1

Landström L, Márton Z, Arnold N, Högberg H, Boman M, Heszler P (2003) In-situ monitoring and characterization of nanoparticles during w ablation in N_2 atmosphere. J Appl Phys 94 (3):2011

Lowndes DH, Geohegan DB, Puretzky AA, Norton DP, Rouleau CM (1996) Synthesis of novel thin-film materials by pulsed laser deposition. Science 273(5277):898–903. doi:10.1126/science.273.5277.898

Matsumura H (1991) Formation of polysilicon films by catalytic chemical vapor deposition (Cat-CVD) method. Jpn J Appl Phys 30(8B):L1522

Pant A, Russell TWF, Huff MC, Aparicio R, Birkmire RW (2001) Hot-wire chemical vapor deposition of silicon from silane: Effect of process conditions. Ind Eng Chem Res 40(5):1377–1385. doi:10.1021/ie000423+

Seol KS, Tsutatani Y, Fujimoto T, Okada Y, Takeuchi K, Nagamoto H (2001) New in situ measurement method for nanoparticles formed in a radio frequency plasma-enhanced chemical vapor deposition reactor. J Vac Sci Technol B 19(5):1998–2000. doi:10.1116/1.1404979

Seto T, Nakamoto T, Okuyama K, Adachi M, Kuga Y, Takeuchi K (1997) Size distribution measurement of nanometer-sized aerosol particles using dma under low-pressure conditions. J Aerosol Sci 28(2):193–206. doi:http://dx.doi.org/10.1016/S0021-8502(96)00071-7

Waman VS, Funde AM, Kamble MM, Pramod MR, Hawaldar RR, Amalnerkar DP, Sathe VG, Gosavi SW, Jadkar SR (2011) Hydrogenated nanocrystalline silicon thin films prepared by hot-wire method with varied process pressure. J Nanotechnol 2011:10. doi:10.1155/2011/242398

Yoo S-W, Hong J-S, Lee S-S, Kim C-S, Kim T-S, Hwang N-M (2014) Nonclassical crystallization in low-temperature deposition of crystalline silicon by hot-wire chemical vapor deposition. Cryst Growth Des 14(12):6239–6247. doi:10.1021/cg5008582

Ziemann PJ, Liu P, Rao NP, Kittelson DB, McMurry PH (1995) Particle beam mass spectrometry of submicron particles charged to saturation in an electron beam. J Aerosol Sci 26(5):745–756. doi:http://dx.doi.org/10.1016/0021-8502(95)00009-2

Chapter 11
Deposition Behavior of Charged Nanoparticles

From Chaps. 6–10, it can be said that the gas phase generation of CNPs and their deposition into films and nanostructures are quite general in the CVD and some PVD processes. The electrostatic interaction, which is expected in the deposition of CNPs, is clearly revealed in the growth of nanowires but less clearly revealed in the growth of thin films. Considering that the film growth is simply a result of high temperature colloidal crystallization of CNPs, however, it can be inferred that the electrostatic self assembly must have taken place actively during the film growth. Selective deposition is also the result of the electrostatic effect of CNPs. In this chapter, the deposition behavior of CNPs will be examined with a focus on the electrostatic interaction.

11.1 Comparison of Deposition Behavior of Si Between Floating and Grounded Substrates

In Chap. 6, the deposition behavior of diamond was compared between floating and grounded Fe substrates. On the floating substrate, diamond crystals grew on the initially formed soot whereas on the grounded substrate, porous soot continued to grow. Youn et al. (2014) made similar experiments by comparing the deposition behavior between floating and grounded Si substrates during the atmospheric CVD of Si. First, they experimentally confirmed the generation of CNPs in the gas phase using a nano DMA-FCE system as shown in Fig. 11.1.

The number concentrations and size distributions of negative and positive CNPs at N_2 gas flow rate of 1000 sccm are larger and broader, respectively, than those at 500 sccm. At 1000 sccm of N_2, the number concentrations of negative and positive CNPs are almost the same. At 500 sccm of N_2, however, the number concentration of negative CNPs is larger, below 80 nm, and smaller, above 80 nm, than that of

© Springer Science+Business Media Dordrecht 2016

N.M. Hwang, *Non-Classical Crystallization of Thin Films and Nanostructures in CVD and PVD Processes*, Springer Series in Surface Sciences 60,
DOI 10.1007/978-94-017-7616-5_11

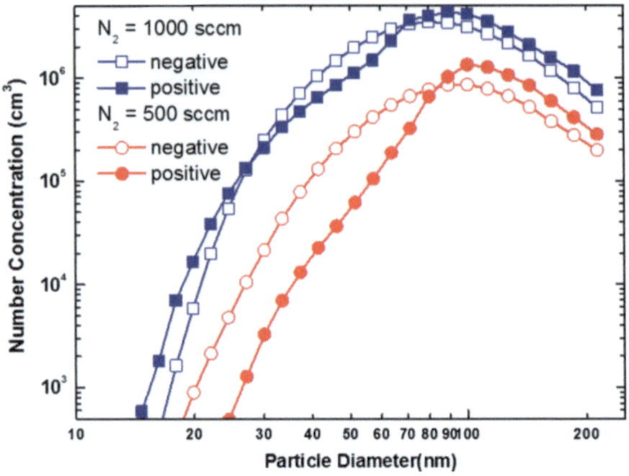

Fig. 11.1 Number concentrations and size distributions of negative (open) and positive (closed) CNPs at the reactor temperature of 900 °C with N_2 flow rates of 500 and 1000 sccm at a SiH_4 flow rate of 5 sccm and H_2 flow rate of 50 sccm (Youn et al. 2014)

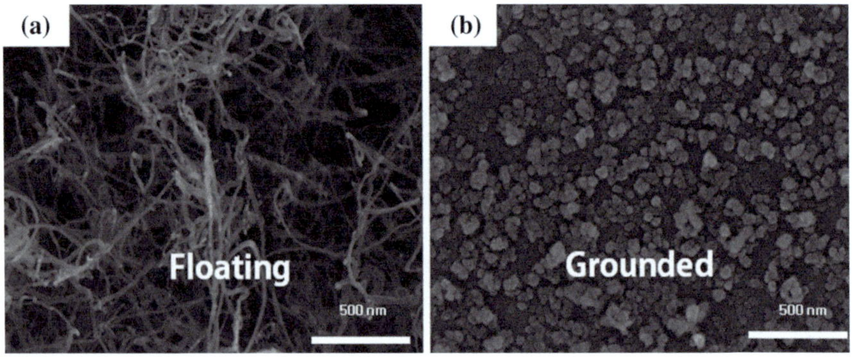

Fig. 11.2 SEM images of microstructures evolved on **a** floating and **b** grounded silicon substrates at a N_2 flow rate of 500 sccm (Youn et al. 2014)

positive CNPs. When the N_2 gas flow rate is 500 sccm, CNPs start to be measured at ~20 nm.

Under the condition where CNPs were generated abundantly in Fig. 11.1, the deposition behavior of silicon was compared between the electrically floating and grounded substrates on a stainless substrate holder plate. To ground the silicon substrate, the stainless holder was connected to the external ground via a stainless rod. The deposition was done for 2 h without a catalyst at the substrate temperature of 900 °C with gas flow rates of 5 sccm 10 % SiH_4–90 % He, 50 sccm H_2, and 500 sccm N_2. Figure 11.2a, b show SEM images respectively on floating and grounded substrates.

Silicon nanowires grew extensively on the floating substrate (Fig. 11.2a) whereas no silicon nanowire grew but only silicon nanoparticles were deposited on the grounded substrate (Fig. 11.2b). The nanowires in Fig. 11.2a had diameters of about 10−30 nm. Since neither catalytic metal nor seed of silicon oxide was used in these experiments, the growth of silicon nanowires in Fig. 11.2a cannot be explained by the vapor–liquid–solid (VLS) (Wagner and Ellis 1964) or oxide-assisted growth (OAG) (Zhang et al. 2003) mechanism. Charge build-up would be maximized on the floating substrate and minimized on the grounded substrate. Since the only difference in the processing condition between Fig. 11.2a, b was floating and grounding of the substrate, charge build-up should be responsible for the growth of silicon nanowires in Fig. 11.2a. This means that the electrostatic interaction between the CNPs and the substrate was responsible for the growth of the silicon nanowires.

To examine the detailed crystalline structure of the silicon nanowires, the silicon nanowires in Fig. 11.2a were observed by TEM as shown in Fig. 11.3. The nanowire has a core–shell structure with an inner crystalline silicon core and an outer amorphous oxide (SiO_x) layer. It should be noted that the outer amorphous oxide layer was not formed during the growth in the reactor but formed in an ambient atmosphere after the sample was taken out of the reactor.

The diameter of the silicon core is 5 nm, and the thickness of the outer oxide layer is 7 nm. The inner silicon core in Fig. 11.3b has the same lattice fringe, indicating that it is a single crystal with $\langle 220 \rangle$ growth direction. The inset in the upper right corner of Fig. 11.3b shows the image of the lattice, which is magnified in the box marked in the lower right. The spacing between the parallel lattice fringes was measured to be 0.19 nm, which is close to the spacing of the (220) planes of silicon.

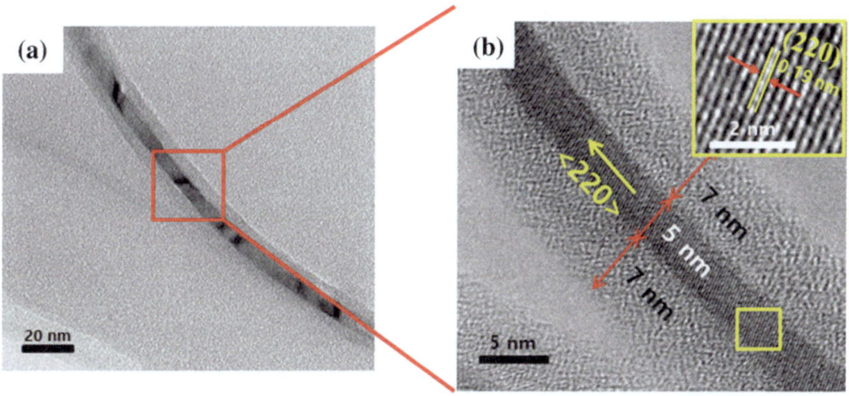

Fig. 11.3 TEM images of silicon nanowires deposited on a floating silicon substrate at a N_2 flow rate of 500 sccm: **a** low-magnification and **b** high-magnification TEM images with the inset of an enlarged HRTEM image of the square-enclosed area (Youn et al. 2014)

Although the building block of the silicon nanowires in Fig. 11.2a is believed to be CNPs, each nanowire is a single crystal with a smooth surface as shown in Fig. 11.3a, b. This result again supports that CNPs are liquid-like. If CNPs are not liquid-like, the self assembly and the oriented attachment of CNPs would produce a pearl necklace-like or chain-like structure instead of smooth nanowires.

This comparison of the deposition behavior between floating and grounded substrates was further studied with a N_2 flow rate of 1000 sccm, under which condition silicon films were deposited. The result is shown in Fig. 11.4.

Figure 11.4a, b shows SEM images of the surface and the cross section of a film, respectively, deposited on a floating silicon substrate; Fig. 11.4c, d shows those of a film deposited on a grounded substrate. The deposition condition was the same as that of Fig. 11.2 except the gas flow rate of N_2. The film on the grounded substrate in Fig. 11.4c, d was much more porous than that on the floating substrate in Fig. 11.4a, b. These results show that drastically different microstructures evolved between the floating and grounded substrates, indicating that the dense film in Fig. 11.4a, b resulted from the electrostatic interaction between the CNPs and the

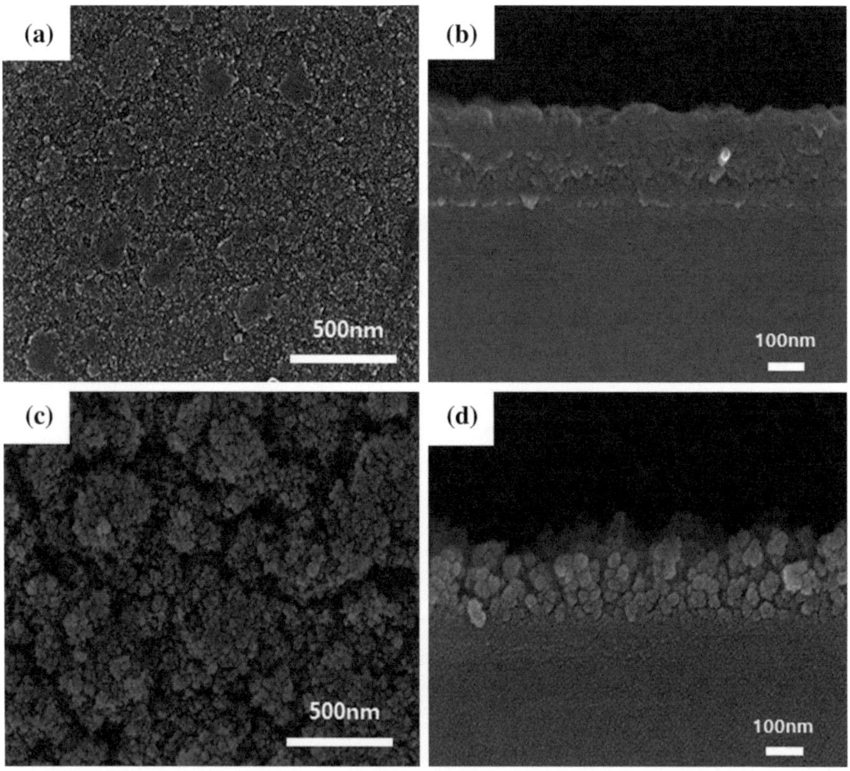

Fig. 11.4 SEM images of **a** plane view and **b** cross section of films deposited on a floating silicon substrate and **c** plane view and **d** cross section of films deposited on a grounded silicon substrate at a N_2 flow rate of 1000 sccm (Youn et al. 2014)

growing surface. The film thicknesses of Fig. 11.4b, d were ~220 and ~190 nm, respectively. Besides, the film in Fig. 11.4b is much denser than that in Fig. 11.4d. Therefore, the growth rate of the film on the floating substrate in Fig. 11.4b is much higher than that on the grounded substrate in Fig. 11.4d.

Then, why is the film growth rate on the floating substrate higher than that on the grounded one? This result may be related with the image force between the CNPs and the growing surface. The electrostatic interaction is very difficult to estimate. Nonetheless, let's approach this problem using (7.6) by assuming that the floating and grounded substrates are regarded as a large sphere with r_2 carrying $n_2 = 10$ and 1 excess charges, respectively. In (7.6), the first term is proportional to n_2 and the third term is proportional to the square of n_2. When the incoming particle carries the opposite sign of charge, it will be attracted more strongly to the floating substrate than to the grounded substrate. When the incoming particle is neutral, n_1 is zero and then the first and the second terms in (7.6) become zero and only the third term remains. In this case also, the incoming neutral particle will be attracted more strongly to the floating substrate. When the incoming particle carries the same sign of charge, the first term in (7.6) is repulsive but the second and the third terms are attractive. In this case, the incoming particle will be less repelled from the floating substrate. Thus, under the condition that a majority of the incoming particles of the same sign of charge are repelled from both of the floating and ground substrates, the deposition rate on the floating substrate can be higher than that on the grounded substrate.

Then, why is the film on the floating substrate more dense than that on the grounded one? This question would be also related to the evolution of dense films on the substrate with a low charge transfer rate and the evolution of porous soot-like structures on the substrate with a high charge transfer rate described in Chaps. 6 and 7. For dense films to be evolved, nanoparticles should undergo the self-assembly and have a liquid-like property. Since these properties are enhanced by the presence of charge and the build-up of charge on the floating substrate is larger than that on grounded substrate, the floating substrate would produce the denser film.

11.2 Evolution of Puzzling Microstructures During Si CVD

As various ZnO nanostructures are evolved presumably by CNPs generated in the gas phase (Wang 2006b), so various silicon nanostructures are often evolved under certain processing conditions. These various nanostructures appear to result from the electrostatic interaction of CNPs. Here, some peculiar microstructure evolutions presumed to be deposited by CNPs will be presented without detailed explanation as to why. A more advanced treatment on the electrostatic interaction would be needed to understand the detailed mechanism of these microstructure evolutions.

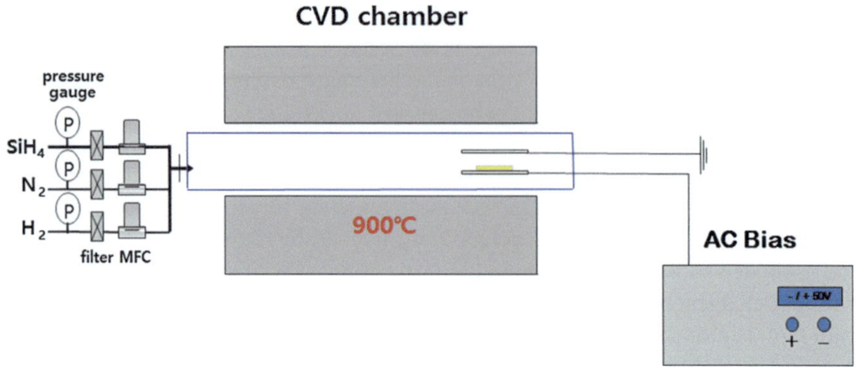

Fig. 11.5 Schematic of the CVD chamber showing the location of the substrate with a power supply for applying the bias

Let me explain briefly the motivation of the experiment, which would help understand the experimental procedure. If CNPs are generated in the gas phase and become the building block of films, the crystalline silicon film can be deposited at the low substrate temperature. Then crystalline silicon films, which can be applied to thin film transistor (TFT) for displays and solar cells, can be deposited on the chief glass substrate. Based on this idea, crystalline Si CNPs are made to generate near the hot zone at 900 °C of the tube CVD reactor and the glass substrate is placed at the low temperature zone of 500 °C. The experimental setup is shown in Fig. 11.5.

What actually happened is that the deposition did not take place on the glass substrate at all. If a metallic substrate is used, it is easily silicided. In other words, a kind of selective deposition occurs. What makes the deposition even more difficult on the glass substrate is the upward drag force in the reactor zone where the substrate is placed as shown in Fig. 11.5. The drag force will be treated in more detail in Chap. 12.

When a DC bias is applied to the substrate holder, the deposition rate on the glass substrate was negligibly small. The reason might be that when the positive bias is applied to the substrate holder, negative CNPs will be attracted to the glass substrate and once they land, their negative charge on the glass substrate will exert the electric field, which will cancel the electric field made by the bias. Therefore, AC bias was applied to the substrate holder. Flow rates of He-diluted SiH_4 (10 % SiH_4–90 % He), H_2 and N_2 were respectively 10, 50, and 300 sccm.

After 2 h of deposition under the AC bias of ±50 V at 1 Hz, an appreciable amount of silicon was deposited on the glass substrate but it still had a porous structure as shown in Fig. 11.6a. The Raman spectrum shows that it is crystalline silicon. An appreciable amount of silicon deposited under the AC bias indicates that CNPs exist in the gas phase in abundance but do not deposit on the glass substrate. There are two possible reasons as to why CNPs do not deposit. One is the weak image force between CNPs and the glass substrate. The other is the drag force.

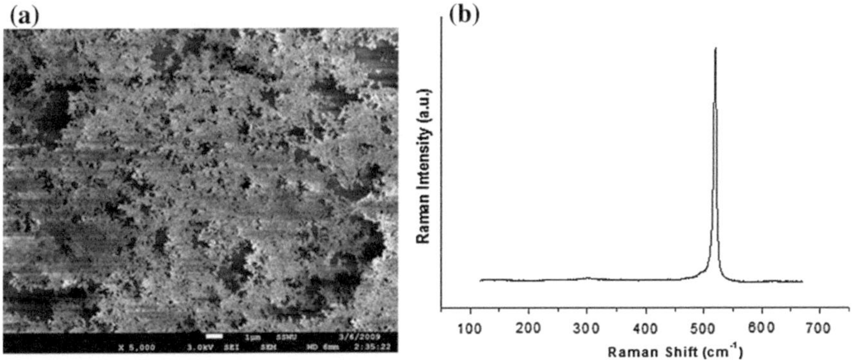

Fig. 11.6 **a** FESEM image of the porous silicon deposited at 500 °C for 2 h under the AC bias of ±50 V at 1 Hz and **b** its Raman spectrum

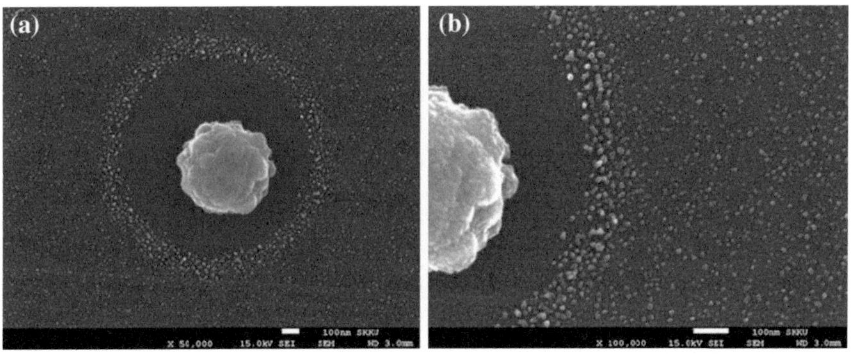

Fig. 11.7 FESEM images of microstructures evolved after 24 h deposition with 5 min of the bias treatment at flow rates of He-diluted SiH_4 (10 % SiH_4–90 % He), H_2 and N_2 being respectively 11, 50, and 300 sccm. **b** is the higher magnification image of (**a**). The *scale bar* is 1 μm

In order to deposit a dense film instead of a porous structure, AC bias of ±50 V at 1 Hz was applied for 5 min to initiate the deposition. After this treatment, the deposition was done for 24 h. It was expected that once silicon nanoparticles are deposited on the glass substrate, they will act as seeds for further deposition. However, the microstructure evolved on the substrate was different from the expectation.

The microstructure is shown in Fig. 11.7. In some area a large silicon grain grows but there is a depletion zone near the large silicon grain as shown in Fig. 11.7a. Outside the depletion zone, fine silicon particles were deposited. The size of particles near the periphery of the depletion zone is larger than that away from the periphery as revealed in Fig. 11.7b.

The microstructure evolution of Fig. 11.7 seems to be related with that of Fig. 7.1 of Chap. 7, where one silicon particle grows exclusively while other small particles are being etched away. Figure 7.1 was explained by the image force of CNPs using (7.5) and (7.6), which shows that a larger charged particle has a strong attractive force with CNPs of opposite sign and a weak attractive force with CNPs of like sign. This effect is accelerated as the size become bigger. Comparing the size of the large particle at the center of Fig. 11.7a with other small particles, a huge amount of CNPs nearby must have been attracted to the large particle. This would be why the depletion zone is formed around the large particle.

Considering Figs. 7.1 and 11.7, under certain conditions where CNPs exist abundantly, one charged nanoparticle can grow exclusively into a very large one. This phenomenon might be related with the initial stage of planet formation in the universe. Dust particles grow at the beginning of planet formation (Blum and Wurm 2008). The growth mechanism of dust is not clearly established. Especially, the growth of dust into gravitationally active planetesimals should be fast. As one of possible mechanisms, electric runaway growth was suggested (Blum and Wurm 2008). Runaway growth is a process by which the mass of a single dust aggregate increases much faster than the masses of all other particles in the system (Blum 2006). In relation to this possibility, Ivlev et al. (2002) showed that runaway growth can occur if dust aggregates are charged. Previously, it was suggested that electric charging of dust can be introduced through tribo-electric effects in collisions, through which electrons are exchanged between the particles (Desch and Cuzzi 2000; Poppe et al. 2000; Poppe and Schräpler 2005). Figures 7.1 and 11.7 clearly show that the electric runaway growth can occur in the presence of charged dust.

The microstructure evolution turned out to be so senstive to the processing condition that a slight change of the condition produced a quite different microstructure. For example, when the gas flow rate was changed to He-diluted SiH_4 (10 % SiH_4–90 % He), H_2 and N_2 being respectively 5, 50, and 1000 sccm with other conditions being the same as those of Fig. 11.7, another peculiar microstructure was evolved as shown in Fig. 11.8.

Figure 11.8a shows three concentric circles by contrast. At the center, there is a small whitish circle surrounded by the dark circle, which is surrounded again by a large whitish circle. Figure 11.8b shows the magnified image of Fig. 11.8a and designates the circles as regions I, II and III. In region I, the silicon nanowires grow most densely as shown in Fig. 11.8c. In region II, the silicon nanowires are less dense but the diameter of nanowires is a little larger than that in region I as shown in Fig. 11.8d. In region III, nanoparticles are densely deposited with very rare nanowires as shown in Fig. 11.8e. This microstructure evolution is difficult to explain. But one thing for sure is that the electrostatic energy of CNPs should be responsible for the evolution.

When the deposition was done for 6 h with the pretreatment of the AC bias of ±50 V at 1 Hz for 10 min with other conditions being the same as those of Fig. 11.7, another peculiar microstructure was evolved as shown in Fig. 11.9.

In Fig. 11.9a, there are dark circles inside a whitish area. When the whitish area is magnified, it consists of relatively dark and relatively bright circles with a black background as shown in Fig. 11.9b. When the dark and bright circles are magnified, they look as Fig. 11.9c, d respectively. In Fig. 11.9c, relatively short nanowires

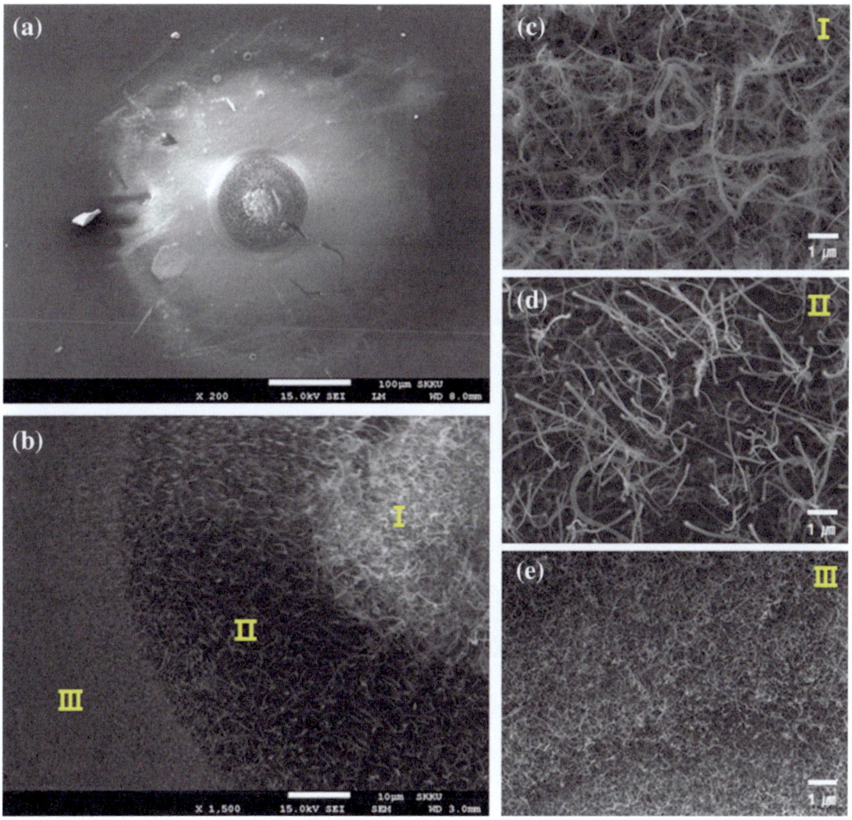

Fig. 11.8 FESEM images of microstructures evolved at flow rates of He-diluted SiH_4 (10 % SiH_4–90 % He), H_2 and N_2 being respectively 5, 50, and 1000 sccm with other conditions being the same as those of Fig. 11.7. **b** is the higher magnification image of **a** and **c**, **d** and **e** are the higher magnification images respectively of the areas I, II and III of **b**

with small diameters are densely grown whereas in Fig. 11.9d, relatively longer wires with larger diameters are grown less densely.

These peculiar structures are observed mainly on an insulating substrate, where some excess charge would build up from the deposition of CNPs and thereby there exists a strong electrostatic interaction between incoming CNPs and the pre-deposited structure on the substrate. When a silicon substrate is used, there is a thin native oxide layer, which has a somewhat insulating effect. If the silicon substrate is annealed in O_2 atmosphere, the oxide layer becomes thicker and the insulating effect is increased.

Swain et al. (2015) prepared such a substrate by thermal annealing of Si (100) substrate in O_2 atmosphere, which produced 100 nm thick SiO_x layer on Si. By elemental analysis by X-ray photoelectron spectroscopy, they determined the stoichiometry of the substrate to be $SiO_{1.53}$. Using such substrates, they could

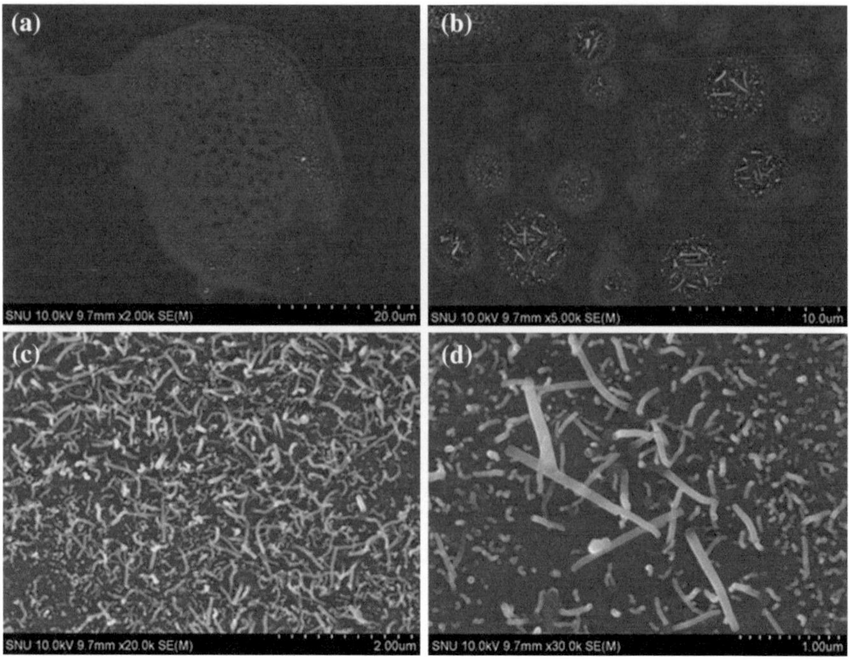

Fig. 11.9 FESEM images of the microstructure evolved after the deposition for 6 h with the pretreatment of the AC bias of ±50 V at 1 Hz for 10 min with other conditions being the same as those of Fig. 11.7. **b** is a higher magnification of (**a**). **c** and **d** are higher magnifications respectively of *dark* and *bright circles* in (**b**)

obtain various microstructure evolutions during CVD using sources vaporized from the solid powder.

0.5 g of fused SiO_2 powder (100 nm) was mixed with 0.5 g Si powder (2–4 μm) and placed in the alumina boat at the hot zone (1100 °C) of the horizontal tube CVD reactor. N_2 or H_2 gas was flown during the CVD reactor. The substrates were placed at 1000 °C in the reactor whereas the hot zone was at 1100 °C.

Figure 11.10a–d shows the microstructure evolved after annealing for 30 min using Si and SiO_2 powders as precursor under flowing N_2. Figure 11.10e, f shows the microstructure evolved after annealing for 90 min. Figure 11.10a–d shows a sunflower-shaped structure. Figures 11.10c, d are the higher magnification images of the square area in Fig. 11.10a, b, respectively. In the middle of Fig. 11.10c, there is a circular area with a diameter of ~4 μm in which the nanoparticles are closely packed and assembled in a random manner. However, outside the circle, it looks as if nanoparticles are emanated from the center in a regular way in all directions.

Figure 11.10f is the higher magnification image of the square area in Fig. 11.10e. After 90 min of annealing, individual nanoparticles in Fig. 11.10a–d changed to nanowires. It appears that nanoparticles in Fig. 11.10a–d act as seeds for nanowire growth.

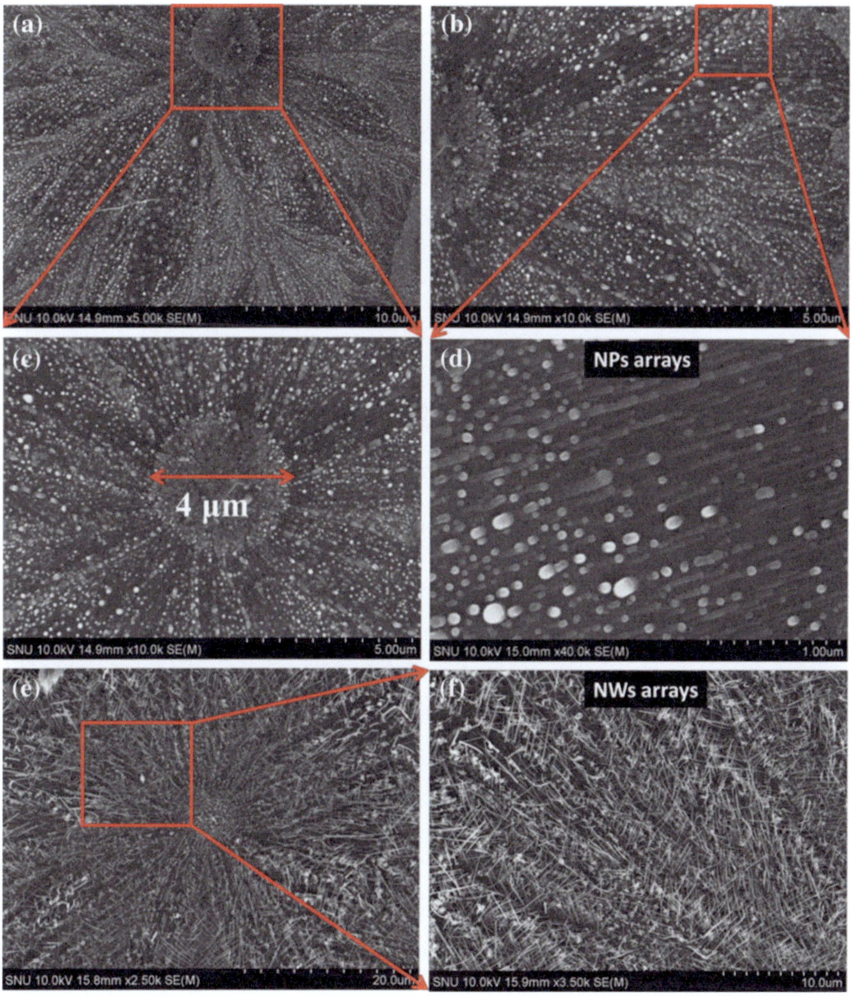

Fig. 11.10 FESEM images of a silicon sunflower structure evolved after annealing for 30 min (**a**–**d**) and for 90 min (**e**, **f**) using Si and SiO_2 powders as precursors under flowing N_2 (Swain et al. 2015)

Figure 11.11 is lower magnification FESEM image of Fig. 11.10e showing the overall view of the sun flower structures on the SiO_x/Si substrate. Several sun flower structures are observed in a regular pattern.

Figure 11.12 shows microstructures evolved when a mixture of Si and SiO_2 powders was annealed for 90 min in H_2 at the flow rate of 200 sccm with the other conditions being the same as those of Figs. 11.10 and 11.11. Deposition is quite non uniform; only in some circular areas silicon was densely deposited. In some

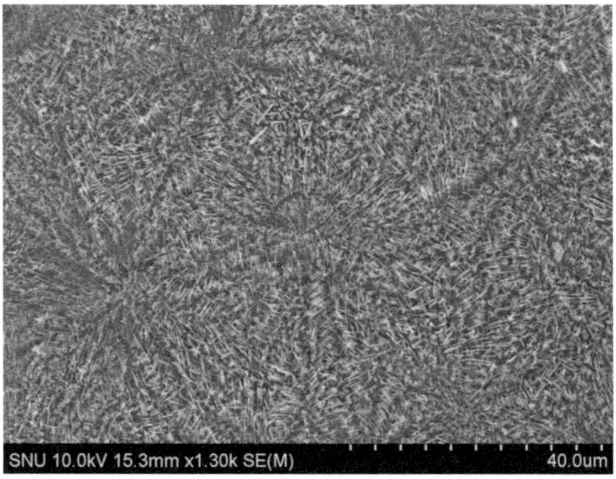

Fig. 11.11 A low-magnification FESEM image of sun flower microstructures observed after 90 min of deposition (Swain et al. 2015)

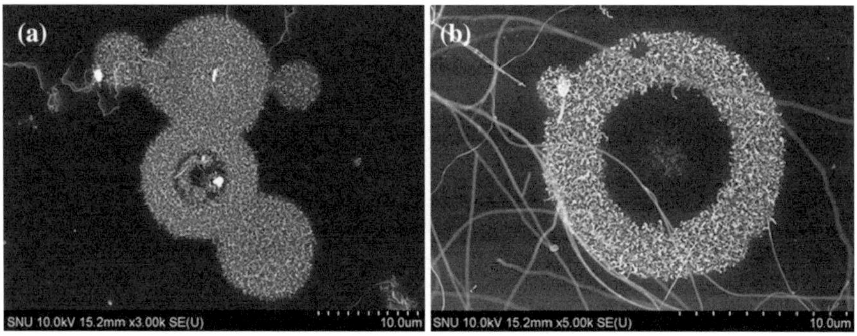

Fig. 11.12 FESEM images of microstructures evolved after 90 min of deposition by evaporating a mixture of Si and SiO_2 power in flowing H_2 (Swain et al. 2015)

case, a depletion zone, where the amount of deposition is much less or non-deposition occurs, exists inside a circle as shown in Fig. 11.12b. As a result, the wreath-like microstructure was evolved. At the center of the depletion zone, however, there are a small circular area, where some deposition occurs.

A wreath-like structure was also observed in the well-known carbothermal synthesis of $SrTiO_3$ as shown in Fig. 11.13 (Swain et al. 2015). The circular area inside the wreath-like structure is a depletion zone with some deposition at the center of the depletion zone. As a result, the depletion zone inside the wreath shows a sun flower structure, which is revealed in Fig. 11.13b–d. Figure 11.13b, c, d are respectively magnified images of the square areas in Fig. 11.13a, b, c.

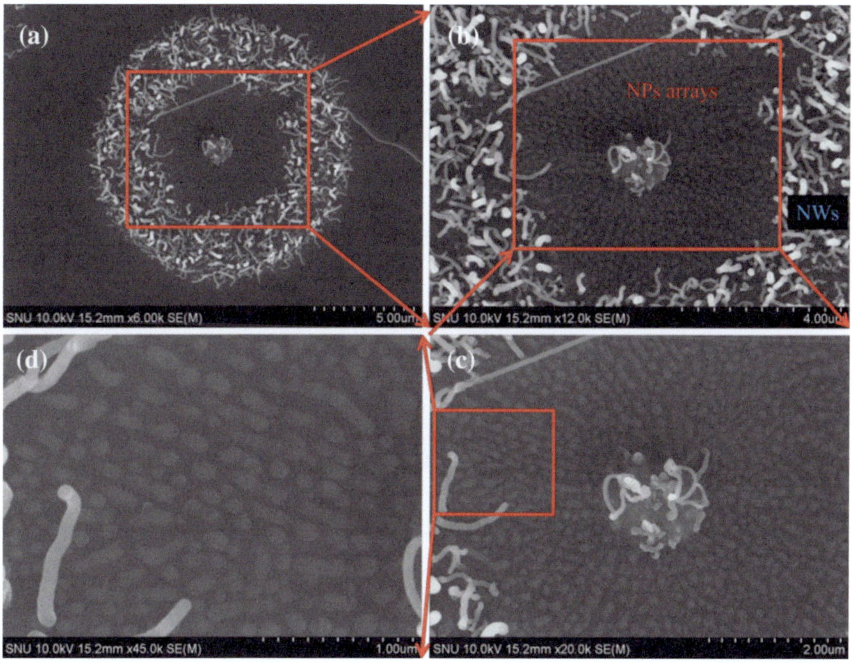

Fig. 11.13 a–d FESEM images of microstructures evolved after 90 min by carbothermal deposition of $SrTiO_3$. **b**, **c** and **d** are respectively magnified images of the squares in **a**, **b** and **c** (Swain et al. 2015)

11.3 Growth of Silicon Nanostructures by Thermal Evaporation of SiO Powders

Pan et al. (2001) investigated the nanomaterials produced from thermal evaporation of SiO powders as a function of the local temperature. Although they did not mention the involvement of CNPs in their microstructure evolution, it would be worth to introduce their results which are believed to be related with the electrostatic interaction of CNPs. Figure 11.14 shows the schematic of their experimental setup.

A 3–5 g sample of 99.9 % pure SiO powder was placed in an alumina crucible and located at the center of the alumina tube. Several strip-like Si wafers (60 mm in length and 10 mm in width) were placed one by one on a long alumina plate (15 cm in length and 20 mm in width) to act as the deposition substrate. SiO powder was heated at 1350 °C for 5 h under a pressure of 300 Torr with H_2–Ar gas mixture kept flowing through the tube at a flow rate of 50 sccm. SiO is dark brown and SiO_2 is transparent white. Products of different colors were formed in the temperature

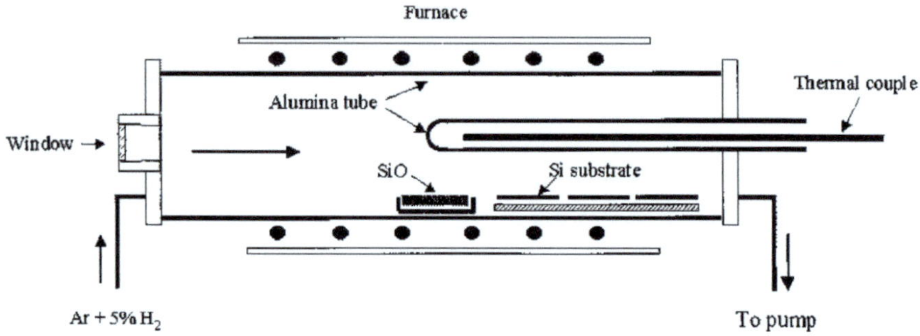

Fig. 11.14 Schematic of the experimental setup for thermal evaporation of SiO powders (Pan et al. 2001)

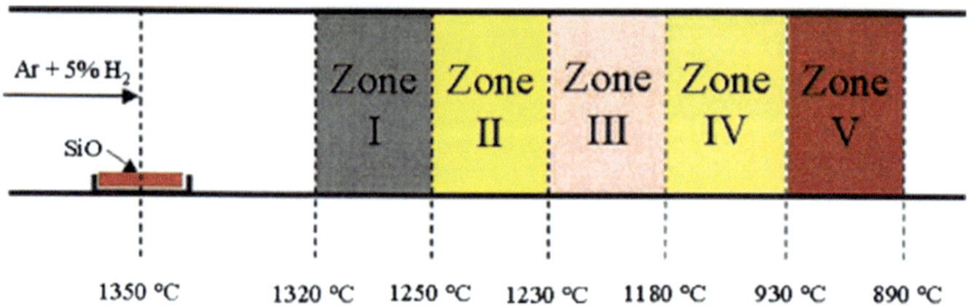

Fig. 11.15 Schematic of the five distinctive growth zones inside the alumina tube. The *colors* of zones I–V were respectively *dark gray*, *bright yellow*, *yellow mixed with pink*, *light yellow* and *brown* (Pan et al. 2001)

range of 890–1320 °C. From the highest temperature of 1320 °C to the lowest temperature of 890 °C, the color changed from dark gray to bright yellow, yellow mixed with pink, light yellow and finally brown. Figure 11.15 schematically depicted the position and the corresponding temperature of the five zones inside the alumina tube.

Various nanostructures were evolved in each zone. Figure 11.16 shows the low-magnification SEM images of the sample in zone I. It shows a large quantity of round-tip rods with 50–200 μm diameters and of 0.5–2 mm length (Fig. 11.16a); octopus-like structure (Fig. 11.16b), pin-like structure (Fig. 11.16c), and nanowires (Fig. 11.16d).

Another worth mentioning is the microstructure evolved in zone III. Products with a color of yellow mixed with pink were formed in zone III. Usually, a few pink stripes and a few yellow stripes, perpendicular to the alumina tube axis, coexist one after the other. Under the naked eye and optical microscopy, these yellow and pink

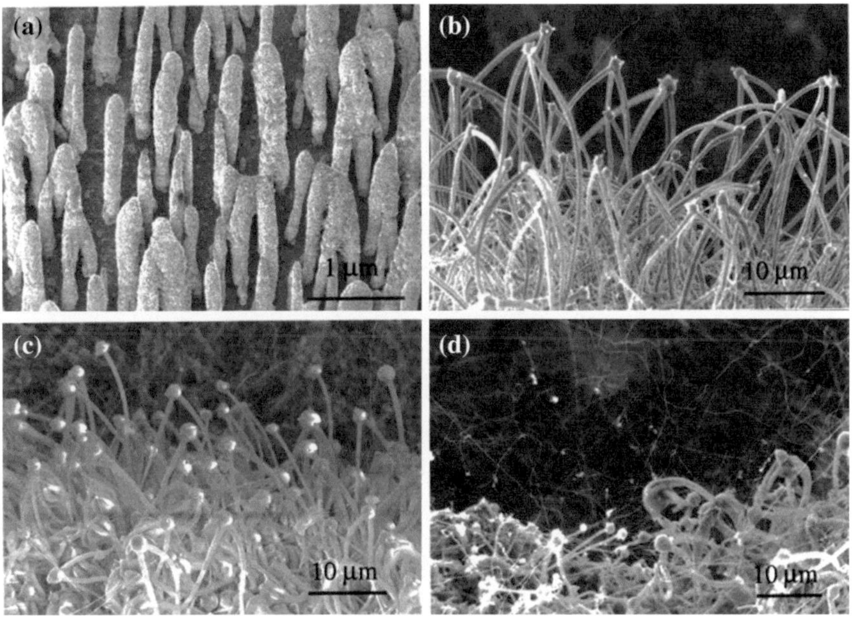

Fig. 11.16 SEM images of the nanomaterials formed in zone I. A large quantity of *dark gray* aligned round-tip rods were formed in this zone (**a**). These rods were mainly composed of octopus-like (**b**), pin-like (**c**), and wire-like (**d**) structures (Pan et al. 2001)

product are powder-like but under SEM, the yellow product mainly consists of fine pin-like structures (Fig. 11.17a), while the pink product mainly consists of tadpole-like structures (Fig. 11.17b).

Figure 11.18a shows a low magnification TEM image of the tadpole-like structure from the pink stripes. It is clear that the tadpole-like structure is actually a chain of pins, in which several pins connect together with the head of one connecting to the tail of another. In Fig. 11.18b, 20 pins connect one by one to form a close circle. Higher magnification TEM examinations show that the particles in the tadpole-like structures are crystalline Si, while the bars between the particles are either amorphous silicon oxide (Fig. 11.18c) or a core-shell structure (Fig. 11.18d).

11.4 Various ZnO Nanostructures

Various ZnO nanostructures could be synthesized by thermal evaporation or carbothermal reductions. These include ZnO nanowires (Yao et al. 2002), nanobelts (Kong and Wang 2004; Kong et al. 2004), nanocoils (Korgel 2004), nanorings

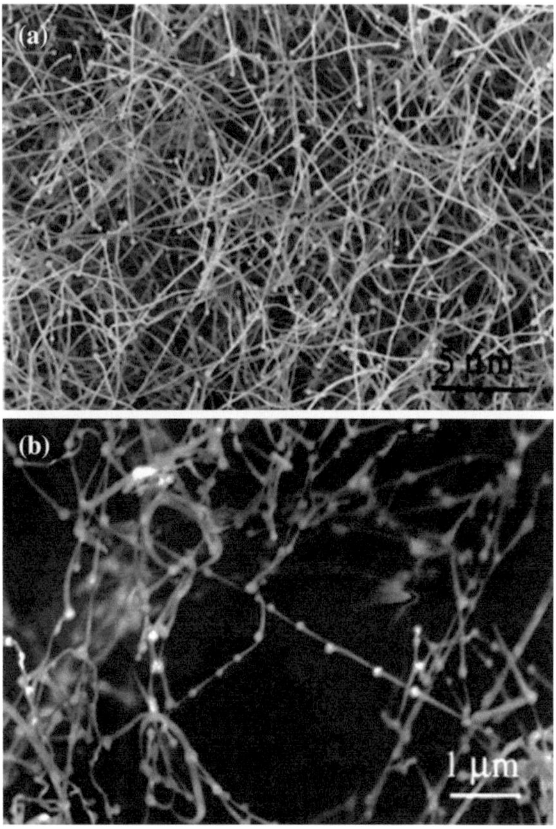

Fig. 11.17 SEM images of the products formed in zone III. The color of the products formed in this zone is yellow mixed with pink. The yellow product is composed of very fine pin-like nanowires (**a**), while the pink product is composed of tadpole-like nanostructures (**b**) (Pan et al. 2001)

(Kong et al. 2004; Kong and Wang 2004), and nanocombs (Wang et al. 2003). Figure 11.19 shows the examples of various ZnO nanostructures (Wang 2006a). As a growth mechanism of ZnO nanostructures, various models such as vapor-liquid-solid (VLS) growth, vapor-solid (VS) growth and redox growth were proposed (Fu et al. 2007). However, many microstructural features such as spontaneous coiling could not be understood from the conventional concept of crystal growth. To explain the growth of the ZnO nanobelts and nanorings, Kong et al. (2004) and Wang (2004) suggested the charging model, where the electrostatic energy plays a critical role in the growth of ZnO nanorings by a spontaneous self-coiling process of polar nanobelts. Their polar charge model is based on the polar nature of ZnO.

Various ZnO shapes shown in Fig. 11.19 strongly imply the involvement of the electrostatic energy in their growth. However, the origin of the electrostatic energy is uncertain. Considering that CNPs were shown to be generated during the

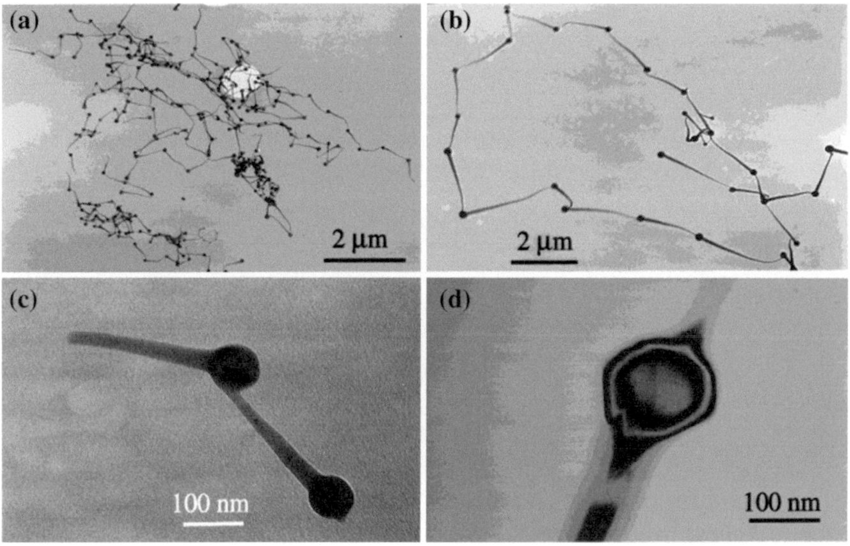

Fig. 11.18 TEM images of the tadpole-like nanostructures (**a** and **b**). The particles in the tadpole-like structures are crystalline Si, while the bars between the particles are either amorphous silicon oxide (**c**) or a Si core sheathed with amorphous silicon oxide layer (**d**) (Pan et al. 2001)

synthesis of ZnO nanostructures in Chap. 9, it is highly probable that the electrostatic energy should come from CNPs.

In conclusion, all the peculiar microstructures shown in this chapter would never be evolved by classical crystallization based on the atomic and molecular growth unit, which is also electrically neutral. Only non-classical crystallization with the building block of CNPs could produce those microstructures. Although very peculiar and unique microstructure evolutions can be made by CNPs, they are synthesized in an almost unpredictable and uncontrollable way. However, the microstructure evolution appears to be reproducible under the similar processing condition. If the technique is developed to control the deposition behavior of CNPs in a predictable and controllable way for example by using the computer simulations, very useful nanostructures such as quantum dots, nanoelectromechanical systems (NEMS), and biomimic materials can be produced. It should be noted that various morphologies are evolved by bio-mineralization, which turns out to occur by non-classical crystallization (Cölfen and Antonietti 2008). In this sense, understanding the electrostatic interaction of CNPs is important not only for the better control and prediction of the thin film growth by CVD and some PVD in fabrications of microelectronic devices and other applications but also for the future technology.

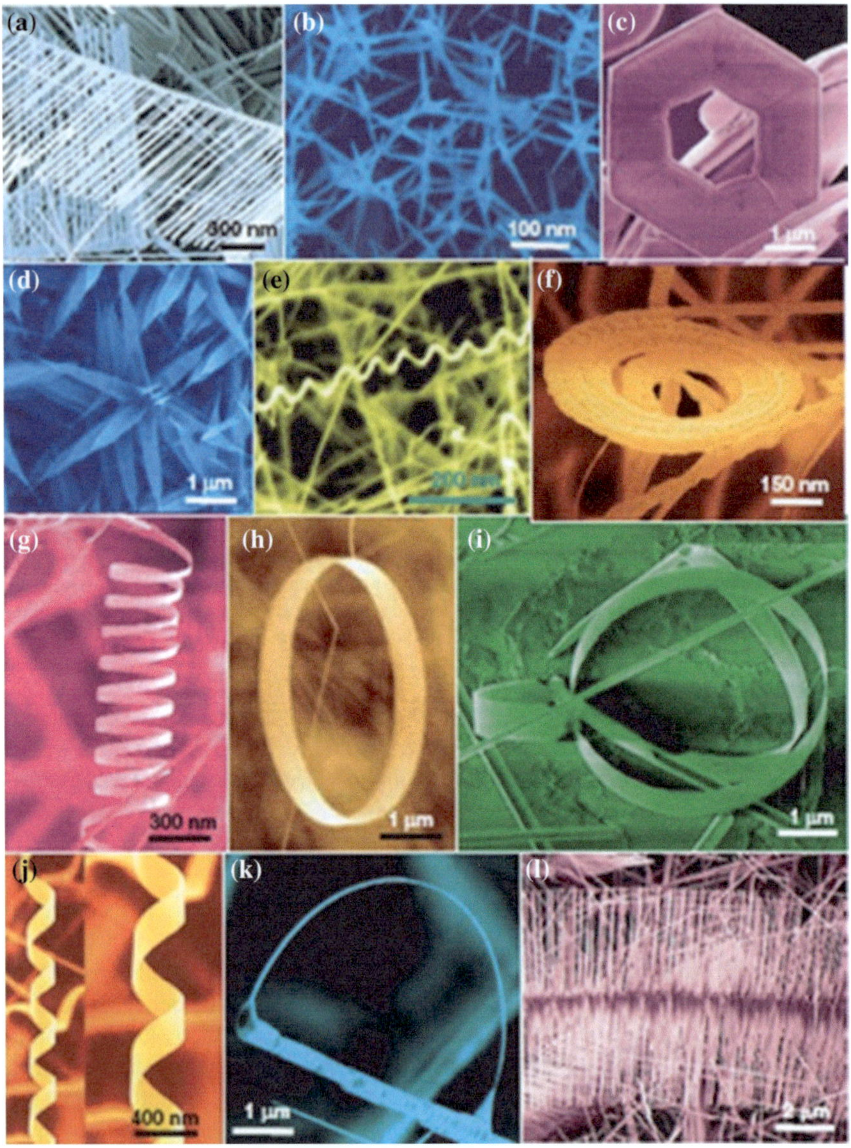

Fig. 11.19 Various ZnO nanostructures synthesized by thermal evaporation of solid powders unless notify otherwise: **a** nanocombs, **b** tetralegs, **c** hexagonal disks/rings synthesized by solution-based chemical synthesis, **d** nanopropellers, **e** deformation-free nanohelixes, **f** spiral of a nanobelt, **g** nanosprings; **h** single-crystal seamless nanoring, **i** a nanoarchitecture composed of a nanorod, nanobow and nanoring, **j** double-sided nanocombs, **k** nanobow structure, and **l** rigid helix (Wang 2006a)

References

Blum J (2006) Dust agglomeration. Adv Phys 55(7–8):881-947. doi:10.1080/00018730601095039
Blum J, Wurm G (2008) The growth mechanisms of macroscopic bodies in protoplanetary disks. Annu Rev Astron Astr 46(1):21–56. doi:10.1146/annurev.astro.46.060407.145152
Cölfen H, Antonietti M (2008) Mesocrystals and nonclassical crystallization. Wiley, p 73–101
Desch SJ, Cuzzi JN (2000) The generation of lightning in the solar nebula. Icarus 143(1):87–105. doi:http://dx.doi.org/10.1006/icar.1999.6245
Fu H, Li H, Jie W, Zhang C (2007) The O_2-dependent growth of ZnO nanowires and their photoluminescence properties. Ceram Int 33(6):1119–1123. doi:http://dx.doi.org/10.1016/j.ceramint.2006.03.024
Ivlev AV, Morfill GE, Konopka U (2002) Coagulation of charged microparticles in neutral gas and charge-induced gel transitions. Phys Rev Lett 89(19):195502
Kong XY, Wang ZL (2004) Polar-surface dominated ZnO nanobelts and the electrostatic energy induced nanohelixes, nanosprings, and nanospirals. Appl Phys Lett 84(6):975–977
Kong XY, Ding Y, Yang R, Wang ZL (2004) Single-crystal nanorings formed by epitaxial self-coiling of polar nanobelts. Science 303(5662):1348–1351
Korgel BA (2004) Self-assembled nanocoils. Science 303(5662):1308–1309
Pan ZW, Dai ZR, Xu L, Lee ST, Wang ZL (2001) Temperature-controlled growth of silicon-based nanostructures by thermal evaporation of SiO powders. J Phys Chem B 105(13):2507–2514. doi:10.1021/jp004253q
Poppe T, Schräpler R (2005) Further experiments on collisional tribocharging of cosmic grains. A&A 438(1):1–9
Poppe T, Blum J, Henning T (2000) Experiments on collisional grain charging of micron-sized preplanetary dust. Astrophys J 533(1):472
Swain B, Park J-W, Yang S-M, Mahmood K, Swain B, Lee J-G, Hwang N-M (2015) Alignment of nanoparticles, nanorods, and nanowires during chemical vapor deposition of silicon. Appl Phys A 120(3):889–895. doi:10.1007/s00339-015-9310-1
Wagner R, Ellis W (1964) Vapor-liquid-solid mechanism of single crystal growth. Appl Phys Lett 4(5):89–90
Wang ZL (2004) Zinc oxide nanostructures: growth, properties and applications. J Phys: Condens Matter 16(25):829-858
Wang ZL (2006a) Chapter 10 - Novel nanostructures and nanodevices of ZnO. In: Jagadish C, Pearton S (eds) Zinc oxide bulk, Thin films and nanostructures. Elsevier, Oxford, pp 339–370. doi:http://dx.doi.org/10.1016/B978-008044722-3/50010-5
Wang ZL (2006b) Novel nanostructures and nanodevices of ZnO. Amsterdam, Elsevier,
Wang ZL, Kong XY, Zuo JM (2003) Induced growth of asymmetric nanocantilever arrays on polar surfaces. Phys Rev Lett 91(18):185502.
Yao BD, Chan YF, Wang N (2002) Formation of ZnO nanostructures by a simple way of thermal evaporation. Appl Phys Lett 81(4):757.
Youn W-K, Lee S-S, Lee J-Y, Kim C-S, Hwang N-M, Iijima S (2014) Comparison of the deposition behavior of charged silicon nanoparticles between floating and grounded substrates. J Phys Chem C 118(22):11946–11953. doi:10.1021/jp5001144
Zhang RQ, Lifshitz Y, Lee ST (2003) Oxide-assisted growth of semiconducting nanowires. Adv Mater 15(7–8):635-640. doi:10.1002/adma.200301641.

Chapter 12
Bias Effect on Deposition Behaviour of Charged Nanoparticles

If CNPs are generated in the CVD process, it is expected that the deposition behaviour would be affected by the electric field, which can be achieved by applying the bias to the substrate. In the bias experiment, a grounded conducting plate is normally made to face the substrate to define the electric field, which is the voltage difference between the two conducting electrodes divided by the distance. Positive bias will repel positive CNPs and attract negative CNPs; negative bias vice versa. Besides, the bias is expected to increase the deposition rate because it will impose the directional motion on CNPs. In some cases, the bias effect is pronounced in increasing the deposition rate whereas in other cases, the bias effect is not so pronounced.

12.1 Dependence of Bias Effect on the Conductivity of Depositing Materials and Substrates

The bias effect on the deposition rate depends on whether the depositing material (CNP) is a conductor, semiconductor or insulator and also on whether the substrate is a conductor, semiconductor or insulator. Sometimes, the bias effect also depends on whether the conducting or semiconducting substrate is floating or grounded. Even when the substrate is an insulator, the bias effect as well as the deposition behaviour depends on the thickness of the insulating substrate on the conducting holder. It should be noted that in case of an insulating substrate, the bias is applied to a conducting substrate holder. In this case, the electric field is diminished by the insulating substrate and thereby the bias effect is also diminished.

The bias will increase the deposition rate most pronouncedly when both substrate and depositing material are conducting with the substrate being grounded. The bias effect will be least pronounced when both substrate and depositing material are insulating. Considering that a charge transfer rate (CTR) of materials is

© Springer Science+Business Media Dordrecht 2016

N.M. Hwang, *Non-Classical Crystallization of Thin Films and Nanostructures in CVD and PVD Processes*, Springer Series in Surface Sciences 60,
DOI 10.1007/978-94-017-7616-5_12

also important in deposition behaviour of CNPs, metallic CNPs with a high CTR would have a self-catalytic effect in deposition. Therefore, metallic CNPs with a high CTR on a grounded metallic substrate with a high CTR would be a combination to produce the highest deposition rate. In this case, it should be worried that a porous structure might be evolved.

When a conducting material is deposited on an insulating substrate, the electric charge will build up on the substrate surface, producing some electrostatic effect. Since the image force for incoming CNPs is stronger on the conducting surface than that on the insulating surface, CNPs tend to land on the pre-deposited conducting area, resulting in non-uniform deposition. This is a kind of selective deposition. Once the insulating surface is entirely covered with the conducting film, the further deposition behaviour would be like that on the floating conducting substrate.

If a depositing material or CNP is insulating and a substrate is conducting, CNPs would deposit more easily than on an insulating substrate. This aspect was observed in the diamond CVD process, where diamond deposition is more difficult on insulating substrates than on conducting or semiconducting substrates. In some cases, however, such as diamond and GaN, CNPs have difficulty in landing even on conducting substrates. CNPs of diamond and GaN deposit more easily on conducting substrates with a high CTR than that with a low CTR. Once the conducting substrate is entirely covered with an insulating film, the charge will build up on the growing surface, which will cancel the applied bias. As a result, the bias effect will be diminished pronouncedly.

The effect will also be affected by the reactor pressure. When the pressure is high, the mean free path of atoms and molecules will be short and because of the collisions between atoms and molecules, which will exert a drag force, the bias effect will be diminished. When the pressure is low, the vice versa. The bias effect will also depend on the size of CNPs and the relative concentration of positive and negative ions and CNPs.

The effect of bias was most extensively studied in the Si CVD from the viewpoint of the TCN. The bias effect turned out to be not so simple.

12.2 Bias Effect on Deposition Behavior of Si on Conducting, Insulating and Semiconducting Substrates

Lee and Hwang (2008) applied electric biases of +150, 0, and −150 V to the substrates of stainless steel and polyethylene terephthalate (PET) maintained at ~90 °C during silicon hot wire CVD (HWCVD) under 0.6 Torr at a filament temperature of 1700 °C for a deposition time for 30 min with gas mixtures of 5–20 % SiH_4 and 80–95 % H_2. Either a stainless steel or a PET substrate was placed 15 cm below the wire.

In order to produce the uniform electric field from the bias, the electrode is designed as a metal plate of twice larger surface area than the substrate. For each deposition, two substrates of the identical material and size were placed in the chamber with one substrate biased and the other unbiased in order to compare the deposition behaviour between biased and unbiased substrates under the same processing condition. The mass increase of the substrate after deposition was measured to determine the growth rate of the films, using a microbalance with a scale in 10^{-5} g.

Each sample was weighed 5 times consecutively and the averaged value with a standard deviation less than 3×10^{-5} g was obtained. In order to rule out the possible variation from experiment to experiment in comparing the deposition rate or the weight increase between biased and unbiased substrates, both biased and unbiased substrates were placed in the same chamber and the deposition was done simultaneously under the same processing conditions. The mass increase or decrease of the biased sample after deposition was obtained by subtracting the mass increase of the unbiased sample from the mass increase of the biased sample. Therefore, the obtained mass data can be expressed by an equation, $M(\text{mass}) = (M_{ba} - M_{bb}) - (M_{ua} - M_{ub})$, where M_{ba} and M_{bb} are respectively the mass of the biased sample after and before deposition and M_{ua} and M_{ub} are respectively the mass of the unbiased sample after and before deposition.

Figure 12.1 shows the mass difference of silicon films between biased and unbiased stainless steel substrates. For 20 % SiH_4, the mass was positive both for positively and negatively biased substrates, indicating that the mass of the films deposited on a biased substrate was higher than that on an unbiased substrate. In this case, the mass increase was higher for +150 V than for −150 V. For 10 % SiH_4, however, the mass was positive for a positively biased substrate but negative for a negatively biased substrate, indicating that the growth rate of the films on a

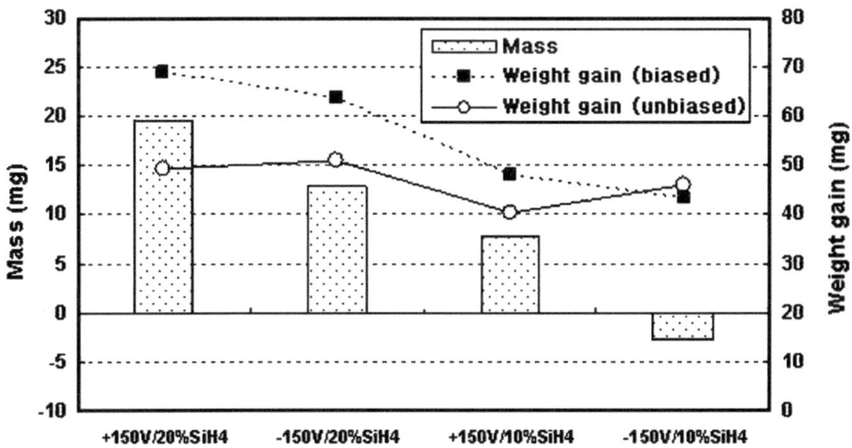

Fig. 12.1 Masses of silicon thin films deposited on a stainless steel substrate with biases applying to the substrate or the conducting substrate holder (Lee and Hwang 2008)

negatively biased substrate was lower than that on an unbiased substrate. For the positive bias of +150 V, the mass increase was higher for 20 % SiH_4 than for 10 % SiH_4.

The previous current measurements during silicon HWCVD showed that the negative CNP were dominant (Kim et al. 2006). Therefore, the mass decrease for the negative bias for 10 % SiH_4 can be explained by the fact that the electric field induced by the negative bias repels the incoming negative CNPs. The mass increase for the positive bias for 20 and 10 % SiH_4 can be explained similarly by the electric field attracting the negative CNPs. However, the mass increase for the negative bias for 20 % SiH_4 cannot be explained by this logic. One possibility is that once the surface of the stainless steel substrate is completely covered with silicon, the surface would not be conducting like stainless steel anymore, and then the bias effect will be changed. Another possibility is that if the amount of positive CNPs is large enough for 20 % SiH_4, the electric field induced by the negative bias will attract the positive CNPs to the growing surface, leading to the deposition rate higher than that for the unbiased substrate.

Figure 12.2 shows the mass difference of silicon thin films between biased and unbiased PET substrates. Since PET is insulating, the bias was applied to a copper plate under the substrate. The bias effect of Fig. 12.2 is quite different from that of Fig. 12.1. In Fig. 12.2, the mass was negative for the positive bias, indicating that the growth rate of the films deposited on a positively biased PET substrate was lower than that on an unbiased PET substrate. However, the mass was positive for the negative bias, indicating that the growth rate of the films deposited on a negatively biased PET substrate was higher than that on an unbiased PET substrate.

This different behavior of the bias effect between Figs. 12.1 and 12.2 would come from the conducting and insulating property of the substrates. The electric charge carried by the nanoparticles would not be discharged on the insulating PET

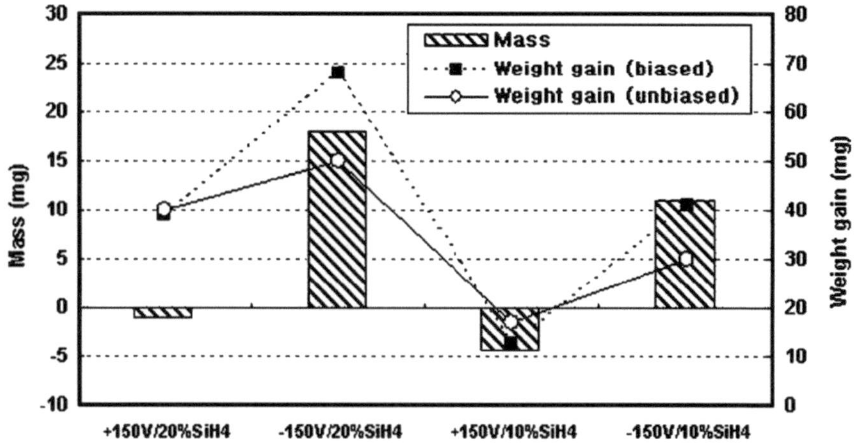

Fig. 12.2 Masses of silicon thin films deposited on a PET substrate with applying biases to the substrate or the conducting substrate holder (Lee and Hwang 2008)

substrate. Note that the negative CNPs are dominantly generated in the gas phase over positive CNPs. In an unbiased substrate, the surface charge will be negative and can be discharged by the positive charge carriers in the gas phase. In this case, the supply of the positive charge carrier to the growing surface would be a rate determining step of the film growth. For the positively biased PET substrate, the negative charge will be attracted to the surface. This negative charge would repel the incoming negative CNPs, which are the dominant flux, thereby decreasing the growth rate compared with that of the unbiased substrate. For the negatively biased PET substrate, the positive charge will be attracted to the surface. This positive charge would attract the incoming negative CNPs, thereby increasing the growth rate compared with that of the unbiased substrate. In other words, the negative bias increases the film growth rate by collecting the positive charge carriers on the surface, which are the rate-determining steps.

In order to further investigate the deposition behavior of charged nanoparticles through the bias effect, the biases of −150, 0, and +150 V were applied to a copper mesh which was placed 2 mm over the PET substrate. Figure 12.3 shows the mass difference on a PET substrates between biased and unbiased cases. For the four samples, the mass of silicon films deposited under the bias was decreased compared with the unbiased condition. The mass decrease was higher for +150 V than for −150 V and for 20 % SiH_4 than for 10 % SiH_4. For the bias applied on the copper mesh, the CNPs will be attracted to the mesh and become depleted near the PET substrate, resulting in the decrease of the growth rate. This effect will be higher for the positive bias since the positive bias would attract the negative CNPs, which are dominant over the positive CNPs. Similar effects on the deposition in the condition of surface electrical charges were reported (Levchenko and Ostrikov 2007; Qijin and Xu 2007).

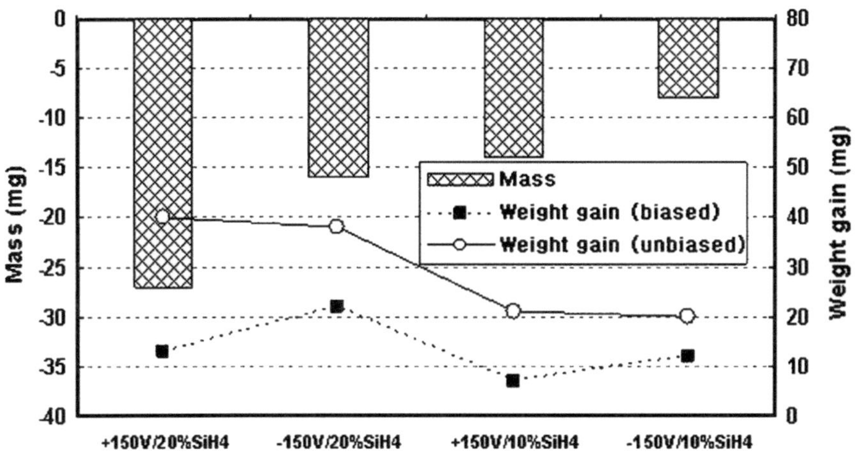

Fig. 12.3 Masses of silicon thin films deposited on a PET substrate with biases applying to the copper mesh 2 mm above the substrate (Lee and Hwang 2008)

Lee et al. (2008) investigated the bias effect of the deposition behaviour on the insulating glass substrate. The stainless steel substrate holder had a diameter of 10 cm, and the size of a glass substrate (Corning 1737 of 1 mm thickness) was $2.5 \times 2.5\ \text{cm}^2$. The substrate heater was placed behind the substrate holder to control the substrate temperature. Silicon films were deposited on the glass substrate under 0.1 Torr at the filament temperature of 1700 °C. The distance between the filament and the substrate was 12 cm. With gases flowing and filament turned on, the substrate temperature was controlled to stabilize at 300 °C during deposition using the controller with a thermocouple. The gas flow ratio of $R_H = [H_2]/([SiH_4] + [H_2])$ was 0.95 and the total gas flow rate was 36 sccm. The DC electric bias voltages of −200, 0 and +200 V were applied on the substrate holder.

The film growth rate was evaluated from the thickness measurements of cross-sectional observation of films using FESEM (JEOL, JSM-6330F). The surface roughness and the morphology of films were observed by FESEM.

Figure 12.4a–c shows the FESEM images of the cross-sections of silicon films deposited at the biases of −200, 0 and +200 V on the stainless substrate holder, respectively, at the filament temperature of 1700 °C and the substrate temperature of 300 °C under 0.1 Torr for 10 min. The film thicknesses at the biases of −200, 0 and +200 V were respectively 384, 284 and 250 nm. The growth rate of silicon films was changed drastically by applying an electric bias on the substrate holder. The measured growth rates at the biases of −200, 0 and +200 V were respectively 6.4, 4.7 and 4.1 Å/s. The silicon film grew much faster at the negative bias than at zero and positive biases. The growth rate under the positive bias decreased slightly compared with that under the zero bias.

The dependence of the growth rate on the applied bias is consistent with Figs. 12.1 and 12.2 reported by Lee and Hwang (2008): the negative bias on the insulating polyethylene terephthalate (PET) substrate showed a higher growth rate than the positive and the zero biases whereas on the conducting stainless steel substrate the positive bias showed a higher growth rate than the negative and the zero biases.

Figure 12.4 can be explained similarly based on the previous report that negative CNPs are more dominant than positive CNPs during HWCVD. If negative CNPs, which existed dominantly in the gas phase, deposited on the glass substrate, the negative electric charge would build up on the insulating glass substrate and then the additional deposition of negatively charged nanoparticles would become difficult due to the repelling electric force. Then, positive-charge carriers including nanoparticles, which existed in a minor amount in the gas phase, would be attracted to the negatively charged glass surface by electric attraction and start to discharge the negative charge built up on the glass. Under the condition where negatively charged nanoparticles are dominant in the gas phase, the rate-determining step for deposition would be the supply of positive charge on the insulating surface.

Therefore, the negative bias, which will collect the positive charge on the glass substrate, would increase the growth rate compared with the zero and the positive bias. Similarly, the positive bias, which would retard the supply of positive charge carriers, would have the lowest growth rate.

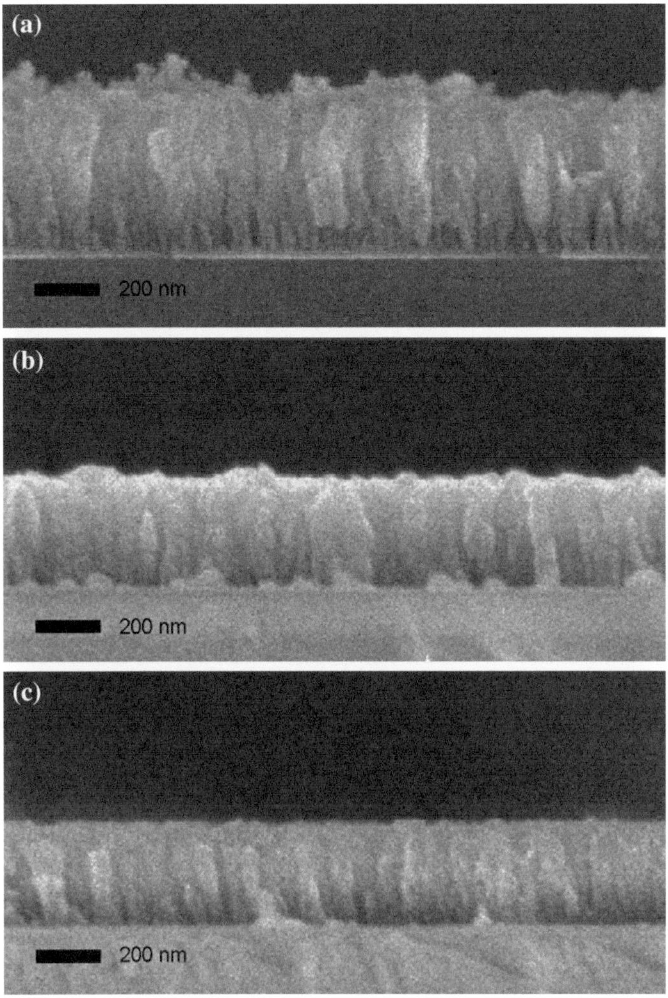

Fig. 12.4 Cross-sectional FESEM images of silicon films deposited at electric biases of **a** −200, **b** 0 and **c** +200 V (Lee et al. 2008)

Figure 12.5a–c shows the FESEM images of the surface of silicon films respectively, for the specimens of Fig. 12.4a–c. Quite different surface morphologies were evolved under different electric biases. At the negative bias, the surface of the film had a very porous skeletal structure as shown in Fig. 12.5a. At the zero bias, the surface of the film had a relatively compact structure as shown in Fig. 12.5b. At the positive bias, the surface of the film had the smoothest structure and the grain size distribution as well as the film thickness was rather uniform as shown in Fig. 12.5c.

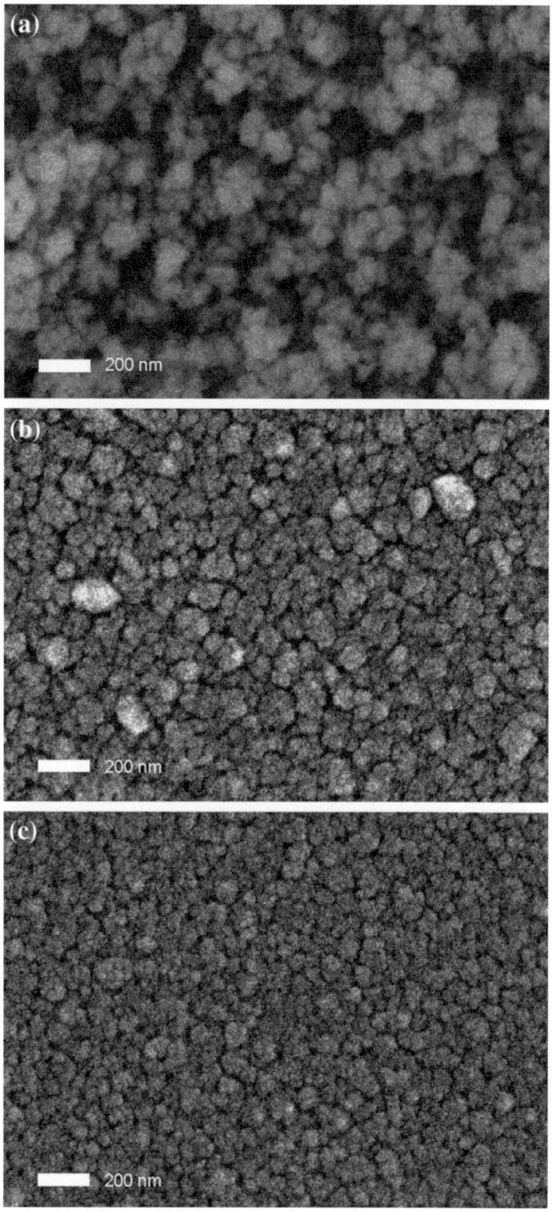

Fig. 12.5 FESEM images of the surface of silicon films deposited at electric biases of **a** −200, **b** 0 and **c** +200 V (Lee et al. 2008)

12.3 Comparison of Charge Generation from Hot Wires Between AC and DC Power Supply

Park et al. (2009) showed that the electric field around the hot wires heated by the DC power supply interacted with the charged carriers, which affected the film quality and the deposition rate. However, the electric field around the hot wires heated by the AC power would be different from that by the DC power. This difference might be related with the different deposition behavior between DC and AC powered HWCVD systems. Park et al. (2010) compared the generation behavior of charged carriers between DC and AC powered HWCVD systems, where hot wires produce a different electric field.

A gas mixture of 10 sccm of hydrogen-diluted silane of 30 % SiH_4–70 % H_2 and 50 sccm of H_2 was supplied. The total percentage of hydrogen was 0.95 %. The reactor pressures were 0.3 Torr. Figure 12.6 shows measured currents for AC and DC powers with varying wire temperature, when the electrodes were placed 1.5 cm below the input and output positions. Two facts should be noticed. First, the currents measured for the AC power were far more negative than those for the DC power at both input and output positions. For example, when the wire temperature was 1800 °C, the currents measured 1.5 cm below the input position were -3.65×10^4 and -0.11 nA/cm^2, respectively, for AC and DC powers. Second, for the AC power, the current below the input position was more negative than that below the output position as shown in Fig. 12.6.

This is opposite to the case for the DC power. Although the current below the input position was more positive than that below the output position for the DC power, the difference was so small compared with that for the AC power that it is not distinguishable in Fig. 12.6.

The previous explanation as to why the current below the input position was more positive for the DC power was that the electric field generated around the input position of the wire at 15 V would repel positively charged species and attract negatively charged species. However, the electric field around the output position at

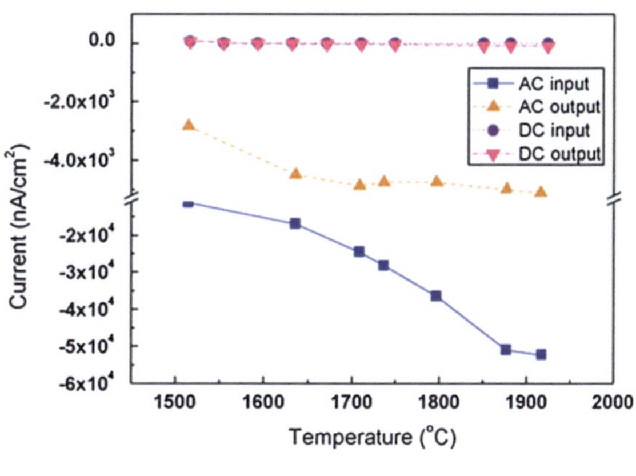

Fig. 12.6 Temperature dependence of current measured 1.5 cm below the input and output positions of the hot wire for AC and DC powers (Park et al. 2010)

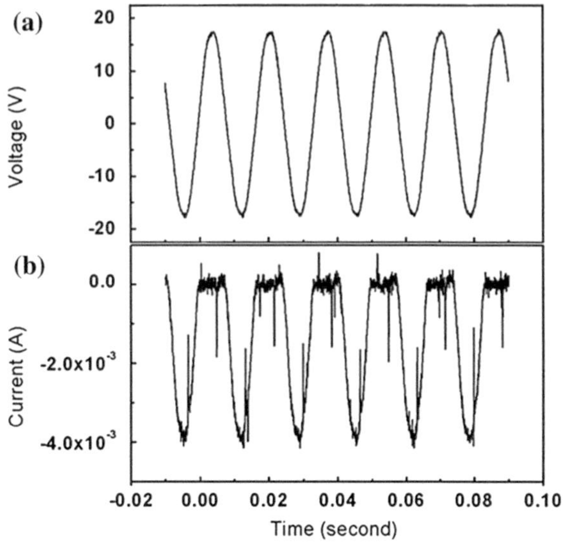

Fig. 12.7 **a** Voltage oscillation with time measured by an oscilloscope at the input position of the hot wire under the AC power and **b** current oscillation with time measured by an oscilloscope on the electrode placed 1.5 cm below the input position of the hot wire (Park et al. 2010)

0 V would have a negligible effect on the movement of positively charged or negatively charged carriers.

The current generation behavior for the AC power, which is quite different from that for the DC power as shown in Fig. 12.6, may be also explained by considering the electric field around the hot wire. However, the AC power generates an oscillating electric field, which makes the situation a little bit more complicated. Figure 12.7a, b, respectively, shows the voltage of the AC power and the current measured on the electrode by an oscilloscope at the input position. In Fig. 12.7a, the AC voltage oscillates at 60 Hz in the range of −18 to +18 V. At the same time, the current in Fig. 12.7b also oscillates at 60 Hz as the AC voltage does. It is obvious that the current is affected by the oscillating electric field of the AC power. However, the amplitude of the current is not symmetric, oscillating between 0 and -4×10^{-3} μA although the AC voltage oscillates symmetrically with 0 V as a center.

This result can be explained by the unbalanced generation of positive and negative charges from the hot wire. A relatively large amount of negative charges or electrons would be generated by thermal emission from the hot wire but a much less amount of positive charges would be generated by the positive surface ionization on the hot wire (Zhang 1999). If the silicon clusters, which may or may not be hydrogenated, are formed, they would have relatively low ionization potentials and high electron affinity, both of which approach the work function of silicon as the cluster size increases.

Therefore, the positive and negative surface ionization of a silicon cluster on the hot wires would have much lower activation energy than those of a silicon atom. When the hot wire exerts a positive potential, it will attract the negative charges and repel the positive ones. And the positive current measured on the electrode would be small because the amount of positive charges generated by positive surface ionization would be small. When the hot wire exerts a negative potential, however, it will attract the positive charges and repel the negative ones. The measured negative current would be large because the amount of negative charges generated by the thermal emission would be large. Another factor that is responsible for the non-symmetric oscillation of the current in Fig. 12.7b would be the difference in the mobility between positive and negative charges. Most of the negative charges would be electrons, which must have much higher mobility than positive charges, which would be molecular or cluster ions. Electrons would respond to the given electric field much faster than molecular and cluster ions. This analysis explains why the current for the AC power in Fig. 12.6, which is the average value of such oscillating current as shown in Fig. 12.7b, is much more negative than that for the DC power. This analysis also explains why the current is more negative below the input position than below the output position of the hot wire for the AC power, considering that the AC voltage is oscillating in the range of −18 to +18 V at the input position of the hot wire whereas the voltage is zero at the output position.

Si thin films deposited by HWCVD were compared between the samples prepared by AC and DC powered hot wires. Figure 12.8 shows Raman spectra of thin films deposited 1.5 cm below the wire at the input and output positions at the wire temperature of 1800 °C. The spectra from top to bottom in the order of increasing crystallinity are for DC output, DC input, AC input, and AC output positions. To determine the volume fraction of crystalline silicon in the films, the Raman spectra were decomposed into three peaks for a monocrystalline region centered at about 520 cm^{-1}, a polysilicon region around 505–517 cm^{-1}, and an amorphous region around 480 cm^{-1} (Chung et al. 2009). The crystalline volume fraction of the films was determined to be 0.517, 0.532, 0.541, and 0.548, respectively, for DC output, DC input, AC input, and AC output by using $I_c/(I_c + I_a)$, where I_c and I_a represent the integrated intensity of crystalline and amorphous peaks, respectively (Chung et al. 2009). Thin films deposited under the AC power had higher crystalline fraction than those under the DC power. This result is consistent with the result in

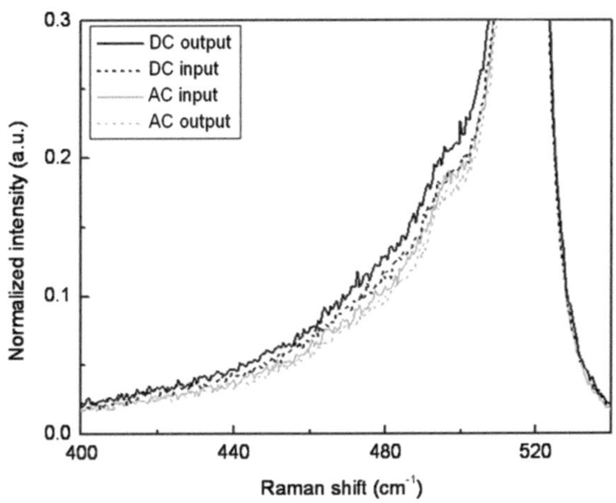

Fig. 12.8 Raman spectra of films, which were deposited under DC and AC powers and 1.5 cm below input and output positions of the hot wire (Park et al. 2010)

the previous section for Si HWCVD under the DC power that the crystalline fraction was increased under the deposition condition where more negative current was measured. Because of high mobility and small cross-section of electrons, a significant percentage of negative charges appears to be an electron although its exact percentage has not been determined. However, it is certain that the amount of negatively charged clusters (or nanoparticles) would increase with increasing amount of negative charges. Therefore, the amount of negative charges affects the amount of negative CNPs, which actually affects the film growth and the film quality. On the other hand, the difference in crystallinity of thin films between input and output positions is smaller under AC power than that under DC power. Therefore, it can be said that the AC power in the Si HWCVD process would produce higher crystalline thin films than the DC power.

12.3.1 Comparison of the Effect of Wire Bias Between AC and DC Powers

When the additional wire bias was applied to the output position of the wire, the measured currents were changed. Figure 12.9 shows measured currents 1.5 cm below the input and output positions of the wire when AC and DC powers were used with the additional wire bias applied from −20 to +40 V with respect to the ground. As expected, the current under AC power at the input position showed the highest negative value. The AC voltage with the negative wire bias oscillates with the applied negative wire bias as the center.

For example, when the wire bias is −25 V with its root mean square value of AC voltage being 15 V, the AC voltage oscillates at the center of −25 V, in the range of −46 and −4 V. When the wire bias is −25 V for the DC power, the lowest voltage is −25 V at the output position with −10 V at the input position due to the potential

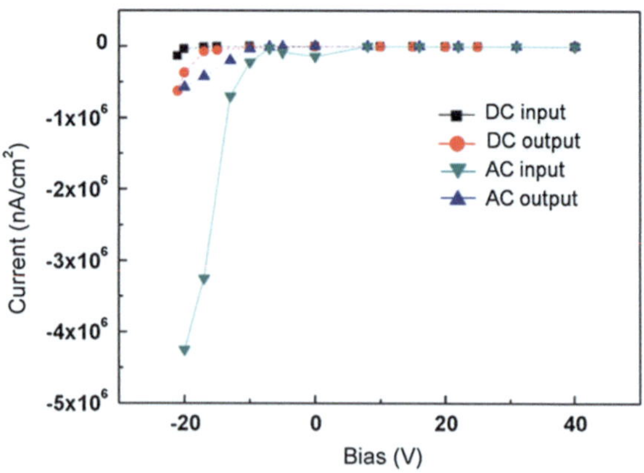

Fig. 12.9 Dependence of the current on the additional wire bias for AC and DC powers. The current was measured at 1.5 cm below the input and output positions of the hot wire (Park et al. 2010)

difference of 15 V between input and output positions. Therefore, although the same magnitude of the wire bias is applied, more negative current would be measured on the substrates for hot wires heated by AC power than that by DC power.

12.4 Non-uniform Deposition Using AC Power Supply in Si HWCVD

In the previous section, the generation behaviour of charged species was quite different between DC and AC powered HWCVD systems. Considering that CNPs can be a major building block for thin film growth, the deposition behaviour would be different between DC and AC powered HWCVD systems. In relation to this possibility, Song et al. (2007), who used an AC powered HWCVD system, observed that the silicon deposition behaviour on the glass substrate was quite different from a DC powered HWCVD system with respect to the uniformity of deposition.

In order to establish uniform gas flow, a cylindrical gas showerhead with a diameter of 12 cm was placed beneath 4 helical tungsten wires (99.99 %). Each wire was 8 cm long with a diameter of 0.3 mm. The substrate holder, made of stainless steel, had a diameter of 10 cm, and the substrate size was 2.5×2.5 cm^2. The substrates used in the experiments were Corning 1739 glass. On the Corning 1739 glass substrate, 40 nm-thick silicon oxide was deposited by plasma enhanced CVD (PECVD). The deposition was performed under 0.5 Torr at the substrate temperature of 450 °C with the distance of 6.5 cm between wires and substrate. A 20 % SiH_4–80 % H_2 gas mixture was supplied at 30 sccm. Since the deposition is relatively uniform on a silicon substrate, the deposition on the glass substrate will be shown.

Figure 12.10 are photographs of film surfaces that were deposited on the glass on which 40 nm thick silicon oxide was deposited. Noticeably, the deposition behavior depended on the wire temperature.

Although the thickness of the film deposited at the wire temperature of 1590 °C (Fig. 12.10a) was uniform, at the wire temperatures of 1670 °C (Fig. 12.10b) and 1800 °C (Fig. 12.10c), the thickness was not uniform between the center, the edge and the corner. The thickness of the center, the edge and the corner at the wire temperature of 1670 °C (Fig. 12.10b) was measured as 70, 125, and 157 nm, respectively, which indicated that the deposition rate was highest at the corner and lowest at the center, with the edge being medium. At 1800 °C (Fig. 12.10c), the silicon was deposited mainly at the corner and hardly deposited at the center.

The non-uniform deposition shown in Fig. 12.10 would come from the charge build-up on the insulating glass substrate. The dependence of the non-uniform deposition on the wire temperatures would be related to the relative fraction of the negatively-charged depositing flux; the amount of the negatively charged flux would increase with increasing wire temperature, which is indirectly confirmed by the observation that the amount of negative current increased with increasing wire temperature (Kim et al. 2006). Therefore, the fraction of the negatively charged flux

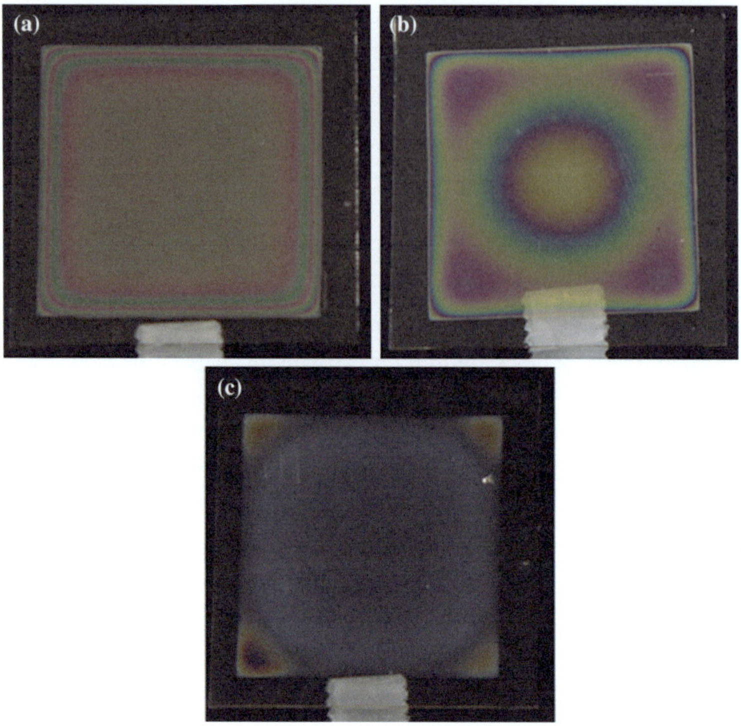

Fig. 12.10 Photographs of silicon films deposited on a glass for 4 min at the wire temperatures of **a** 1590 °C **b** 1670 °C and **c** 1800 °C. The non-uniform deposition becomes more pronounced with increasing wire temperature (Song et al. 2007)

would be maximal at the wire temperature of 1800 °C, in which case the center of the substrate was hardly deposited.

Once the entire insulating glass is deposited by silicon, which is a semiconductor, the surface conductivity increases. Then, the deposition rate would become uniform because the charge build up would be decreased. This effect is confirmed in Fig. 12.11, which show the deposition behavior on an insulating glass, at the wire temperature of 1670 °C, according to deposition times ranging from 2 to 6 min. Up to 4 min (Fig. 12.11a, b, c), the non-uniformity in thickness was pronounced, but with increasing time and further deposition, the film thickness became relatively uniform (Fig. 12.11d, e). Up to 4 min, the glass substrate was not entirely deposited by silicon, resulting in non-uniform deposition, and after 4 min, the glass substrate was entirely deposited by silicon, resulting in relatively-uniform deposition.

Between the center and the corner of a glass substrate, not only the deposition rate but also the surface roughness was different. After deposition for 4 min at the wire temperature of 1670 °C, the roughness at the center and at the corner was 10.6 and 7.5 nm, respectively. The roughness was higher at the center than at the corner whereas the deposition rate was lower at the center.

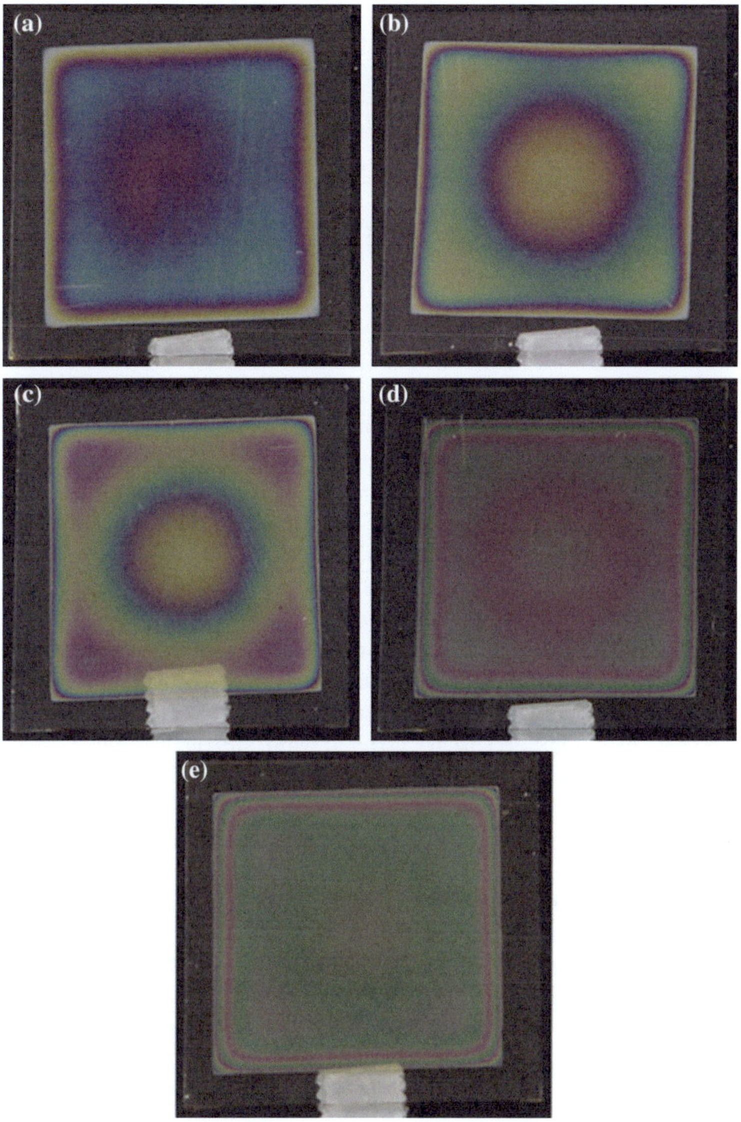

Fig. 12.11 Photographs of silicon films deposited on a glass at the wire temperature of 1670 °C for the deposition times of **a** 2 min, **b** 3 min, **c** 4 min, **d** 5 min, and **e** 6 min. The non-uniform deposition was maintained up to 4 min and then tended to disappear as the silicon films grew (Kim et al. 2006)

In Fig. 12.12, the roughness, with varying deposition times and at the wire temperature of 1670 °C, is compared between the center and the corner. For all of the deposition times, the roughness at the center was higher than that at the corner. The initial roughness (0 min) corresponds to the surface roughness of as-deposited

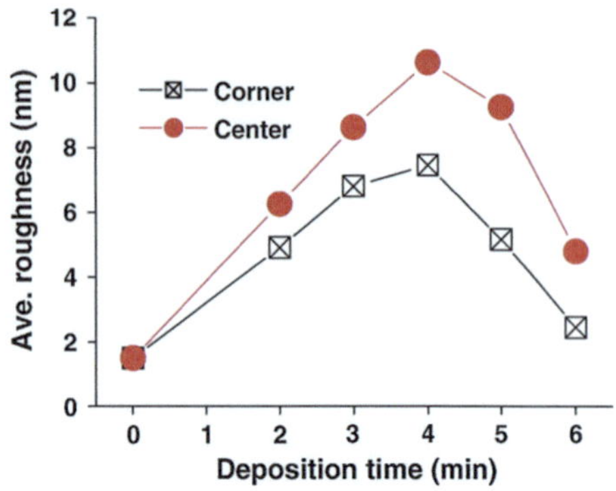

Fig. 12.12 The variation of average surface roughness of silicon films with deposition time at the center and at the corner at the wire temperature of 1670 °C. The roughness at 0 min is for deposited silicon oxide on a glass substrate (Kim et al. 2006)

silicon oxide on a glass. The roughness initially increased with deposition times and then decreased after reaching the maximum at 4 min both at the center and at the corner. The roughness variation with deposition time had some correlation with the non-uniformity in film thickness; up to 4 min, the non-uniformity between the center and the corner was pronounced, but tended to disappear after 4 min (Fig. 12.11a–e). In Fig. 12.12, the roughness was higher at the center, where the film growth rate was lower than at the corner.

Comparing the tendency of the roughness variation with that of the thickness non-uniformity, it seems that the roughness tended to be higher where a higher charge was accumulated. It should be noted that after 4 min of deposition, when the glass substrate was expected to be covered with silicon and the charge to be accumulated much less because of the relatively easy charge removal of the semi-conducting silicon film, the roughness decreased.

One possible explanation for the roughness difference between the center and the corner and between the deposition times is that when the charge builds up, the charged flux is repelled and mainly the neutral flux is deposited. If a neutral flux consists mainly of clusters or nanoparticles instead of atoms or molecules, they tend to land on the surface in a random way by Brownian coagulation, which increases the roughness, whereas the CNPs tend to land in a highly regular, self-assembly way, which decreases the roughness. This might be related with the previous observation that during diamond HWCVD porous and skeletal soot of high roughness was evolved on an iron substrate by the neutralization of negative CNPs just before landing whereas smooth diamond crystal was evolved on a silicon substrate by the self-assembly of CNPs.

Although the results in this study are for the substrate of $2.5 \times 2.5\ \text{cm}^2$, the non-uniformity was more pronounced when the substrate of $6 \times 6\ \text{cm}^2$ was used. The non-uniformity would increase with the size of the substrate. Non-uniformity in deposition can be a big issue for a large area deposition by HWCVD.

12.5 Effect of Alternating Bias on Deposition in Si CVD

So far the DC bias was applied to the substrate. When the depositing material is not conducting but insulating or semiconducting, the electric field made by the DC bias tends to be diminished with deposition, because the charge would build up on the growing surface. For example, When + 100 V is applied to the substrate, negative charges will build up on the growing surface, which will diminish the electric field. As a result, the bias effect would not be so effective.

If alternating biases are applied to the substrate with a proper frequency, the charge build-up would not occur; then the bias effect would continue. Motivated by this idea, Youn et al. (2012) examined the alternating current (AC) bias effect on the microstructure evolution and the growth rate. Applying the AC bias turned out to be a new processing parameter in the thermal CVD process where charged nanoparticles are generated.

Silicon films were deposited on a quartz substrate by conventional atmospheric-pressure CVD using a quartz tube reactor. The quartz substrate of $10 \times 10 \times 1$ mm^3 was on a stainless substrate holder plate of $12 \times 12 \times 6$ mm^3 to apply the bias. The substrate was placed at the center zone of the quartz tube. Another electrode of a stainless plate was placed above the substrate. The distance between the substrate holder and the upper electrode was fixed at 1 cm. The bias by the AC power supply was applied between the two electrodes with the upper electrode being grounded.

To examine the frequency effect of the alternating bias on the deposition behavior, the ac bias voltage of $\pm$100 V with frequencies of 0.2, 0.5, and 5 Hz was applied to the substrate holder. For comparison, the silicon was deposited on the substrate without the bias under the same processing conditions. Figure 12.13 shows the FESEM images of the surface morphology of silicon films deposited on the quartz substrates for 2 h at the reactor temperature of 900 °C with gas flow rates of 5 sccm 10 % SiH_4–90 % He, 50 sccm H_2, and 1000 sccm N_2.

Figure 12.13a is for the zero bias, and Fig. 12.13b–d are for the bias voltage of $\pm$100 V at frequencies of 0.2, 0.5, and 5 Hz, respectively. At the zero bias (Fig. 12.13a), the film shows appreciable roughness on the surface. At the frequency of 0.2 Hz, the surface is mainly covered with flake-like structures (Fig. 12.13b). At 0.5 Hz, a flake-like structure disappeared, and the surface became smoother than that of 0.2 Hz (Fig. 12.13c). At 5 Hz, the surface had a very porous structure (Fig. 12.13d). The surface morphology of the film deposited at the zero bias (Fig. 12.13a) is similar to that of films deposited by PECVD (Cheng et al. 2009) and HWCVD (Chung et al. 2011).

Care must be taken of when the AC bias is applied in PECVD and HWCVD processes, where the amount of negatively charged nanoparticles is much larger than that of positively charged ones. When the amount of positive and negative CNPs is pronouncedly unbalanced, the unbalanced AC bias would be necessary to produce the result similar to Fig. 12.13. These results indicate that the surface morphology of films can be changed considerably by changing the frequency of the alternating bias.

Fig. 12.13 FESEM images of silicon films deposited on a quartz substrate at **a** the zero bias and the ac bias voltage of ±100 V with frequencies of **b** 0.2, **c** 0.5, and **d** 5 Hz applied to the substrate holder (Youn et al. 2012)

To examine the effect of the bias voltage, the bias was applied to the stainless substrate holder at the biases of ±50, ±100, ±150, and ±200 V with a frequency of 1 Hz. Figure 12.14 shows the FESEM images of the surface morphology of silicon films deposited under the same processing condition of Fig. 12.13. Under the bias of ±50 V as shown in Fig. 12.14a, the film microstructure tended to have a flake-like structure. When the bias was increased to ±100 V, the microstructure tended to have round nodules as shown in Fig. 12.14b. When the bias was increased to ±150 V, the microstructure became porous as shown in Fig. 12.14c. The nanoparticles of ~100 nm were aggregated as a chain-like structure with their surface covered with numerous nanorods. Finally, as the bias was increased to ±200 V, the nanoparticles deposited on the surface were ~50 nm as shown in Fig. 12.14d, being smaller than those deposited at ±150 V. The microstructure deposited at ±200 V was much more porous than that deposited at ±150 V, whose aspect was shown more clearly in FESEM images of lower magnification.

To have an overall view of the microstructure evolution, the lower magnifications of Fig. 12.14 are shown in Fig. 12.15. The microstructure became more porous as the bias voltage increased. The film deposited at ±50 V is relatively dense as shown in Fig. 12.15a. Although the porosity difference between Fig. 12.15a, d appeared not to be appreciable, the microstructure of lower

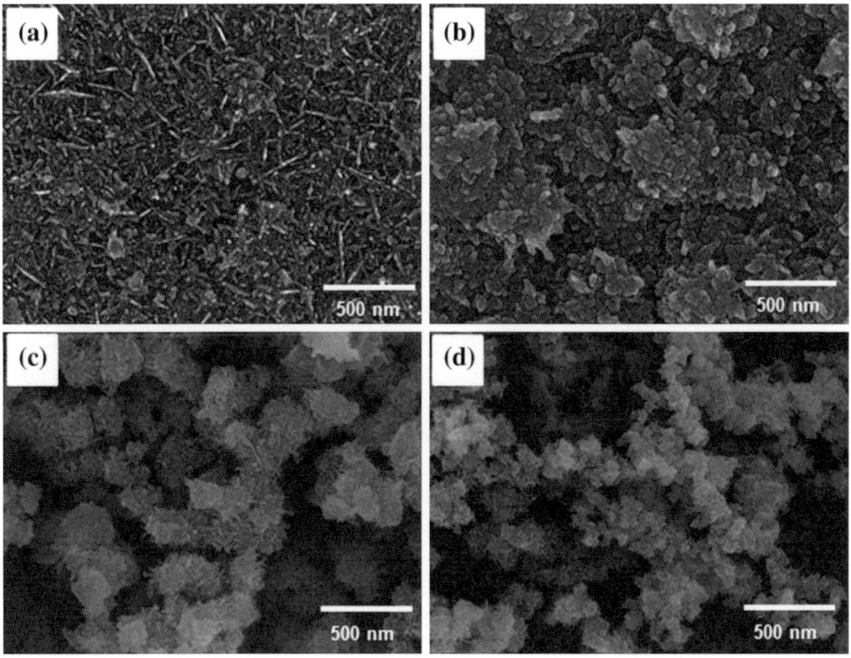

Fig. 12.14 FESEM images of silicon films deposited on a quartz substrate at the ac bias voltages of **a** ±50, **b** ±100, **c** ±150, and **d** ±200 V with the frequency of 1 Hz applied to the substrate holder under the same processing conditions as Fig. 12.13 (Youn et al. 2012)

magnification showed a drastically different porosity as shown in Fig. 12.15c, d. These results show that drastically different microstructures were evolved by changing the intensity of the alternating bias.

To observe the magnified image of nanorods attached to nanoparticles in Fig. 12.14d, the sample of Fig. 12.14d was immersed in ethanol with ultrasonic treatment. The suspended particles, which had been captured on a TEM grid membrane of holey carbon, were observed by HRTEM. Figure 12.16a shows the low-magnification TEM morphology of the silicon nanoparticles with their surface covered with numerous nanorods. Individual nanorods were single crystalline as revealed by the HRTEM image of Fig. 12.16c. Figure 12.16c shows an enlarged-lattice image of Fig. 12.16b. The spacing between the parallel fringes of the crystalline was measured to be 0.31 nm. It is equal to the spacing of the {111} planes of crystalline silicon.

These results imply that controlling the AC bias frequency or the magnitude of bias voltages may produce various microstructures. Besides, the AC bias technique can provide a promising method to produce highly porous microstructures. CNPs are a kind of nanosized colloidal particles, which undergo self-assembly and produce highly ordered three-dimensional arrays. This phenomenon is called "colloidal crystallization" (Shaw and Costello 1993; Velev et al. 2000). In addition to the

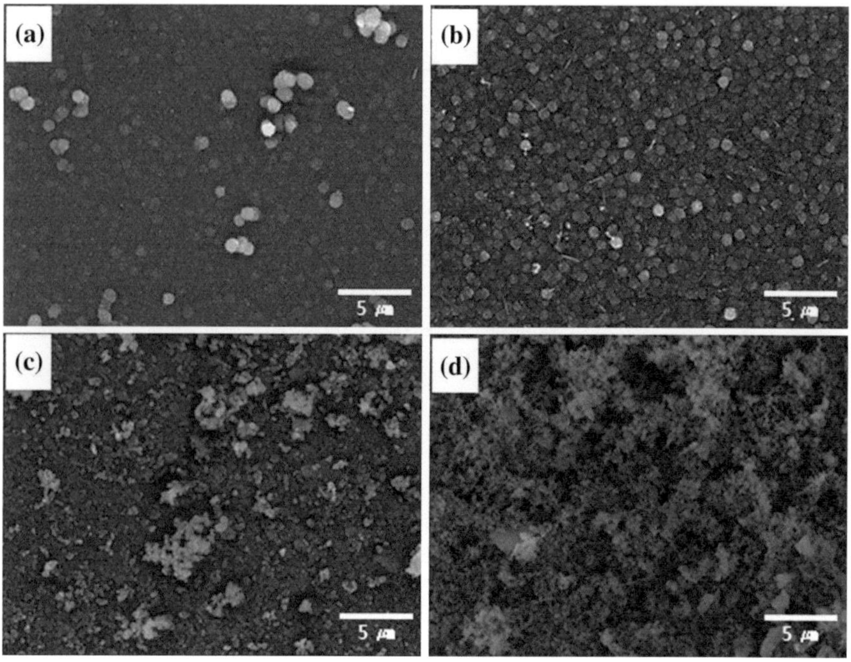

Fig. 12.15 Low-magnification FESEM images for silicon films deposited on a quartz substrate at the AC bias voltages of **a** ±50, **b** ±100, **c** ±150, and **d** ±200 V with the frequency of 1 Hz applied to the substrate holder under the same processing conditions as Fig. 12.13 (Youn et al. 2012)

property of self-assembly, CNPs have the property of charge enhanced diffusion, resulting in the epitaxial colloidal crystallization without voids. Therefore, the self-assembly of CNPS tends to produce a dense structure. The AC bias would prevent CNPs from undergoing the self-assembly and produce a porous structures. The structure grown by CNPs without the AC bias would resemble the dense deflocculation of colloidal particles whereas that with the AC bias would resemble the porous flocculation of them. The highly porous structure has a high surface area, which is favorable for a low dielectric constant (Miller 1999), catalytic effect (Leofanti et al. 1998) and electrode materials (Simon and Gogotsi 2008).

The bias frequency and the magnitude of bias voltages can be changed during deposition, producing more diverse microstructure. This technique could also produce films of a layered structure with each layer of different microstructures. For example, the first layer is dense, the second layer is porous, and the third layer is dense again. These results provide a significant implication in the CVD process, showing clearly that the electric bias, which can be either AC or DC, can be a new processing parameter in the deposition process, where the CNPs are generated.

To examine the bias effect on the film growth rate, the mass increase in the substrate after deposition was measured using a microbalance with a scale in 10^{-5} g. Figure 12.17a shows the effect of the bias frequency on the mass change of silicon

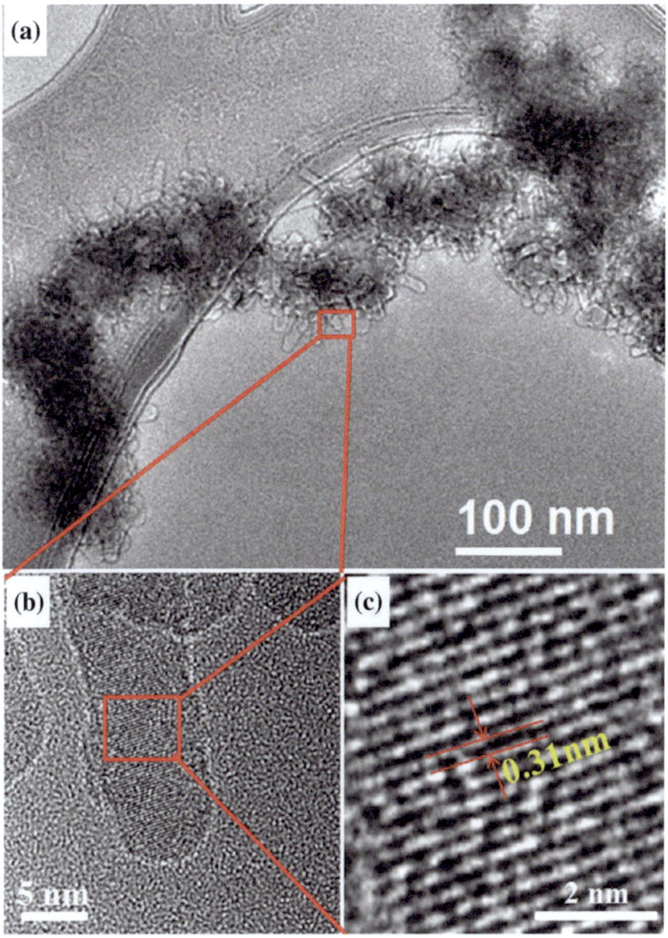

Fig. 12.16 TEM images of silicon nanoparticles with numerous nanorods deposited on a quartz substrate at the AC bias voltages of ±200 V: **a** a low-magnification TEM image, **b** a high-magnification TEM image, and **c** an enlarged HRTEM image from the square enclosed area (Youn et al. 2012)

films grown on the substrate. As the frequency was increased, the mass continued to increase. As the frequency was increased from 0.2 to 0.5, 1, and 5 Hz, the mass of the film was increased by 1.1, 2.0, 3.0, and 3.5 times, respectively, compared with that of the film deposited under the zero bias.

The slope of the mass increase with respect to bias frequency in Fig. 12.17a tended to decrease with increasing frequency. Figure 12.17b shows the effect of the bias voltage on the mass change of silicon films. The effect of the mass increase by bias voltage is more pronounced than that by bias frequency. As the bias voltage was increased from 0 to ±50, ±100, ±150, and ±200 V, the mass was increased respectively by 1.8, 3, 4.5, and 8.5 times.

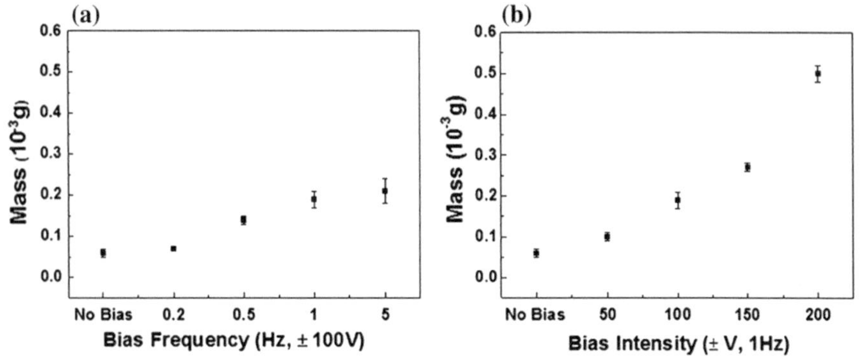

Fig. 12.17 Mass changes of silicon films deposited on a quartz substrate for **a** the ac bias frequencies and **b** the voltages (Youn et al. 2012)

Increasing the bias voltage applied to the substrate holder means the increase of the electric field between the two electrodes, which would increase the electric force to attract the CNPs toward the electrodes. Considering that both positive and negative CNPs exist abundantly in the gas phase as shown in Chap. 9 and that the charge would not build up on the quartz substrate because the AC bias is applied, it is expected that the deposition rate continued to increase with increasing bias voltage as shown in Fig. 12.17b.

The increased deposition rate by the AC bias indicates the increased production yield. Therefore, in the process where the production yield is important, the AC bias can be highly effective. For example, in the Siemens process, which produces polysilicon by decomposing trichlorosilane using CVD, the gas phase nuclei would have an adversary effect on the production yield. The AC bias would efficiently collect the CNPs, increasing the production yield.

12.6 Effect of Alternating Bias During ZnO CVD

As described in Chap. 9.2, the generation of charged ZnO nanoparticles was experimentally confirmed during their syntheses by carbothermal reduction (Kim et al. 2009a). It was also confirmed that charged ZnO tetrapod nanowires are generated in the gas phase (Park et al. 2013). This means that the deposition of CNPs and charged tetrapod nanowires would be affected by the bias applied to the substrate. Motivated by this idea, Park et al. (2015) examined the bias effect during the synthesis of ZnO nanostructures by the carbothermal reduction described in Chap. 9.

The source materials of 4 g of ZnO powder and 0.8 g of graphite powder were loaded on an alumina boat which was placed at the center of the highest temperature zone of 1000 °C in a quartz tube reactor with the inner diameter and length of the

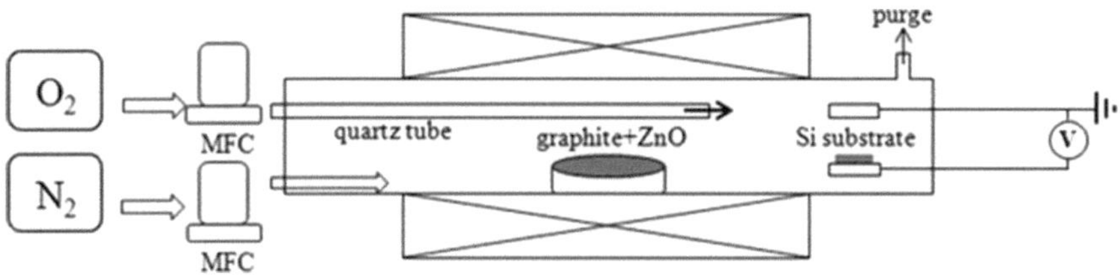

Fig. 12.18 Schematic of the CVD reactor connected to a power supply for applying electric bias (Park et al. 2015)

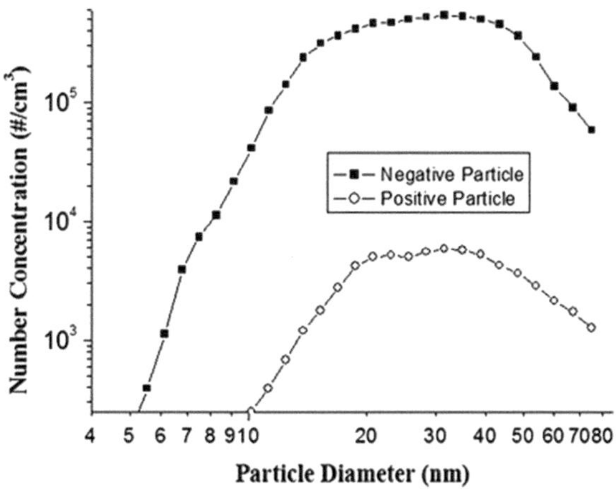

Fig. 12.19 The size distribution of charged ZnO nanoparticles that were generated in the gas phase during the synthesis of ZnO nanostructures (Park et al. 2015)

tube being 5 and 100 cm, respectively. Nitrogen was supplied to the inlet of the reactor as a carrier gas at a flow rate of 500 sccm. In order to minimize the contact of the oxygen gas with the powder mixture of ZnO and graphite, the oxygen gas was supplied at a flow rate of 500 sccm through a small quartz tube of 6 mm in inner diameter, which was placed 2 cm away from the powder mixture toward the outlet of the reactor tube as schematically shown in Fig. 12.18.

For some reason having not been clearly understood at the time of experiment, the deposition did not occur on a silicon wafer substrate at the zone of 450 °C although the abundant generation of CNPs was experimentally confirmed by the DMA-FCE system. This is why a silicon wafer substrate was placed at the zone of 450 °C for bias experiment whereas the temperature of the hot zone of the reactor was 1000 °C. The substrate was on a stainless steel holder that was connected to a power supply for applying electric bias. The bias was applied to the substrate holder with respect to the grounded plate, which is 1 cm away from the holder. Therefore, the bias produced the electric field between the holder and the grounded plate.

For measurements by the DMA, the substrate holder and the grounded plate were taken out of the reactor and a small quartz tube with the inner diameter of 5 mm was placed in the outlet of the reactor at a distance of 11 cm away from the source materials (800 °C) and was connected to the DMA-FCE system at the outside of the reactor.

Figure 12.19 shows the size distribution of ZnO CNPs measured by the DMA-FCE system. Although the number concentrations of both positive and negative CNPs are normally not different much (Kim et al. 2009a, b, 2010; Park et al. 2013), the number concentration of the negative CNPs was about two orders of magnitude higher than that of the positive ones in this synthesis condition as shown in Fig. 12.19. This might be due to the fact that p-type defects such as oxygen interstitials, zinc vacancies and oxygen antisites are generated under the oxygen-rich processing condition and ZnO nanoparticles with p-type defects are more easily negatively charged because the work function of p-type materials is higher than that of the intrinsic and n-type ones (Tuomisto et al. 2003; Janotti and Van de Walle 2007; Liu et al. 2009). This possibility is supported by the observation that the number concentration of negative ZnO nanoparticles was similar to that of positive ones when the oxygen gas was not supplied (Park et al. 2013).

Considering the charge neutrality, an enormous amount of excess positive charges should build up on the quartz wall. It might be suspected that the insulating quartz wall could accommodate such a large amount of excess charges. In relation to the possible accommodation of the charge, McCarty et al. (2007) measured the electrostatic charge on individual microspheres of 50–450 μm and found out that the charge on a microsphere was proportional to its surface area with about 1 elementary charge per 2000 nm^2, which corresponds to about 5×10^8 charges per mm^2. The area of the quartz tube is estimated to be large enough to accommodate the expected excess charges.

Although there exists a huge amount of charged ZnO nanoparticles in the gas phase of the CVD reactor as shown in Fig. 12.19, they do not land on the substrate placed in the reactor as shown in Fig. 12.20a. When +100 V was applied to the substrate holder, ZnO nanoparticles of $\sim$10 nm were deposited in not a little amount as shown in Fig. 12.20b. When the bias was increased to +300 V, a little larger ZnO nanoparticles of $\sim$40 nm were deposited in an appreciable amount as shown in Fig. 12.20c. When the bias was increased even further to +600 V, tetrapod-ZnOs (T-ZnOs) were deposited as shown in Fig. 12.20d. SEM observation of larger magnifications revealed that T-ZnOs were deposited together with smaller ZnO nanoparticles. Figure 12.20b–d show that as the bias voltage was increased, the size of particles that could be deposited was increased.

Figure 12.20 indicates that charged ZnO nanoparticles and T-ZnOs exist in the gas phase. Then, why didn't they deposit on the substrate placed at 450 °C zone without the bias? The reason would be the drag force that is influenced by the relative velocity of particles in the medium (Davies 1966). Considering the gas flow dynamics in the reactor, the hot gas that comes from the heated reactor zone would move upward over the colder gas in the unheated zone due to their density difference arising from the temperature gradient. This gas stream would exert a drag

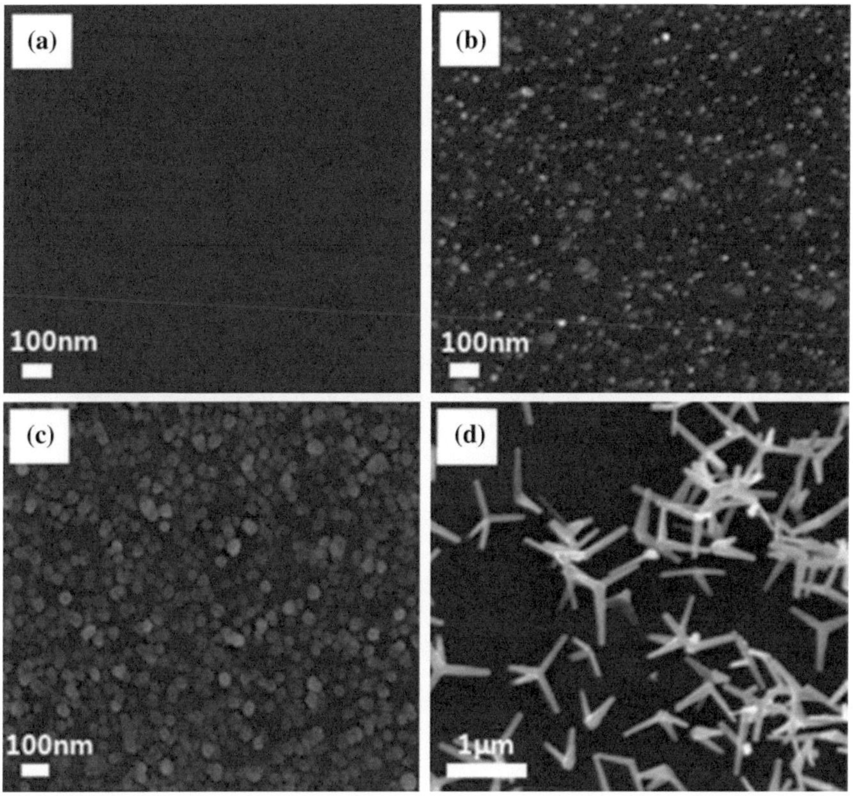

Fig. 12.20 The deposition behavior of ZnO nanoparticles under the positive biases of **a** 0 V **b** +100 V **c** +300 V and **d** +600 V (Park et al. 2015)

force on the nanoparticles that are generated in the hot zone, so that they would move upward over the gas of 450 °C zone.

The bias experiment in Fig. 12.20 reveals clearly that the charged ZnO nanoparticles are under the influence of the drag force and explains why CNPs confirmed in Fig. 12.19 did not land on the substrate in Fig. 12.20a.

Since the bias was positive, deposited particles would be negatively charged. A drag force exerting on particles by the gas flow is expressed as

$$F_D = 1/2 C_D A \rho v^2 \tag{12.1}$$

where F_D = drag force, C_D = drag coefficient, A = cross-sectional area perpendicular to the flow, ρ = density of the medium, v = velocity of the body relative to the medium. This equation reveals that particles with a large area have a larger drag force and thereby a larger resistance to move toward the substrate than those with a smaller area. If particles have the same shape, the area of particles would be proportional to the square of the size. Therefore, in order to deposit large particles, the high bias voltage is needed because of the drag force arising from the upward gas stream caused by the temperature gradient.

In the 450 °C zone in Fig. 12.18, where the carrier gas moves upward, the drag force hinders the ZnO particles from moving toward the substrate and ZnO particles just flew out with the carrier gas. This would be why non-deposition occurs under the zero bias voltage in Fig. 12.20a. And under the bias of +100 V in Fig. 12.20b, small particles of ~10 nm were deposited but large particles flew out with carrier gas. Because the drag force for small charged particles is weak, only small electric force is needed to overcome the drag force of the small particles, which results in the deposition of small charged particles on the substrate. Under the bias of +300 V in Fig. 12.20c, a little larger particles of ~40 nm were deposited with much larger particles being flowed out with a carrier gas. Under the bias of +600 V in Fig. 12.20d, however, even much larger tetrapod particles with the leg length

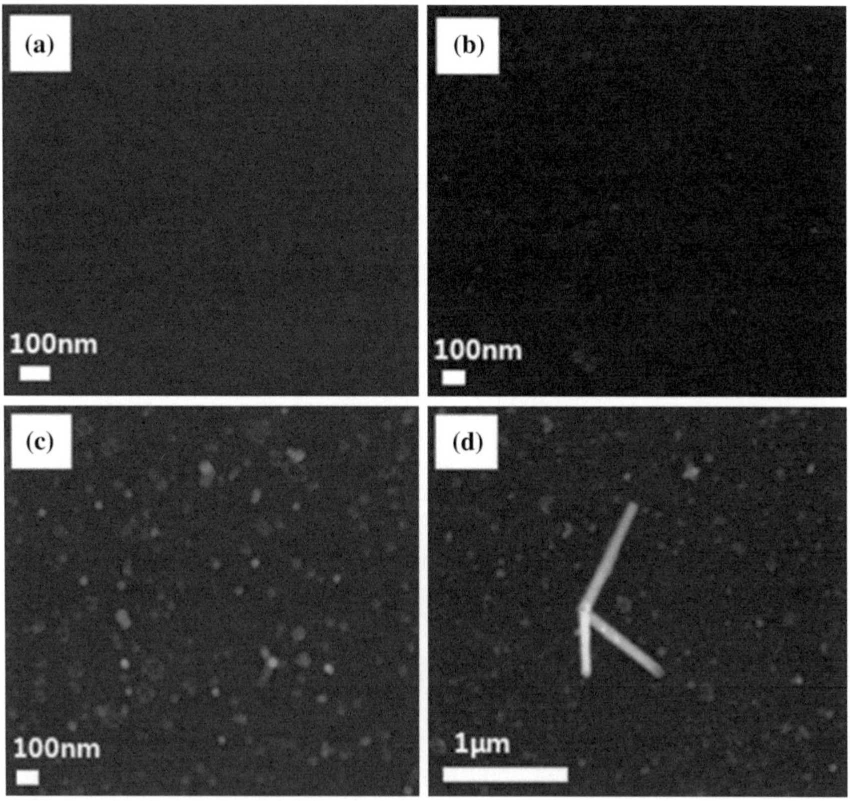

Fig. 12.21 The deposition behavior of ZnO nanoparticles under negative biases of **a** 0 V **b** −100 V **c** −300 V and **d** −600 V (Park et al. 2015)

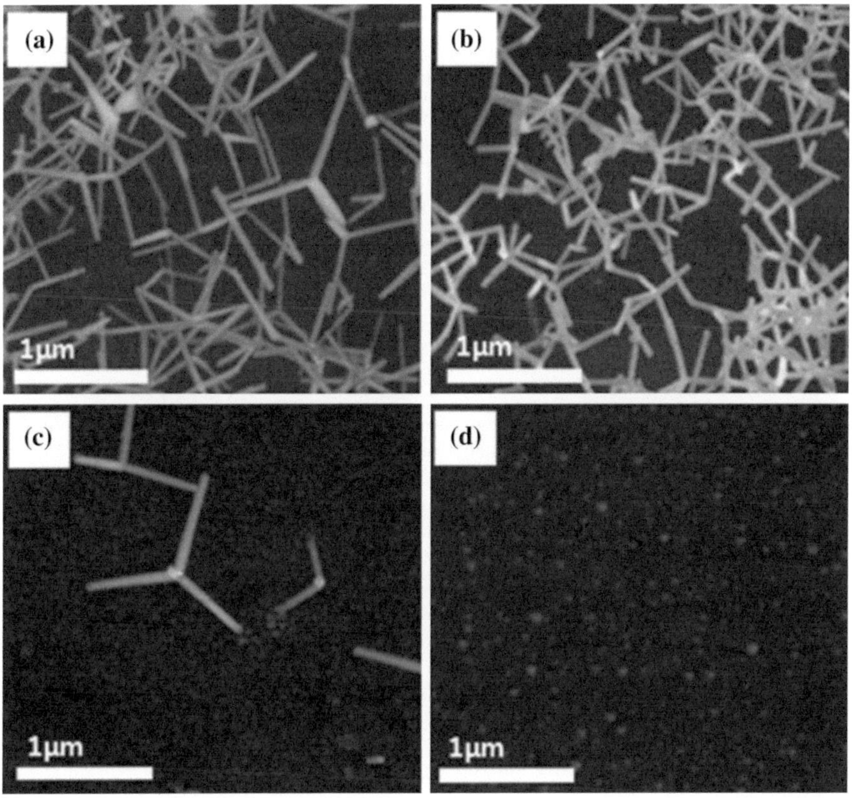

Fig. 12.22 The deposition behavior of ZnO nanoparticles under an alternative bias fixing the bias strength at 600 V with various bias frequencies of **a** 0.2 Hz **b** 1 Hz **c** 3 Hz and **d** 5 Hz (Park et al. 2015)

of ~600 nm could be deposited on a substrate, indicating that the electric force is high enough to overcome the drag force.

When the negative bias was applied, the similar result was obtained but the amount of deposited particles was much less than that of Fig. 12.20 because the number concentration of positive particles was much less than that of negative ones as shown in Fig. 12.21, which is consistent with Fig. 12.19.

When the alternative bias of 600 V was applied, the deposition behavior was affected by bias frequencies as shown in Fig. 12.22a–d. For frequencies of 0.2 and 1 Hz (Fig. 12.22a, b), an appreciable amount of T-ZnOs was deposited similarly to the deposition behavior for positive 600 V bias (Fig. 12.20d). However, the amount of T-ZnOs was markedly decreased for 3 Hz (Fig. 12.22c) and T-ZnOs were hardly deposited for 5 Hz (Fig. 12.22d). This dependence of deposition behavior on the bias frequency might be related with the distance travelled by charged T-ZnOs until the bias was reversed.

More specifically, the distance travelled by charged T-ZnOs during 5 s for 0.2 Hz or 1 s for 1 Hz is expected to be longer than the distance (1 cm) between the two electrodes, which results in deposition of T-ZnOs. On the other hand, the distance travelled by charged T-ZnOs during 0.33 s for 3 Hz or 0.2 s for 5 Hz is expected to be shorter than ~ 1 cm, which results in non-deposition. In this case, charged T-ZnOs would fluctuate above the substrate and flow out eventually with a carrier gas. Thus, non-deposition under the alternating bias of high frequency is attributed to the drag force that hindered the movement of the T-ZnOs.

References

Cheng Q, Xu S, Ostrikov K (2009) Rapid, low-temperature synthesis of nc-Si in high-density, non-equilibrium plasmas: enabling nanocrystallinity at very low hydrogen dilution. J Mater Chem 19(29):5134–5140. doi:10.1039/B904227J

Chung Y-B, Lee D-K, Kim C-S, Hwang N-M (2009) Effect of HCl addition on the crystalline fraction in silicon thin films prepared by hot-wire chemical vapor deposition. Vacuum 83 (12):1431–1434. doi:http://dx.doi.org/10.1016/j.vacuum.2009.05.004

Chung Y-B, Lee D-K, Lim J-S, Hwang N-M (2011) Reduction of amorphous incubation layer by HCl addition during deposition of microcrystalline silicon by hot-wire chemical vapor deposition. Sol Energ Mat Sol C 95(1):211–214. doi:http://dx.doi.org/10.1016/j.solmat.2010.04.021

Davies CN (1966) Aerosol science, Academic Press London

Janotti A, Van de Walle CG (2007) Native point defects in ZnO. Phys Rev B 76(16):165202

Kim J-Y, Kim D-Y, Hwang N-M (2006) Spontaneous generation of negatively charged clusters and their deposition as crystalline films during hot-wire silicon chemical vapor deposition. Pure Appl Chem 78(9):1715–1722. doi:http://dx.doi.org/10.1351/pac199870040789

Kim C-S, Chung Y-B, Youn W-K, Hwang N-M (2009a) Generation of charged nanoparticles during synthesis of ZnO nanowires by carbothermal reduction. Aerosol Sci Tech 43(2): 120–125. doi:10.1080/02786820802499068

Kim C-S, Chung Y-B, Youn W-K, Hwang N-M (2009b) Generation of charged nanoparticles during the synthesis of carbon nanotubes by chemical vapor deposition. Carbon 47(10): 2511–2518. doi:http://dx.doi.org/10.1016/j.carbon.2009.05.005

Kim C-S, Kwak I-J, Choi K-J, Park J-G, Hwang N-M (2010) Generation of charged nanoparticles during the synthesis of silicon nanowires by chemical vapor deposition. J Phys Chem C 114 (8):3390–3395. doi:10.1021/jp910242a

Lee J-I, Hwang N-M (2008) Effect of electric bias on the deposition behavior of silicon on flexible polyethylene terephthalate (PET) during hot-wire chemical vapor deposition. J Cryst Growth 310(15):3668–3673. doi:http://dx.doi.org/10.1016/j.jcrysgro.2008.05.022

Lee D-K, Chung Y-B, Kim J-K, Hwang N-M (2008) Effects of the electric bias on the deposition behavior of silicon films on glass during hot-wire chemical vapor deposition. J Cryst Growth 310(19):4368–4372. doi:http://dx.doi.org/10.1016/j.jcrysgro.2008.07.041

Leofanti G, Padovan M, Tozzola G, Venturelli B (1998) Surface area and pore texture of catalysts. Catal Today 41(1–3):207–219. doi:http://dx.doi.org/10.1016/S0920-5861(98)00050-9

Levchenko I, Ostrikov K (2007) Nanostructures of various dimensionalities from plasma and neutral fluxes. Journal of Physics D: Appl Phys 40(8):2308. doi:10.1088/0022-3727/40/8/S11

Liu Y-Y, Jin H-J, Park C-B, Hoang GC (2009) Analysis of photoluminescence for N-doped and undoped p-type ZnO thin films fabricated by RF magnetron sputtering method. T Electr Electr 10(1):24–27

McCarty LS, Winkleman A, Whitesides GM (2007) Ionic electrets: electrostatic charging of surfaces by transferring mobile ions upon contact. J Am Chem Soc 129(13):4075–4088

Miller RD (1999) In Search of Low-k Dielectrics. Science 286(5439):421–423. doi:10.1126/science.286.5439.421

Park SY, Yang SM, Kim CS, Hwang NM (2009) Effect of bias applied to hot wires on generation of positive and negative charges during silicon hot-wire chemical vapor deposition. J Phys Chem C 113(39):17011–17017. doi:10.1021/jp906196b

Park S-Y, You S-J, Kim C-S, Hwang N-M (2010) Comparison of charge generation from hot wires between AC and DC power supply and its effect on deposition behavior during hot-wire chemical vapor deposition of silicon. J Cryst Growth 312(16–17):2459–2464. doi:http://dx.doi.org/10.1016/j.jcrysgro.2010.05.023

Park S-H, Park J-W, Kim C-S, Hwang N-M (2013) Formation of tetrapod-shaped nanowires in the gas phase during the synthesis of ZnO nanostructures by carbothermal reduction. J Nanosci Nanotechno 13(10):7198–7201. doi:10.1166/jnn.2013.7697

Park S-H, Park J-W, Yang S-M, Kim K-H, Hwang N-M (2015) Effect of electric bias on the deposition behavior of ZnO nanostructures in the chemical vapor deposition process. J Phys Chem C 119(44):25047–25052

Qijin C, Xu S (2007) Low-temperature synthesis of homogeneous nanocrystalline cubic silicon carbide films. J Appl Phys 102(5):056101. doi:http://dx.doi.org/10.1063/1.2776155

Shaw DJ, Costello B (1993) Introduction to colloid and surface chemistry, Butterworth-Heinemann, Oxford

Simon P, Gogotsi Y (2008) Materials for electrochemical capacitors. Nat Mater 7(11):845–854

Song J-H, Jo W, Hwang N-M (2007) Non-uniform deposition in the early stage of hot-wire chemical vapor deposition of silicon: The charge effect approach. Thin Solid Films 515 (19):7446–7450. doi:http://dx.doi.org/10.1016/j.tsf.2006.11.170

Tuomisto F, Ranki V, Saarinen K, Look DC (2003) Evidence of the Zn vacancy acting as the dominant acceptor in n-type ZnO. Phys Rev Lett 91(20):205502

Velev OD, Lenhoff AM, Kaler EW (2000) A class of microstructured particles through colloidal crystallization. Science 287(5461):2240–2243. doi:10.1126/science.287.5461.2240

Youn W-K, Kim C-S, Lee J-Y, Lee S-S, Hwang N-M (2012) Generation of charged nanoparticles and their deposition behavior under alternating electric bias during chemical vapor deposition of silicon. J Phys Chem C 116(47):25157–25163. doi:10.1021/jp310705p

Zhang H (1999) Ion sources. Springer, Berlin

Chapter 13
Charge-Enhanced Kinetics

The most important presumption in the TCN is that small CNPs should be liquid-like. This presumption was derived from Figs. 6.17 and 6.18, which show that small CNPs undergo epitaxial recrystallization and produce highly crystalline diamond crystals with well-defined facets whereas large CNPs undergo non-epitaxial recrystallization and produce ball-like nanostructured diamonds. It appears that this liquid-like property changes abruptly with the size of CNPs. No theory is available now to explain such a size effect of CNPs on the liquid-like property, but weakening of the bond strength by charge and the dependence of the electrostatic energy on the size of nanoparticles mentioned in Chap. 1 seem to be the best explanation. In this chapter, experimental results supporting that charge enhances atomic diffusion or chemical kinetics will be presented.

13.1 Magic Size

In relation to the size effect of nanoparticles on the liquid-like property, the phenomenological explanation was given by Fujita (1994, 1991) who suggested that the property of nanoparticles changes abruptly at a specific size which was called the 'magic size'. In the order of increasing size, particle coalescence, epitaxial recrystallization and diffusional sintering were suggested to take place. Fujita defined 'atom clusters' as ultrafine particles whose size is equal to or smaller than the magic size. As one way to determine the magic size of atom clusters, he used the minimum crystalline size in the transition from crystal to amorphous phase induced by electron irradiation at 2 MV.

Figure 13.1 shows an example of crystal nucleation in an amorphous Al_2O_3–ZrO_2 composite, in which the lattice fringes appear in crystallized regions of ZrO_2. In the figure, the partly crystallized region of ZrO_2 and the magic size, i.e., the crystal nucleation, are indicated with dotted and solid circles respectively. It is noted that each of the lattice fringes appears just within each individual solid circle

© Springer Science+Business Media Dordrecht 2016

N.M. Hwang, *Non-Classical Crystallization of Thin Films and Nanostructures in CVD and PVD Processes*, Springer Series in Surface Sciences 60,
DOI 10.1007/978-94-017-7616-5_13

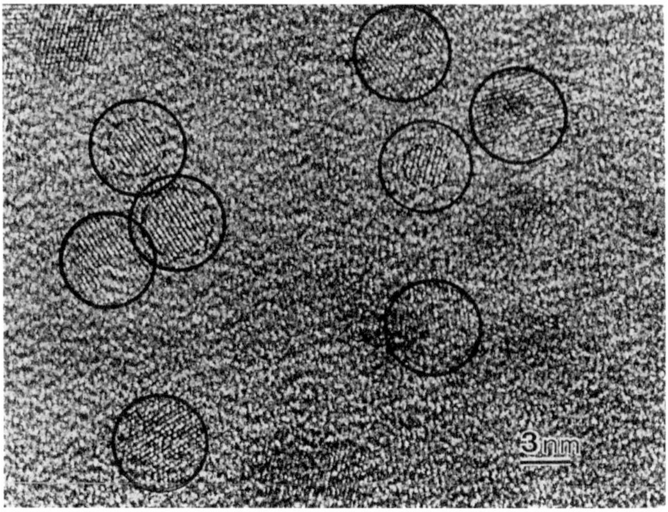

Fig. 13.1 Lattice images appearing in embedded blocks of ZnO_2 in an amorphous Al_2O_3–ZrO_2 composite. *Dotted* and *solid circles* correspond to partly crystallized regions and the magic size of ZrO_2 crystals, respectively. The micrograph was taken at 2 MV (Fujita 1991)

whose size is the very same, i.e., 6 nm Φ, in all the regions. In this case, orientations of the lattice fringes are different from each other when the solid circles are separated from each other, as seen in four solid circles at the right-hand side. On the contrary, their orientation coincide with each other when the solid circles slightly overlap each other, as seen in three solid circles at the middle left side. These facts mean that coalescence among the embedded atom clusters easily occurs when they contact each other, as in the case of sintering among the isolated atom clusters. Therefore, the atom clusters are liquid-like.

The magic size can also be determined by the crystal nucleation in liquids and vapors. Essentially, however, the magic size can be determined as the critical size at which the material behavior, such as sintering speed, alloy speed, catalysis, etc., markedly changes (Fujita and Komatsu 1987). Fujita (1991) estimated the magic size for several materials. The magic size of isolated particles was estimated to be 15–20 nm∅ for FCC metals and alloys, 10–15 nm∅ for body centred cubic (BCC) metals and alloys, 15–20 nm∅ for Al_2O_3, and 3–5 nm∅ for TiC and SiC, where ∅ represents the diameter of nanoparticles. For embedded particles, it is ~2 nm∅ for Si and ~1 nm∅ for diamond. The magic size for an isolated diamond particle is expected to be ~2 nm∅.

Figure 13.2 is a schematic illustration showing the change of the material behavior as a function of the size. Curves A, B and C in Fig. 13.2 show the change in behavior of each individual isolated cluster at the magic size. Deformation and diffusion phenomena also rapidly and easily occur at the magic size just like

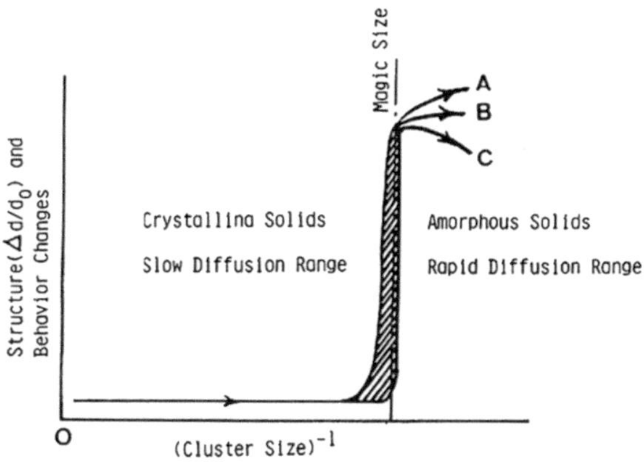

Fig. 13.2 Abrupt change of the material behaviour at the magic size (Fujita 1991)

interaction among liquid drops. Namely, mechanical twining, transformation and plastic or viscous deformation easily occur within each individual cluster even at very low temperatures.

Sintering (Fujita and Komatsu 1987; Komatsu and Fujita 1989) and alloying (Mori et al. 1991) among the isolated atom clusters rapidly occur even at subzero temperatures and/or even when the atomic size ratio is extremely large. Coalescence among the atom clusters in Fig. 13.1 is an example showing such diffusion phenomena.

Although Fujita did not mention the role of charge in inducing the liquid-like behaviour of nanoparticles, their extensive works clearly show that sufficiently small nanoparticles, when charged, exhibit the liquid-like behaviour, which should be the most important concept in understanding the non-classical crystallization.

13.2 Experimental Evidences of Charge-Enhanced Diffusion During In-situ TEM Observation

As in Fujita's study on the magic size in the transition from crystalline to amorphous phase induced by electron irradiation of TEM, charging of the sample is inevitable during TEM observation. Therefore, if charge really enhances atomic diffusion of nanoparticles, this aspect will be revealed during in-situ TEM observations. In some literatures surveyed, the role of charge was noticed and mentioned but in other literatures, it was not noticed. In some other cases, the enhanced kinetics was explained by the kinetic energy of electron bombardment.

So far, no systematic study has been made as to the presumption that charge enhances atomic diffusion in nanoparticles. However, there are many experimental results supporting that charge enhances diffusion. One of them was reported by Iijima and his colleagues (Iijima and Ichihashi 1986; Iijima 1987; Ajayan and Iijima 1991; Iijima and Ajayan 1991). More recent results, which are believed to be related with charge-enhanced diffusion, will be introduced after Iijima et al's works.

13.2.1 Au Nanoparticles Under Electron-Beam Irradiation

Iijima and Ichihashi (1986) observed morphologies and internal structures of ultrafine particles of gold with a high resolution TEM (HRTEM) newly developed at that time. The specimens, clusters of gold, were prepared by in situ vacuum deposition onto fine and spherical particles of crystalline silicon or a percolated amorphous carbon film (Iijima 1984; Iijima and Ichihashi 1985). The silicon particles were less than several hundred angstroms in size and were usually covered with SiO_2 layers 10–30 Å in thickness. The micrographs were reproduced from pictures each of which corresponds to a single frame of the video-recording tape, which has a time resolution of 1/60 s. The sequence of evolution of a particular gold cluster was studied by examination of a set of picture frames.

A typical series of electron micrographs of the same particle of gold is reproduced in Fig. 13.3. These were selected from a video tape recording (VTR) over a period of 5 min. The particle sat on a SiO_2-covered Si substrate and contained about 460 gold atoms. The shape of the particle itself constantly changed approximately every few tenths of a second. The change was often accompanied by rotational and translational motion of the particle. The center of gravity moved over distances of 30–60 Å on the substrate. The internal structure also changed from a single crystal to a twinned crystal and vice versa. With an increase of the particle size, the movement became slow and no rapid change was observed in the particle if it was more than 100 Å in size.

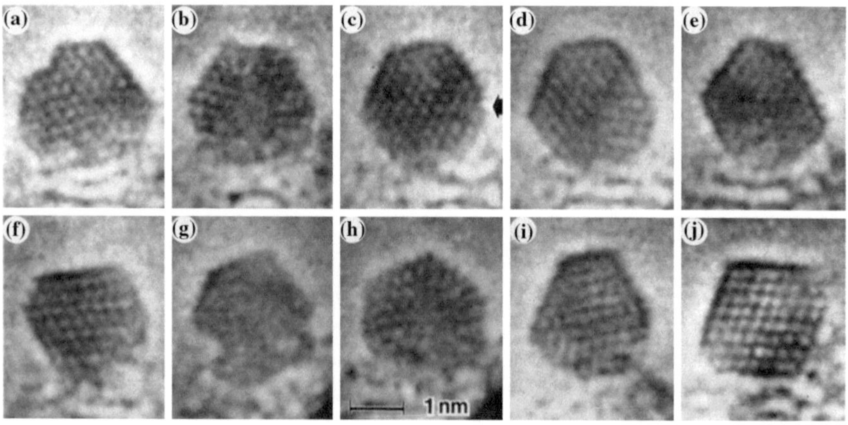

Fig. 13.3 Electron micrographs showing various shapes of an ultrafine particle consisting of about 460 gold atoms reproduced from a video tape recorder. The shape of the particle itself was changing continually under electron-beam irradiation. The lattice fringes appearing in the particles correspond to d_{111} =2.35 Å. The particles in (**a**), (**d**), and (**i**) are single twins. Single crystals with cuboctahedral shape are seen in (**e**), (**f**), and (**i**). From the size of the cuboctahedron (**j**), the particle theoretically contains 459 gold atoms. The particle also transforms into a multiply twinned icosahedral particle, (**b**) and (**h**) (Iijima and Ichihashi 1986)

By taking into account the beam intensity, the stopping power of the substrate material (amorphous carbon), and its thermal conductivity, they estimated the substrate temperature to be roughly 100 °C. They tried to estimate the cluster temperature by other methods (Iijima 1987). Firstly, Bi clusters were prepared and observed in exactly the same procedure as the Au clusters. The results confirmed a solid state of the Bi clusters, since they exhibited lattice fringe images. This means that the substrate temperature during the observation in the microscope would not exceed the melting temperature of Bi which is 271 °C. Secondly, to examine an effect of the electron beam heating, conventional and micro-beam modes in electron microscopic observations were compared in terms of the probe size. There was no major change in cluster activities. A cluster temperature was determined by a balance of heat dissipation through conduction and heat gain through inelastic scattering events of the incident electron beam in the cluster. The cluster temperature was estimated to be almost the same as that of the substrate, which is room temperature. The result supported their assumption that the instability of the clusters may not be caused by a rise in specimen temperature.

The rate of movement was increased by a decrease in the area of the particle in contact with the substrate, which is located at the lower portion of each micrograph. Most importantly, the evolution of the particles, however, became sluggish when an electrically good conductors such as graphite or amorphous carbon was used as a substrate (Iijima 1987). Alumina substrates gave the same results as the SiO_2-covered Si but almost no activities of the clusters were observed when a substrate of α-Fe_2O_3 was used. Also gold particles supported on a gold film remained inactive without undergoing translational motion or structural fluctuation (Ajayan and Iijima 1991).

Considering the fact that the evolution rate of particles depends on the conductivity of the substrate, the active movement comes from the amount of charge built up rather than from the electron bombardment. Iijima and Ichihashi (1986) proposed that the state of such a small particle should be called "quasi solid state," which is neither solid nor liquid according to the conventional concepts of matter.

Further, Iijima and Aiayan (1991) observed the liquid-like coalescence of small gold particles supported on silica, using HRTEM and real time video recording. Particles, which were a few nanometers in size, and decoupled from substrate interactions, coalesced extremely fast, fusing together like viscous droplets in fractions of a second. It is seen from the images that the coalescence of particles of 1–2 nm is completed in a time interval of one frame or 1/60 s, which is the video time resolution available. Direct fusion of still larger particles of 2–3.5 nm occurred in about 1/20 s.

When two particles of 2–3 nm sit firmly on the substrate, the rate of the coalescence is at least two orders of magnitude slower (order of seconds) than that of particles of 1–2 nm not sitting firmly on the substrate. In this case, the lattice fringes of both particles were seen to align even before the particles actually made contact.

This alignment means that particles rotate into a lower energy misorientation. The interaction energy between two particles depend on their misorientation when they are close enough. The derivative of the interaction energy with respect to a misorientation angle is a torque, which will rotate the nanoparticles in such a direction as to lower the interaction energy. Since epitaxial misorientation is expected to have the minimum interaction energy, the rotation would be in a direction to align the lattice fringes of both particles. This rotation would be related with the oriented attachment of particles suggested by Banfield et al. (2000).

As the particle size increases, coalescence occurs during a time interval of minutes and the process occurs in a very similar fashion as in the sintering of micron size particles. A neck is first developed at the plane of contact which is slowly filled by diffusion from the bulk and the surface to the neck region. As the particle sizes get larger, grain boundaries are formed due to the random orientation of the particles before contact. In this case, any fast plastic deformation or structural rearrangements do not occur.

13.2.2 Melting Point Depression

Occasionally, the liquid-like behaviour of small nanoparticles is attributed to the melting point depression (Iijima and Ichihashi 1986). It may be related indirectly but not directly related to the liquid-like behaviour of CNPs, which is drastically different from that of neutral nanoparticles. To make it clear, however, it might be necessary to describe briefly the concept of the melting point depression.

The melting point depression is originated from the fact that the liquid-vapor interface energy (σ_l) is smaller than the solid-vapor interface energy (σ_s). As the size gets smaller, the contribution of the interface free energy to the total free energy of the particle becomes more important. Then, a phase with the low interface energy tends to be more stable than that with the high interface energy. The Gibbs free energy of a nanoparticle consists of the bulk and interface free energies and expressed by (3.2). At the melting point, the solid and the liquid have the same Gibbs free energy. If it is assumed that the molar volumes of solid and liquid are the same for simplicity, the Gibbs free energy of the spherical nanoparticles of solid with radius r should be the same as that of liquid at the melting temperature

$$\Delta G_s = \Delta G_l = \frac{4\pi r^3}{3}\Delta f_s + 4\pi r^2\sigma_s = \frac{4\pi r^3}{3}\Delta f_l + 4\pi r^2\sigma_l, \tag{13.1}$$

where the subscripts, s and l, represent respectively solid and liquid.

If the following approximation valid for small undercooling is substituted in (13.1)

$$\Delta f_l - \Delta f_s = \frac{\Delta H \Delta T}{T_m}, \tag{13.2}$$

where T_m is the melting point, ΔH the heat of fusion, and ΔT undercooling, which can also be the melting point depression. Then, the melting point depression, ΔT, is derived as

$$\Delta T = \frac{3T_m(\sigma_s - \sigma_l)}{r\Delta H} \tag{13.3}$$

The melting point depression is inversely proportional to the size and proportional to the difference of the interface energy between solid and liquid. As the size of nanoparticles gets small, the melting point depression may contribute additionally to the liquid-like behaviour, which should be distinguished from the effect of charge on the liquid-like behaviour.

13.2.3 Electron-Assisted Superplasticity of Amorphous Silica

Zheng et al. (2010) observed the superplastic deformation behaviour of nanoscale amorphous silica near room temperature using in situ experiments inside a TEM with low beam intensities without obvious rise in sample temperature. They called the phenomenon electron-assisted superplasticity. Figure 13.4 shows the typical deformation response of amorphous silica nanoparticles in compression tests inside a TEM. Figure 13.4a displays the centered dark-field image of a silica particle with a diameter of 510 nm before the compression test.

After imaging for the positioning of the sample and the diamond flat punch, the beam was blocked off with the condenser lens aperture, and the silica particle was compressed with the Hysitron PicoIndenter (Shan et al. 2008a, b). The particle is

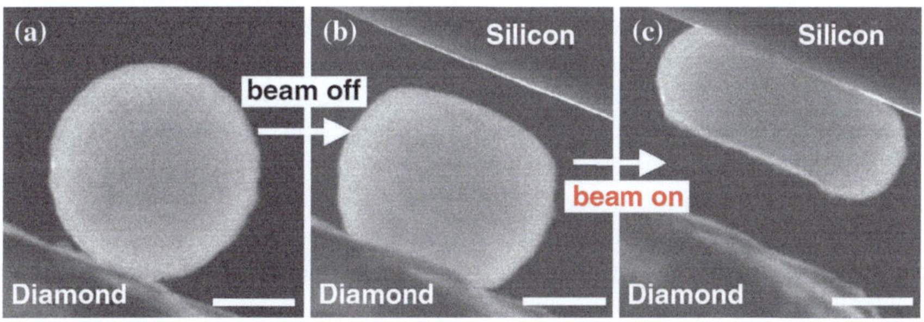

Fig. 13.4 Nanocompression of amorphous silica particle inside a TEM. Two consecutive compression runs were performed with the electron beam being off and on, respectively. Panels **a**–**c** show the centered dark-field images. In a, the particle is adhered to the diamond punch and the silicon has not moved into the picture. In **b**, the particle is imaged after it has been pressed with the beam off. Subsequently, with the beam on, the particle can be compressed into a pancake with a moderate force, as seen in **c**. The *scale bars* are for 200 nm (Zheng et al. 2010)

plastically deformed as shown in Fig. 13.4b because of the e-beam irradiation it had experienced during the imaging. Three load-partial unload cycles were performed under this beam-off condition. After ~40 percent compression (Shan et al. 2008a), where the contact pressure is estimated to be 9.2 GPa and the force is approaching the capacity limit of the Hysitron Picoindenter, the beam was brought on the sample to image the particle (Fig. 13.4b). On unloading, the total plastic percent compression was 27 %. On compressive loading again with beam on, surprisingly, the plastic flow of the glass was continuous and smooth, with no sign of shear banding or cracking. Much easier flow was observed, with the ensuing beam-on deformation in a second set of in situ experiments, resulting in a pancake shape (Fig. 13.4c). Although the contact area kept increasing, the forces required to deform was at levels considerably lower than those in the beam-off condition.

Similar superplastic behaviour of Li_2O under e-beam irradiation was reported by Zheng et al. (2012). They could form the polycrystalline Li_2O nanowires in situ by touching and pulling lithium hydroxide under e-beam irradiation at room temperature inside the TEM. They compared the elongation behaviour of Li_2O nanowires with and without e-beam irradiation. The Li_2O nanowires sustained an enhanced elongation (from 80 % to 176 %) under low dose e-beam irradiation near room temperature as compared with that (from 51 % to 57 %) without e-beam irradiation. The extremely high deformability could be understood by the fast Li_2O diffusion under e-beam irradiation and tensile stress condition.

13.3 Comparison of Deposition Behaviour Between Neutral and Charged Nanoparticles

The best analogy for the difference in the deposition behaviour between neutral and charged nanoparticles would be the difference in the sedimentation behaviour between flocculation and deflocculation of colloid particles as shown in Fig. 6.6. The neutral nanoparticles (NNPs) have only van der Waals attraction without repulsive interaction. As a result, NNPs tend to undergo diffusion-limited aggregation (DLA), leading to a porous fractal structure. Bardotti et al. (1995) studied the deposition behaviour of large Sb clusters on graphite as shown in Fig. 13.5. Figure 13.5a, b show typical structures generated by random deposition of Sb clusters on graphite respectively at 25 and 100 °C.

The metallic vapor obtained from a heated crucible is condensed in Ar and cooled at liquid nitrogen temperature. This leads to the formation of the beam of incident clusters, which are neutral and have low kinetic energy less than 10 eV/cluster. The size distribution of the incident antimony clusters is centered on 2300 atoms, which corresponds to a mean diameter d = 5 nm assuming a spherical shape for the clusters.

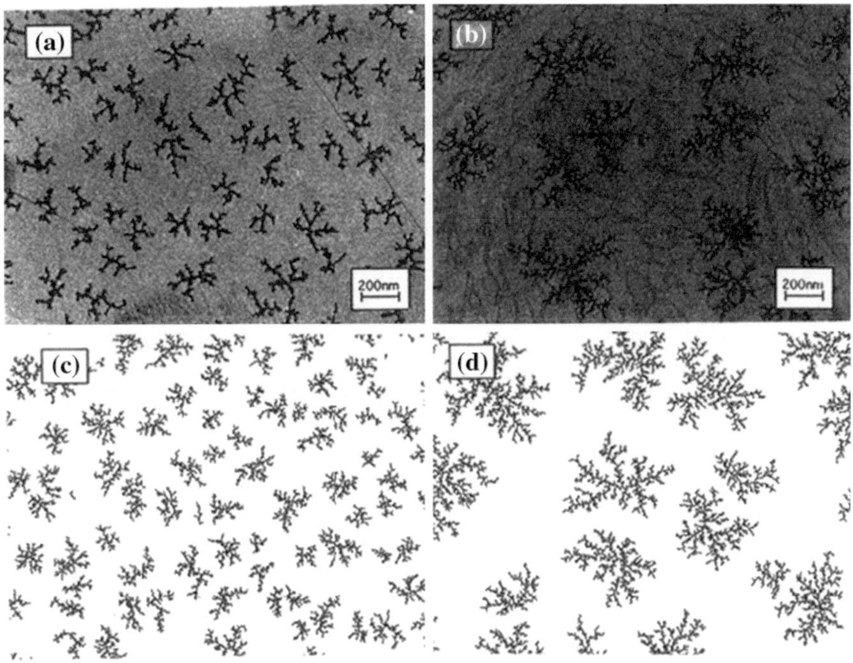

Fig. 13.5 TEM images of typical islands formed by low energy deposition of neutral Sb clusters on highly oriented (0001) pyrolitic graphite (HOPG) at (**a**) 25 °C and (**b**) 100 °C, (**c**, **d**) Predicted island morphologies using the deposition-diffusion-aggregation (DDA) model (Bardotti et al. 1995)

To simulate the growth of the ramified islands in Fig. 13.5a, b, they used the deposition-diffusion-aggregation (DDA) model, which assumes as follows.

(1) Deposition: Clusters are deposited at randomly chosen positions of the surface at a flux F per unit surface per unit time.
(2) Diffusion: Each diffusion time τ, all isolated clusters are chosen at random and moved in a random direction by 1 diam.
(3) Aggregation: If two clusters come to occupy neighboring sites, they stick irreversibly and form an island. Islands are assumed to be immobile.

The result of computer simulations based on the DDA model is shown in Fig. 13.5c, d. The fractal or ramified structure is well reproduced. Therefore, the fact that the deposition of NNPs produce a fractal structure is well established experimentally and theoretically.

In contrast to the deposition behaviour of NNPs, CNPs undergo self-assembly, resulting in a highly regular 3D packing. This aspect of self-assembly is well known in colloidal crystallization. Figure 13.6 shows a typical example of hexagonally close packed long-range ordering of monodisperse latex particles. The microstructures in

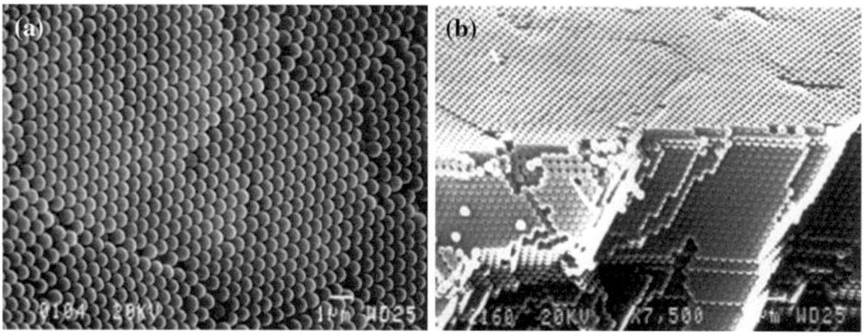

Fig. 13.6 SEM images of (**a**) a typical area on the surface of the assemblies and of (**b**) the edge of a broken particle. *Scale bars*, 1 μm. (Velev et al. 2000)

Fig. 13.6 were evolved from the self-assembly of negatively-charged polystyrene latex microspheres of 630 nm in diameter.

Considering the correlated motion and epitaxial alignment of two charged particles during electron irradiation reported by Yuk et al. (2012) as mentioned in Chap. 1, CNPs has an orientation assembly in addition to self-assembly. The orientation-assembly would result in oriented attachment of nanoparticles. Self-assembly and orientation-assembly are not enough to explain the non-classical crystallization because the hexagonal close packed structure shown in Fig. 13.6 has a packing density of 0.73, which means that the structure has a minimum amount of voids of 23 %. Non-classical crystallization normally result in 100 % dense structure. Note that the structure evolved by coalescence in Figs. 1.5, 1.6 and 1.7 in Chap. 1 is dense without any porosity. Therefore, in order for CNPs to be a building block for non-classical crystallization, CNPs should have a liquid-like property in addition to self-assembly and orientation-assembly. The instantaneous fusion-like coalescence of nanoparticles in Figs. 1.5, 1.6 and 1.7, which was also reported by Ajayan and Iijima (1991) cannot be explained without assuming that CNPs are liquid-like.

Non-classical crystallization of thin films and nanostructures is similar to colloidal crystallization. There are two major differences. One is the size. CNPs in the thin film reactor are normally less than ~100 nm whereas colloidal particles are normally larger than ~100 nm. The other is the temperature. The gas or the substrate temperature in the thin film reactor is high whereas the temperature in colloidal crystallization cannot be made high because colloidal particles are in solution. The small size and the high temperature make CNPs liquid-like. Thereby, CNPs can be a building block to produce dense films with high quality even for epitaxial growth at a moderate rate.

The liquid-like behaviour of small CNPs including Fujita's concept of the magic size can be best explained by the effect of charge on weakening the bond strength shown in Table 1.1 in Chap. 1. The active vibration of atoms in small nanoparticles during TEM observation such as shown in Fig. 13.3 has been explained by electron bombardments. However, Iijima's observation that the pronouncedly active vibration of atoms on the insulating substrate changes to less active or non-active vibration on the conducting substrate indicates that the charge is responsible for the pronounced activity. If this is correct, many effects that have been attributed to electron or ion bombardments might actually come from the charge itself. The effect of charge can be more generally understood if the charge weakens the bond strength as introduced in Chap. 1. Then, the charge would enhance chemical reactions and probably explain the following phenomenon of gas-mediated electron or ion beam induced deposition and etching.

13.4 Gas-Mediated Electron or Ion Beam Induced Deposition and Etching

Recently, beams of electrons and ions are used to modify a surface locally at micron and submicron dimensions or fabricating in three dimensions (Utke et al. 2008), which is known as gas mediated electron beam induced etching (Randolph et al. 2011), electron beam induced chemical dry etching (Lobo et al. 2012) or subtractive 3D printing (Martin et al. 2014). Focused electron and focused ion beams can have diameters below 10 nm and occupy a range between individual atom manipulation with scanning tunneling microscope probes and laser beams, both in dimensions and writing speed (Utke et al. 2008).

The apparatus used in both cases resembles a scanning electron microscope. A schematic of the configuration used for either electron or ion induced processes is shown in Fig. 13.7. The gas is introduced to the close proximity of where the beam is incident on the sample through a needle of ~ 1 mm diameter.

Randolph et al. (2011) studied the kinetics of gas mediated electron beam induced etching (EBIE) using the XeF_2 precursor and SiO_2 substrate. At room temperature, XeF_2 does not give rise to etching of SiO_2 (Loudiana et al. 1984; Joyce et al. 1988; Ibbotson et al. 1984; Hills and Arnold 1991; Coburn and Winters 1979). However, etching is induced by electron irradiation (Loudiana and Dickinson 1985; Rack et al. 2003; Randolph et al. 2006; Utke et al. 2008). EBIE can be carried out at room temperature with sub-5 nm and sub-beam-size lateral resolution but the mechanisms behind etch kinetics are not fully understood (Randolph et al. 2011). Figure 13.8 shows cylindrical etch pits with a diameter (d) of 450 nm together with a schematic illustration of gas mediated etching.

Martin et al. (2014) fabricated the nano-architecture of diamond pillar using gas-mediated EBIE as shown in Fig. 13.9. The technique achieves dry chemical etching at room temperature through the dissociation of surface-adsorbed H_2O molecules by energetic electrons in a water vapor environment.

They also patterned the letters 'UTS' and 'NANO' into individual microparticles as shown in Fig. 13.10 simply by tracing out the letters using an electron beam etching with a 20 keV electron beam, while charging was stabilized using a low

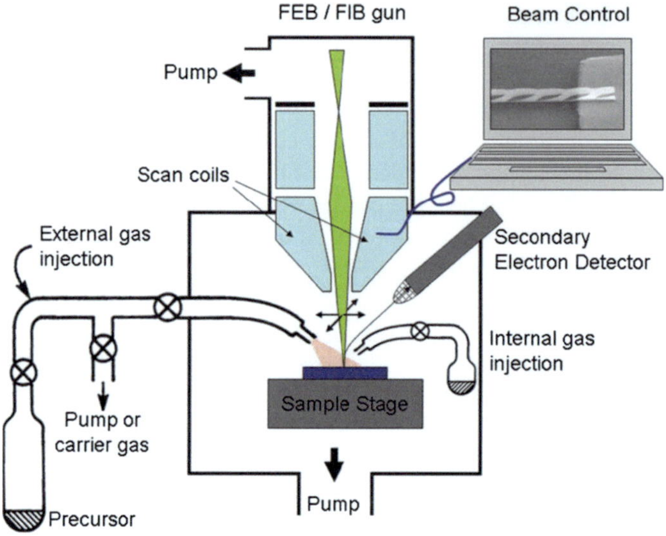

Fig. 13.7 Schematics of field ion beam (FIB) and field electron beam (FEB) nanofabrication systems. Gas injection systems with external and internal precursor reservoirs are shown (Utke et al. 2008)

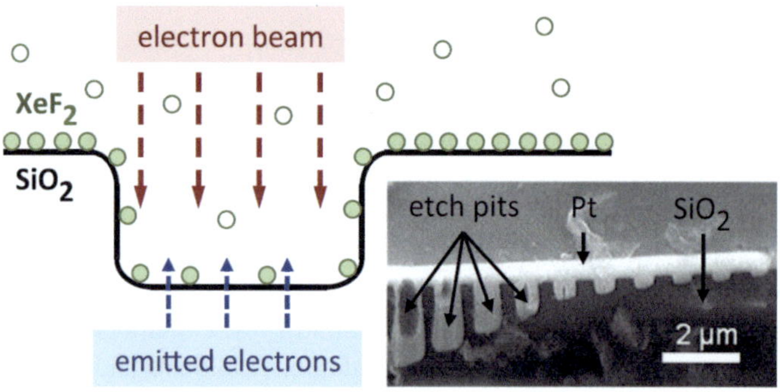

Fig. 13.8 Schematic illustration of gas mediated etching induced by a stationary electron beam with a top-hat flux profile. Empty and filled circles represent precursor gas molecules and adsorbates respectively. *Inset* Cross-sectional electron image of a typical set of Pt-capped etch pits in SiO_2 (Randolph et al. 2011)

vacuum (13 Pa) environment of H_2O. The letters are clearly visible in SEM images (Fig. 13.10a), while atomic force microscope (AFM) maps of the 'UTS' logo show line widths and depths of 100 nm (Fig. 13.10b)).

Figure 13.10c shows an individual diamond microparticle with visible (111) facets and Fig. 13.10d shows the word 'NANO' imprinted in the crystal with the letters 'NA', 'N' and 'O' occupying all three (111) planes, respectively.

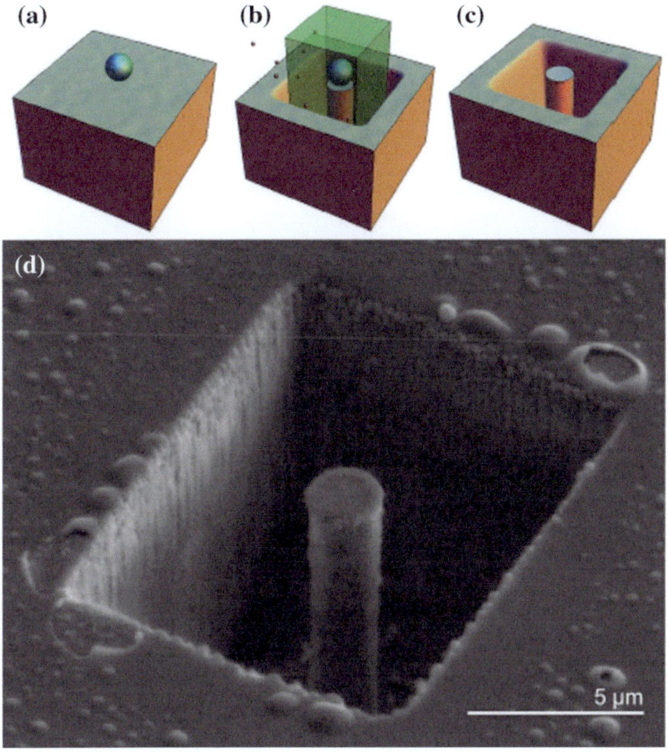

Fig. 13.9 Diamond pillar fabricated by mask-based EBIE. **a–c** Schematic illustration of pillar fabrication by H_2O mediated EBIE. **a** Silica bead on a diamond surface. **b** Diamond volatilization achieved by scanning a 2 keV electron beam over a rectangle repeatedly in a H_2O environment. The silica bead acts as an etch mask that prevents the electrons from reaching the diamond surface. **c** Final pillar geometry after the silica bead was removed from the substrate. **d** Electron image of a pillar fabricated in single crystal diamond by H_2O-mediated EBIE using the process shown in **a–c** (Martin et al. 2014)

However, the etching and deposition mechanism of EBIE is not clearly understood. Definitely the thermodynamic driving force exists for this irreversible process but it is difficult to explain the enhanced kinetics at room temperature. However, considering that the electric charges may weaken the bond strength of the diamond surface as well as the adsorbed H_2O, it is highly probable that the electric charge might be related with the enhanced kinetics of EBIE although the detailed mechanism needs further study.

H_2O cannot etch diamond directly and must be decomposed into H and O to etch diamond. In enhancing the rate of this decomposition, the electric charge will play an important role. Besides, the charged hydrogen and oxygen are expected to be much more reactive than neutral ones and thereby enhance the etching rate of diamond.

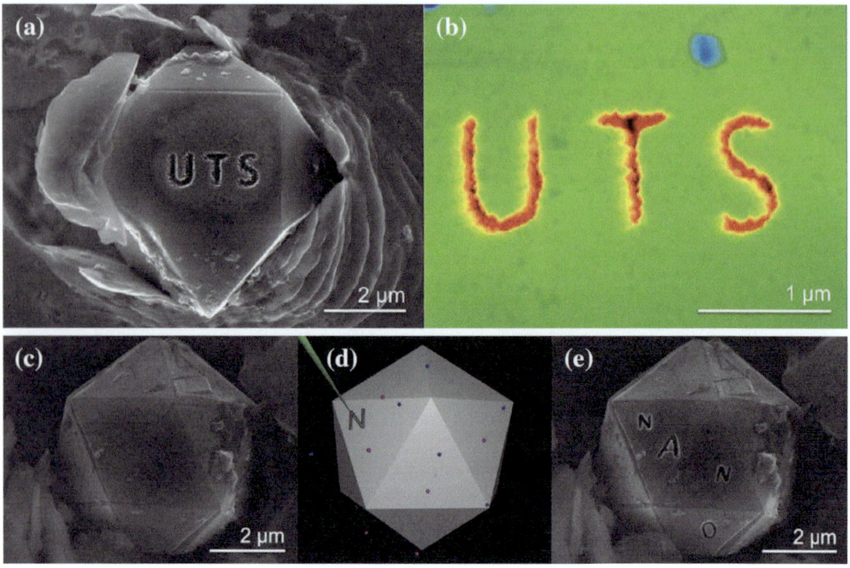

Fig. 13.10 **a** SEM image of the symbol 'UTS' patterned by H_2O mediated EBIE on the 110 plane of a single diamond micro-particle embedded in platinum. **b** AFM image of the symbol 'UTS' shown in a (depth of each letter, 100 nm). **c** SEM image of a diamond micro-particle. **d** Schematic illustration of the process used to pattern the micro-particle shown in c. Each letter of 'NANO' was patterned individually using H_2O mediated EBIE on three different 111 faces of diamond. **e** SEM image of the microparticle shown in c after the letters 'NANO' were patterned by EBIE (Martin et al. 2014)

The weakened bond between carbon atoms of the charged diamond surface might also contribute to the enhanced rate of etching. This mechanism of weakening bond strength by charging may be confirmed by ab initio calculations.

13.5 Catalytic Effect by Charge

The weakening bond strength by charge can also explain numerous catalytic phenomena, including the enzyme catalyzed reaction, where the charge is involved. For example, the reaction of neutral molecules, $H_2 + ½O_2 \rightarrow H_2O$ has a high kinetic barrier but the reaction of charged species $H^+ + (OH)^- \rightarrow H_2O$ has a much lower kinetic barrier, which occurs even at room temperature in the electrochemistry. The same is true for the dissociation of a water molecule, which is a big issue as a way to get hydrogen these days. It can be said that charge is a catalyser.

The catalytic metals that have a high hydrogen evolution rate are listed in Table 6.1. The high charge transfer rate of these transition metals might be related with partially-filled d orbitals. This high charge transfer rate is responsible for the soot growth on the substrate of these metals such as Pd, Pt, Ni and Fe. According to new understanding of the soot formation on the Fe substrate in the diamond CVD process, the soot is a result from the transformation of charged diamond nanoparticles after losing an electron to Fe. Considering the loose skeletal structure of soot, the electron transfer from diamond to Fe should occur by the tunnelling process. This means that Fe has an ability to receive electron from negative CNPs by tunnelling. When CNPs are positive, Fe has an ability to give electrons to CNPs. In other words, the catalytic metals have the ability to exchange charge quickly with others. This ability to exchange charge might be related with the catalytic ability of Pd, Pt, Ni and Fe. This would be why there exists an almost perfect correlation between the hydrogen evolution rate and the soot formation in Chap 6.6.

References

Ajayan PM, Iijima S (1991) Wetting and de-wetting transitions of small metal particles on substrates under electron irradiation. J Colloid Interf Sci 147(1):281–285. doi:http://dx.doi.org/10.1016/0021-9797(91)90155-2

Banfield JF, Welch SA, Zhang H, Ebert TT, Penn RL (2000) Aggregation-based crystal growth and microstructure development in natural iron oxyhydroxide biomineralization products. Science 289(5480):751–754

Bardotti L, Jensen P, Hoareau A, Treilleux M, Cabaud B (1995) Experimental observation of fast diffusion of large antimony clusters on graphite surfaces. Phys Rev Lett 74(23):4694–4697

Coburn JW, Winters HF (1979) Ion- and electron-assisted gas-surface chemistry—An important effect in plasma etching. J Appl Phys 50(5):3189–3196. doi:http://dx.doi.org/10.1063/1.326355

Fujita H (1991) Atom clusters—new applications of high-voltage electron microscopy "micro-laboratory" to materials science. Ultramicroscopy 39(1–4):369–381. doi:http://dx.doi.org/10.1016/0304-3991(91)90217-T

Fujita H (1994) Studies on atom clusters by ultra-high voltage electron microscopy. Mater T JIM 35(9):563–575. doi:http://doi.org/10.2320/matertrans1989.35.563

Fujita H, Komatsu M (1987) In situ experiments of sintering process by high voltage electron microscopy. Sintering'87 1:423–428

Hills MM, Arnold GS (1991) Chemical treatment of fused silica with xenon difluoride. Appl Surf Sci 47(1):77–90. doi:http://dx.doi.org/10.1016/0169-4332(91)90104-R

Ibbotson DE, Flamm DL, Mucha JA, Donnelly VM (1984) Comparison of XeF_2 and F-atom reactions with Si and SiO_2. Appl Phys Lett 44(12):1129–1131. doi:http://dx.doi.org/10.1063/1.94665

Iijima S (1984) Ultra-fine spherical particles of γ-alumina: electron microscopy of crystal structure and surface morphology at atomic resolution. Jpn J Appl Phys 23(6A):L347

Iijima S (1987) Some experiments on structural instability of small particles of metals. In: Sugano S, Nishina Y, Ohnishi S (eds) Microclusters, vol 4. Springer Series Mater. Springer, Berlin Heidelberg, pp 186–199. doi:10.1007/978-3-642-83064-8_25

Iijima S, Ajayan PM (1991) Substrate and size effects on the coalescence of small particles. J Appl Phys 70(9):5138–5140. doi:http://dx.doi.org/10.1063/1.348990

Iijima S, Ichihashi T (1985) Motion of surface atoms on small gold particles revealed by HREM with real-time VTR system. Jpn J Appl Phys 24(2A):L125

Iijima S, Ichihashi T (1986) Structural instability of ultrafine particles of metals. Phys Rev Lett 56(6):616–619

Joyce S, Langan JG, Steinfeld JI (1988) Reactions of XeF_2 with thermally grown SiO_2. Surf Sci 195(1–2):270–282. doi:http://dx.doi.org/10.1016/0039-6028(88)90796-0

Komatsu M, Fujita H HVEM Observation of phase-transformations in ZrO_2 ceramics. In: The forty-fifth annual meeting of the japanese society of electron microscopy, Osaka Shoko Kaigisho, Osaka, May 31–June 2, 1989. vol 38. No. 4, J Electron Microsc, Tokyo, p 265

Lobo CJ, Martin A, Phillips MR, Toth M (2012) Electron beam induced chemical dry etching and imaging in gaseous NH_3 environments. Nanotechnology 23(37):375302

Loudiana MA, Dickinson JT (1985) Summary abstract: Simultaneous exposure of SiO_2 and ThF_4 to XeF_2 and energetic electrons. J Vac Sci Technol B 3(5):1393–1396. doi:http://dx.doi.org/10.1116/1.582998

Loudiana MA, Schmid A, Dickinson JT, Ashley EJ (1984) The chemical sputtering of silica by Ar^+ ions and XeF_2. Surf Sci 141(2–3):409–416. doi:http://dx.doi.org/10.1016/0039-6028(84)90140-7

Martin AA, Toth M, Aharonovich I (2014) Subtractive 3D printing of optically active diamond structures. Sci Rep 4:5022. doi:10.1038/srep05022.http://www.nature.com/articles/srep05022#supplementary-nformation

Mori H, Komatsu M, Takeda K, Fujita H (1991) Spontaneous alloying of copper into gold atom clusters. Phil Mag Lett 63(3):173–178. doi:10.1080/09500839108205987

Rack PD, Randolph S, Deng Y, Fowlkes J, Choi Y, Joy DC (2003) Nanoscale electron-beam-stimulated processing. Appl Phys Lett 82(14):2326–2328. doi:http://dx.doi.org/10.1063/1.1565696

Randolph SJ, Fowlkes JD, Rack PD (2006) Focused, nanoscale electron-beam-induced deposition and etching. Crit Rev Solid State Mater Sci 31(3):55–89. doi:10.1080/10408430600930438

Randolph S, Toth M, Cullen J, Chandler C, Lobo C (2011) Kinetics of gas mediated electron beam induced etching. Appl Phys Lett 99(21):213103. doi:http://dx.doi.org/10.1063/1.3662928

Shan Z, Adesso G, Cabot A, Sherburne M, Asif SS, Warren O, Chrzan D, Minor A, Alivisatos A (2008a) Ultrahigh stress and strain in hierarchically structured hollow nanoparticles. Nat Mater 7(12):947–952

Shan Z, Li J, Cheng Y, Minor A, Asif SS, Warren O, Ma E (2008b) Plastic flow and failure resistance of metallic glass: Insight from in situ compression of nanopillars. Phys Rev B 77(15):155419

Utke I, Hoffmann P, Melngailis J (2008) Gas-assisted focused electron beam and ion beam processing and fabrication. J Vac Sci Technol B 26(4):1197–1276. doi:http://dx.doi.org/10.1116/1.2955728

Velev OD, Lenhoff AM, Kaler EW (2000) A class of microstructured particles through colloidal crystallization. Science 287(5461):2240–2243. doi:10.1126/science.287.5461.2240

Yuk JM, Park J, Ercius P, Kim K, Hellebusch DJ, Crommie MF, Lee JY, Zettl A, Alivisatos AP (2012) High-resolution EM of colloidal nanocrystal growth using graphene liquid cells. Science 336(6077):61–64. doi:10.1126/science.1217654

Zheng K, Wang C, Cheng Y-Q, Yue Y, Han X, Zhang Z, Shan Z, Mao SX, Ye M, Yin Y, Ma E (2010) Electron-beam-assisted superplastic shaping of nanoscale amorphous silica. Nat Commun 1:24. doi:http://www.nature.com/ncomms/journal/v1/n3/suppinfo/ncomms1021_S1.html

Zheng H, Liu Y, Mao SX, Wang J, Huang JY (2012) Beam-assisted large elongation of in situ formed Li_2O nanowires. Sci Rep 2:542. doi:10.1038/srep00542. http://www.nature.com/articles/srep00542#supplementary-information

Chapter 14
Implications and Applications

Charge enhanced kinetics, which is a new discovery derived from the deposition behaviour of CNPs, will have a great scientific and technological impact. Many phenomena in thin film processes as well as other materials processes, which had not been understood clearly, could be understood by charge-enhanced kinetics. From this new discovery, many examples of new understanding and application can be thought of. Among them, some specific examples as well as some speculative ones will be introduced here.

14.1 Summary of the Theory of Charged Nanoparticles

The generation of neutral and charged nanoparticles in the gas phase of the thin film process and their deposition behaviour can be summarized as follows.

- Clustering or nucleation in the gas phase is inevitable under typical processing conditions of thin films by CVD or PVD because the growth rate would be impractically low under the condition without clustering or nucleation in the gas phase.
- Neutral clusters or nanoparticles undergo diffusion-limited aggregation (DLA) producing a fractal-like structure and therefore produce voids in the films whereas charged clusters or nanoparticles undergo not only self-assembly and orientation-assembly but also fusion-like coalescence without leaving any void because they are liquid-like.
- The film quality would be best when the dominant growth flux is an atom or molecule without clustering or nucleation in the gas phase although the growth rate would be impractically low.
- The film quality would be worst with lots of voids when the dominant growth fluxes are neutral clusters or nanoparticles.

© Springer Science+Business Media Dordrecht 2016

N.M. Hwang, *Non-Classical Crystallization of Thin Films and Nanostructures in CVD and PVD Processes*, Springer Series in Surface Sciences 60,
DOI 10.1007/978-94-017-7616-5_14

- The film quality would be moderate without any void with a moderate growth rate when the dominant growth fluxes are charged clusters or nanoparticles.
- Small CNPs are more liquid-like than large CNPs. Therefore, epitaxial growth would be favoured by small CNPs and high substrate temperature whereas non-epitaxial or nanostructure growth would be favoured by large CNPs and low substrate temperature.
- Electrostatic interaction of CNPs is an important factor in the deposition behaviour of CNPs and is related to selective deposition, nanowire growth, interaction with other CNPs and the evolution of a variety of nanostructures. Conducting CNPs would grow much faster than insulating CNPs.
- The motion of CNPs will be normally random or Brownian but will be affected by thermophoretic, drag and electrostatic forces. Applying the electric bias will impose strong electrophoretic force.

14.2 Implications in Thin Film Processes

14.2.1 CVD Process

In most CVD processes, CNPs are generated and frequently become a dominant flux for the growth of thin films and nanostructures. This would be why quantum dots and nanopowders can be synthesized by CVD. The number concentration of CNPs would be an important parameter related with the deposition rate of films. Also, the size of CNPs would be an important parameter related with the film quality. For successful epitaxial growth at the given substrate temperature, the size should be minimized. The size can be made smaller normally by decreasing the flow rate of reactant gases, decreasing the reactor pressure, and increasing the inert gas flow rate. Increasing the reactor temperature normally increases the charging tendency and thereby increases the number density of CNPs.

Increasing the number of charges in the CVD reactor would decrease the size of CNPs. If there is a unipolar charging source, the size of CNPs can be maintained small. This is the case of hot filament CVD (HFCVD), especially for diamond synthesis, where negative CNPs are remarkably dominant. The size of CNPs would be the most important parameter in epitaxial growth because the liquid-like tendency increases with decreasing size.

Multiple charging is another method to make nanoparticles liquid-like and would favour the epitaxial growth. The simplest way to induce multiple charging is to use plasma. From the viewpoint of the TCN, plasma is a very promising tool of deposition because the generation of charges, which is a very important parameter of generating CNPs, can be controlled.

14.2.2 Plasma Enhanced CVD (PECVD)

In the PECVD process, lots of ions, radicals and electrons are generated. Considering that charge weakens the bond strength of molecules, the charged reactants in plasma would decompose at temperature drastically lower than that of thermal CVD. As a result, CNPs would be generated at low temperature. Besides, because of charge-enhanced atomic diffusion of nanoparticles, crystalline CNPs would be formed and deposit as a crystalline film on a substrate at low temperature. This would be why the processing temperature can be much lowered if PECVD is used. Charge-enhanced kinetics, which weakens the bond strength of reactant molecules and CNPs, may play a major role in the enhanced kinetics of plasma CVD.

In addition to charged species, lots of excited species such as radicals are generated in the PECVD process. Excited species have a higher energy than unexcited species which are in their ground state. This energy difference might be related with weakening the bond strength of excited species, which will make them reactive. This presumption is from the analogy that the electrostatic energy of CNPs given by (1.2) might be related with weakening the bond strength. Among these excited or charged species, the species with a sufficiently long life time would contribute to enhancing kinetics.

14.2.3 Low Pressure Synthesis of Diamonds

Chapters 5 and 6 can be summarized as describing thermodynamic and kinetic requirements for the low pressure synthesis of diamond. The thermodynamic requirements that make diamond more stable than graphite are the small size (high capillary pressure) and the negative charge. The kinetic requirement is the sufficiently high mobility of atoms in the nucleation stage of CNPs, which can be achieved by high processing temperature or by high charge density. If the temperature is not high enough, the CNPs may have an amorphous carbon structure. In order for the CNPs to have a crystalline diamond structure, which is necessary for deposition as diamond films, the CNPs should be liquid-like. Considering that the Debye temperature of diamond is $\sim$1930 °C (Spear and Dismukes 1994), the temperature for CNPs to be liquid-like would be rather high.

These conditions are satisfied in the HFCVD process. The small size of CNPs is attributed to the unipolar charging environment of HFCVD. The processing temperature is high enough because the hot filament is higher than $\sim$2000 °C. In the plasma CVD process of a bipolar charging environment, the size is expected to be larger than that in the HFCVD process. Besides, the temperature is not so high as that of HFCVD. These two features would be disadvantageous in forming crystalline diamond nanoparticles. However, one critical advantage would be multiple charging, which not only increases the stability of diamond but also makes CNPs liquid-like as expected from (1.2). Therefore, high density plasma such as microwave can also generate diamond CNPs.

According to this analysis based on the TCN, the charge density of plasma would be critical to synthesizing diamond. The ability to produce diamond CNPs would depend on the plasma density. The plasma density of microwave plasma (MP) is higher by 10–100 times than inductively couple plasma (ICP). Again the plasma density of ICP is higher by 10–100 times than capacitively coupled plasma (CCP) (Chapman 1980). The analysis based on the TCN is supported by the experimental observation that MP is known to produce diamonds whereas there is no report for the diamond synthesis by CCP. In some narrow processing conditions, ICP is known to produce diamonds.

If this analysis is correct, if the gas phase temperature in the CCP and ICP is high enough, CNPs would be liquid-like or the atoms in CNPs would be mobile and diamond CNPs can be formed, which can deposit as diamond films. Then the large area growth of diamond films may be achieved.

14.2.4 Conducting and Insulating CNPs

The collision frequency between CNPs in the reactor would be increased when the reactor pressure is increased because of the increased residence time. Although there may be some exception, it can be said generally that the collision frequency would be higher in atmospheric pressure CVD than in low pressure CVD and the size of CNPs would be larger in atmospheric pressure CVD than in low pressure CVD. For this reason, the film quality is generally better in low pressure CVD.

On the other hand, the magnitude of electrostatic interaction between CNPs described by (7.5) and (7.6) depends strongly on whether CNPs are conducting or insulating. The electrostatic image force between conducting CNPs is much stronger than that between insulating CNPs. This means that the collision frequency of conducting CNPs would be much higher than that of insulating CNPs and thereby conducting CNPs grow much bigger than insulating CNPs. For this reason, the CVD process is not appropriate for depositing conducting films. The ideal reactor to deposit conducting films would be the high vacuum chamber, where the collisions between atoms and between neutral or charged nanoparticles are minimized. This would be why for depositing conducting films, CVD is not preferred but thermal evaporation or sputtering is preferred.

14.2.5 Thermal Evaporation

The generation of charged nanoparticles in the gas phase is least expected during evaporation in vacuum. There are two reasons for this. First, the gas phase nucleation or clustering is not expected, considering that the mean free path is normally larger than the distance between the evaporating source and the substrate. Second, there is no source of charging. Therefore, the thermal evaporation can be a method for ideal atomic deposition, which is also ideal for deposition of conducting metal films. However, the thermal evaporation has a disadvantage of line-of-sight deposition, which results in poor step coverage. Many device fabrication processes require good step coverage.

Nonetheless, there are some evidences indicating the generation of charged nanoparticles. However, the size of nanoparticles is much smaller than that in CVD, the term 'cluster' instead of 'nanoparticle' will be more appropriate. The generation of charged clusters during evaporation of metals in vacuum is worth studying because it will reveal clustering mechanism as well as charging mechanism in vacuum (Barnes et al. 2002; Lee et al. 2002; Jeon et al. 2003, 2005). The generation of clusters during thermal evaporation might be related with the report that the process of the thermal evaporation of metals is irreproducible (Chidsey et al. 1988) or very sensitive to small changes in the evaporation parameters (Poppa et al. 1971). Although the majority of the reports (Chidsey et al. 1988; Buchholz et al. 1991; DeRose et al. 1991; Inukai et al. 1991; Koch et al. 1992) claimed that the film quality was improved by the low evaporation rates, some reports (Levlin et al. 1997; Hwang and Dubson 1992; Baski and Fuchs 1994) claimed that the film quality was improved by the high growth rate. The generation of clusters would be responsible for irreproducibility and poor quality of films often encountered in the thermal evaporation process. The generated clusters may be neutral or charged. If the clusters are neutral, the problem becomes worse.

Barnes et al. (2002) and Jeon et al. (2003, 2005) confirmed the generation of charge by measuring the current during thermal evaporation of metals as shown in Fig. 3.18. The polarity of charge was influenced by the evaporation temperature and by the heating method. For example, when the temperature of copper evaporation was below 1300 °C, the major polarity of charge was positive but when the temperature was above 1300 °C, the major polarity of charge was negative.

The origin of negative charging is attributed to the electron emission from hot metals (Peineke et al. 2006; Peineke and Schmidt-Ott 2008). The origin of positive charging, which is dominant at low evaporation temperature, is less clear than that of negative charging. One possibility for positive charging would be the surface ionization of alkali impurities in the copper, which was suggested by Peineke et al. (2006), Peineke and Schmidt-Ott (2008). Although a systematic study is needed to determine the mechanism of positive charging during evaporation of gold, copper, silver and aluminium, it is worth mentioning some observations made during the experiments. The generation of positive charges during evaporation of metals in the tungsten basket depends on the time. Initially, an appreciable amount of the positive current is measured on the electrode located at the position of the substrate but the positive current decreases with time. After $\sim$10 min of evaporation, a small amount of the negative current is measured. This aspect was observed reproducibly. This means that the source of positive charges disappears with the evaporation time.

From this observation, it is inferred that the tungsten basket has some impurities or oxide on the surface. This surface layer would be responsible for positive charging. Although it is not known whether this layer is intentionally made or not, this layer plays an important role in generating positively charged clusters during evaporation of metals and turns out to be helpful in improving the film quality at a high deposition rate.

Jeon et al. (2003) compared the deposition behaviour of copper between two evaporation temperatures of 950 and 1300 °C because the current measured by an electrometer was below a noise level of ~10 pA/mm^2 at 950 °C compared with ~100 pA/ mm^2 at 1300 °C. Therefore it was expected at the evaporation temperature of 950 °C that clusters would not be generated, resulting in exclusively atomic deposition and the resulting film would have the highest quality although the film growth rate would be much lower than that of the evaporation temperature of 1300 °C. However, quite unexpectedly the surface morphology of the film evaporated at 950 °C was worse than that evaporated at 1300 °C as shown in Figs. 14.1 and 14.2.

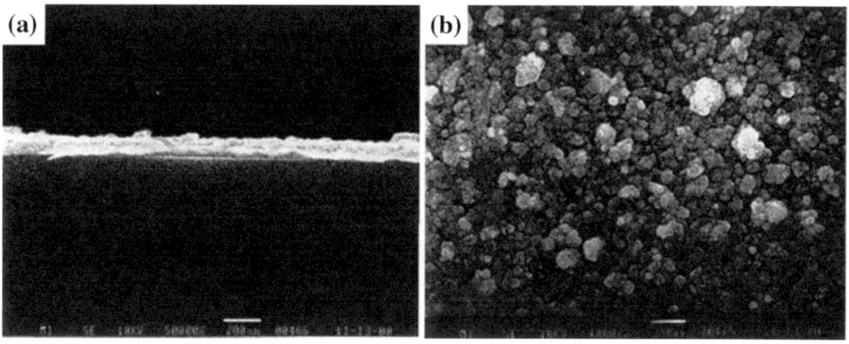

Fig. 14.1 **a** Cross-section and **b** plan views by FESEM of the films deposited by evaporation of copper at 950 °C for 140 min. The magnification bars in **a** and **b** are 200 and 250 nm, respectively (Jeon et al. 2003)

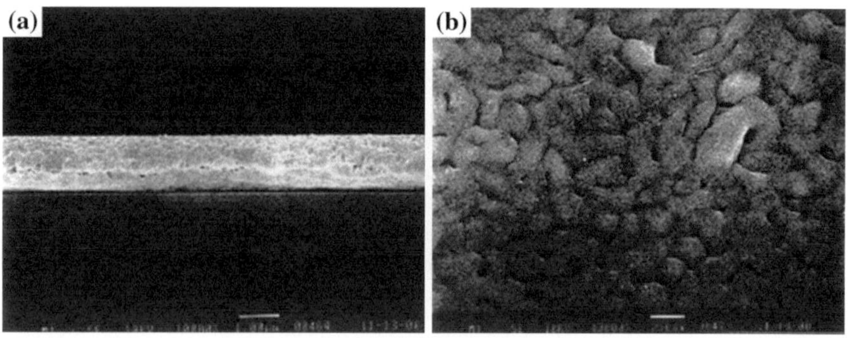

Fig. 14.2 **a** Cross-section and **b** plan views by FESEM of the films deposited by evaporation of copper at 1300 °C for 7 min. The magnification bars in **a** and **b** are 1080 and 250 nm, respectively (Jeon et al. 2003)

This poor film quality is highly in contrast with the film quality deposited at 1300 °C shown in Fig. 14.2.

Figures 14.1a and 14.2a show cross-section images and Figs. 14.1b and 14.2b show plan view images of FESEM of the films. The film growth rate for 950 °C ($\sim$0.01 nm/s) was about 300 times lower than that for 1300 °C ($\sim$3 nm/s). The deposition times for Figs. 14.1 and 14.2 were respectively 140 and 7 min. Although the growth rate was extremely low at the evaporation temperature of 950 °C, the film morphology was much worse than that evaporated at 1300 °C. It would be difficult to explain such microstructure evolutions by classical crystallization. These results are in agreement with previous reports (Levlin et al. 1997; Hwang and Dubson 1992; Baski and Fuchs 1994) claiming that the film quality was improved by the high growth rate.

It is important to understand why the morphology of the film evaporated at 950 °C was worse than that at 1300 °C. Since the surface roughness shown in Fig. 14.1 was not expected to result entirely from atomic deposition, some neutral clusters appeared to have been generated during evaporation at 950 °C. It is known that the soft landing of small neutral clusters results in the random aggregation characteristic of a fractal structure when surface diffusion is not appreciable (Bardotti et al. 1995). In contrast, the smooth morphology of the film evaporated at 1300 °C shown in Fig. 14.2 is attributed to deposition by a major flux of charged clusters. These results imply that charged clusters make a much higher deposition rate and at the same time a much denser and smoother film than neutral ones.

If the clusters happen to be charged and small enough, they will be liquid-like and would not do much harm to the quality of films. But if they are neutral, they will make problems. In order to solve both problems of poor step coverage and generation of neutral clusters, it would be good to introduce the plasma to the thermal evaporation process. In order to generate plasma, the pressure should be increased, which will increase the step coverage. All the clusters will be charged due to plasma and produce the films of relatively high quality at the high deposition rate. Inductively-coupled plasma (ICP)-assisted thermal evaporation, which has been used for deposition of oxides or compounds, would be a good example.

Plasma-assisted thermal evaporation is known to produce higher quality films than normal thermal evaporation (Bunshah 1983; SchÜTz and Hartnagel 1989; Li et al. 2015). It was claimed that plasma-assisted thermal evaporation can produce high quality epitaxial growth of InP and the process does not require any expensive ultrahigh-vacuum system as with the conventional molecular beam epitaxy (MBE) technique and no toxic gases need to be handled as in the metal-organic compound vapour deposition (MOCVD) process (Schütz and Hartnagel 1989).

14.2.6 Sputtering

Another way to solve poor step coverage is the thin film deposition by sputtering. The generation of clusters is well established in sputtering (Katakuse et al. 1985,

1986). If the sputter deposition produces voids in the film, it would be an indirect indication that neutral clusters are generated. Increasing the processing pressure is one way to reduce the percentage of neutral clusters by increasing the collisions between charged species and neutral clusters. However, increasing the pressure results in increasing the cluster size by increasing the collisions, which would be harmful to the film quality. The introduction of the additional plasma would also help to remove neutral clusters and induce multiple charging. Since it is well known that the additional ionization in sputtering improves the film quality as well as the growth rate, the secondary discharge is widely used, which includes inductively coupled plasma (ICP), electron cyclotron resonance (ECR) plasma, high power impulse magnetron sputtering (HIPIMS) and hollow cathode.

14.2.7 Pulsed Laser Ablation

Particle generation during pulsed laser ablation (PLA) is relatively well known and was systematically studied with low pressure DMA by Kawakame et al. (1999) as explained in Chap. 11. PLA has a unique advantage of the stoichiometric material transport above the threshold fluence, which is utilized for deposition of complex compounds such as YBCO high T_c superconductors. From the viewpoint of TCN, the unique advantage of the stoichiometric material transport above the threshold fluence might be related with the generation of CNPs, which needs to be experimentally confirmed.

14.2.8 Aerosol Deposition

Kashu et al. (1984), Hayashi (1997), Kashu and Mihara (1995) reported that dense films could be produced by using a gas jet for the transfer and deposition of metal or ceramic ultrafine particles. This method is known now as aerosol deposition, where submicron ceramic particles are accelerated by gas flow in the nozzle up to velocity of several 100 m/s. During interaction with the substrate, these particles form thick (10–100 μm), dense, uniform and hard ceramics layers at room temperature.

The mechanism of the dense film evolution was investigated by Akedo (2004, 2006, 2008). The densification was attributed to fracture and plastic deformation of the particles on the surface of the substrate, which is called room temperature impact consolidation (RTIC) (Akedo 2008). Akedo claimed that a plasma deformation of starting ceramic particles occur at room temperature based on the observation of the cross-section of the deposited ceramic layer by high resolution TEM (HRTEM) (Akedo 2006).

However, tribocharging or charging by rubbing, which people experience in a daily life, indicates that charging can occur naturally at room temperature. Whenever two dissimilar materials contact each other, electrons will have a greater affinity for one material than the other, and some will move across the interface. This electrostatic charging can be hazardous because the small sparks can easily ignite many flammable materials that are found in industry, often resulting in big explosion in the petroleum industry (Bustin and Dukek 1983). Considering tribocharging or charging by rubbing, it is difficult for the particles to avoid being charged throughout the transfer in the aerosol deposition method.

Actually, the static electricity induced by the friction between power particles was mentioned by the authors based on the observation of luminescence during the fabrication of aerosol gas deposition films (Fuchita et al. 2013). The presence of charge will weaken the bond strength at least of the surface atoms of the particles and contribute to the evolution of the dense films. It is difficult to imagine the evolution of such dense films from the impact of neutral nanoparticles.

14.2.9 Aerosol CVD

Aerosol assisted CVD (AACVD) utilizes the formation of an aerosol precursor solution, which is transported to the reaction chamber by use of a carrier gas, whereby the solvent is evaporated and precursor molecules decompose onto the substrate (Marchand et al. 2013). From precursors dissolved in a solvent, an aerosol is generated by use of an ultrasonic humidifier (nebulizer) (Jones and Hitchman 2009), creating a precursor mist, which is then transported to the CVD reactor by a carrier gas.

Even in this case, the generation of CNPs is indirectly evidenced by the pronounced effect of the electric field on the deposition behaviour (Shaw et al. 2005; Crane et al. 2011; Panjawi et al. 2012). Shaw et al. (2005) applied DC or AC electric bias across the gap of 60 μm while depositing tungsten oxide thin films on gas sensor substrates at 600 °C by AACVD using a precursor $[W(OPh)_6]$ in toluene. Each sensor consisted of a 3 by 3 mm alumina substrate which was screen-printed by gold electrodes of ~10 mm thickness with an interdigitated pattern (finger and gap width of 100 and 60 ± 10 μm respectively).

Deposition of WO_3 on a sensor substrate from the AACVD of $[W(OPh)_6]$ in the absence of an electric field produced material that was localised at the electrode–alumina interface with virtually no growth across the electrode gap. However, if an electric field was applied across the sensor during deposition, it was observed that extended fibres dominated the morphology as shown in Fig. 14.3.

14.2.10 Electrospray Deposition

Electrospraying is a process similar to AACVD but an external electric field is applied to produce charged droplets (Jaworek 2007). The nozzle is usually made in the form of metal capillary, which is biased by a high voltage. If the positive bias is

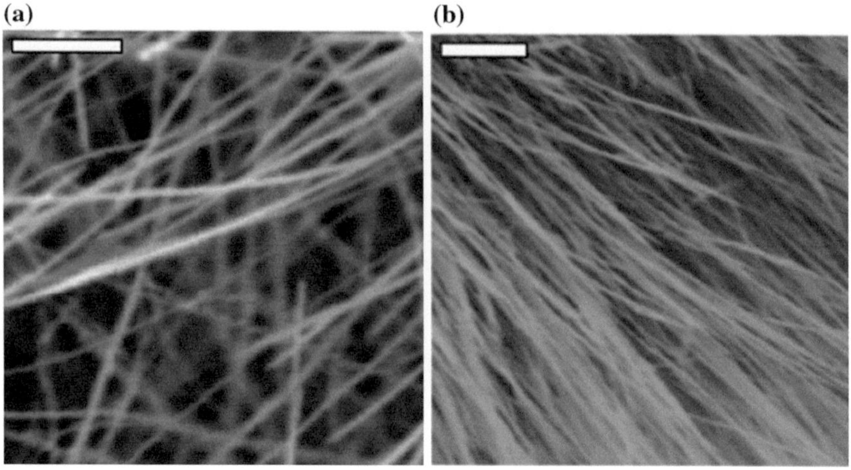

Fig. 14.3 Typical WO_3 fibres deposited by AACVD in the electrode gap under an applied field. **a** Fibres deposited under the DC field strength of 167 V cm^{-1} (scale bar has a width of 1 μm). **b** Fibres deposited under the AC field strength of 167 V cm^{-1} at 50 Hz (scale bar has a width of 2 μm) (Shaw et al. 2005)

applied to the nozzle with respect to the ground, the liquid that comes out of the nozzle has highly multiple positive charges. The shear stress on the liquid surface, due to the established electric field, causes elongation of a jet and its disintegration into charged droplets (Jaworek 2007). When the liquid is evaporated from the droplets, the radius is decreased with charges being maintained. As the evaporation goes on, the charge density continues to increase and eventually Coulomb explosion occurs when the magnitude of charge on a drop overcomes the surface tension, leading to disintegration of the drop. The droplets obtained by this method can be extremely small, in special cases down to nanometers.

Small droplets of a solution or colloidal suspension can provide films of thickness thinner than 1000 nm, depending on the solute concentration or size of suspended particles (Jaworek et al. 2009). When the solvent is evaporated, the remaining charged particles deposit as a film. In addition to the capillary effect of the liquid, the effect of charge on the particle would play an important role in the evolution of dense films.

14.3 Applications in the CVD Process

14.3.1 Epitaxial and Non-epitaxial Growth

Grain size is determined by the ratio of epitaxial and non-epitaxial coalescences of CNPs. If epitaxial coalescence is dominant, the grain size would be large and the structure is columnar whereas if non-epitaxial coalescence is dominant, the grain size would be so small that the resulting structure is called microcrystalline or nanocrystalline. For epitaxial coalescence, the small size of CNPs, multiple charging and high substrate temperature are required. For non-epitaxial coalescence, the large size of CNPs, single charging and low substrate temperature are required. Therefore, depending on what type of film microstructure is needed, the proper processing condition can be designed. The acceleration of CNPs by electric field can be another processing parameter, which will affect the deposition behaviour.

As mentioned earlier, Tagaki et al. (1975, 1976), Takagi (1986) developed the ionized cluster beam deposition (ICBD) technique. The ICBD is known to have a unique advantage to produce high quality or epitaxial films at low substrate temperature. Considering that CNPs are spontaneously generated in most CVD processes, the electric field can be applied in the CVD reactor to accelerate the CNPs toward the substrate. Then a beneficial effect similar to that of ICBD is expected. Lee (2014) tested this possibility in the Si hot wire CVD (HWCVD) process. He applied +200 V to the silicon wafer substrate at 600 °C for 5 min with flow rates of SiH_4, HCl, and H_2 being 2.5, 10 and 10 sccm, respectively, at the wire temperature of 1800 °C and the reactor pressure of 0.1 Torr. Homoepitaxial silicon film could be successfully grown as shown in Fig. 14.4. Although there are some stacking faults, this experiment is just for checking whether the idea work or not and the process can be further optimized.

14.3.2 Low Temperature Deposition of Crystalline Phase

If CNPs are crystalline in the gas phase, they would deposit as a crystalline phase even when deposited at low substrate temperature. This situation occurs in the HWCVD process. Crystalline CNPs will be formed in the high temperature region near the hot filament. If these crystalline CNPs land on the substrate, they may maintain the crystalline structure. According to this logic, crystalline silicon can be deposited even at room temperature or below. This possibility was experimentally confirmed by Lee et al. (2008), who divided the hot-wire reactor into three by two-stage orifices and captured the CNPs on the carbon membrane at room temperature in the third chamber. The working pressures of the first (the HWCVD reactor), the second and the third chambers were maintained respectively at 0.2, 0.01 and 10^{-5} Torr. During the process, the CNPs that were formed in the first chamber of a hot-wire reactor were extracted through the first and the second orifices into the third chamber, where the amorphous carbon membrane of TEM grid was placed for capturing these nanoparticles. Any surface reaction on the TEM grid was not expected to occur because it was at room temperature. The lattice fringes of silicon of 7–8 nm are clearly revealed in Fig. 14.5.

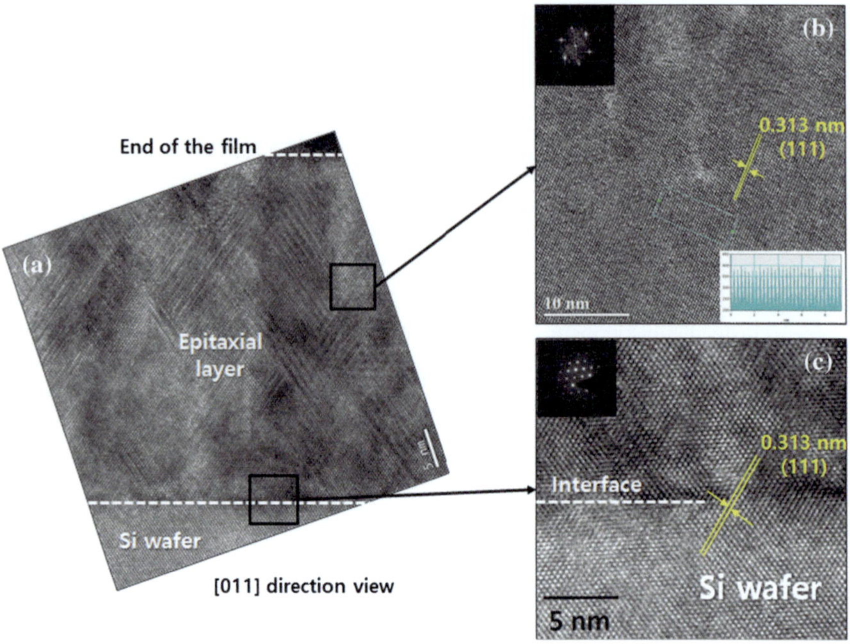

Fig. 14.4 **a** HRTEM image of 50 nm-thick homoepitaxial silicon grown on a Si wafer under the bias of +200 V to the substrate holder during HWCVD. **b** and **c** are the high magnification images of the corresponding *square* areas in (**a**) (Lee 2014)

The diffraction patterns shown in the inset of Fig. 14.5 indicate the silicon lattice. The size of the crystalline silicon nanoparticles decreases with decrease in SiH_4 concentration with 7–8 nm for 20 % SiH_4, 6–7 nm for 15 % SiH_4, 5–6 nm for 10 % SiH_4 and 4–5 nm for 5 % SiH_4. This effect of SiH_4 concentration on the nanoparticle size should be related with the increased amount of silicon precipitation in the gas phase with increasing SiH_4 concentration (Lee et al. 2008).

If these crystalline nanoparticles are deposited on the substrate at low temperature, a crystalline film would be obtained. This is why microcrystalline silicon films can be grown on the glass substrate at such a low temperature (<500 °C) that only amorphous silicon is grown by the thermal CVD process. This concept can be applied to other materials. Based on this concept, Kim et al. (2009) could successfully deposit crystalline silicon nitride at the substrate temperature of 700 °C using HWCVD as shown in Fig. 14.6.

Figure 14.6 shows the TEM images of nanocrystalline silicon nitride films deposited on the Si (100) wafer with a natural silicon oxide surface layer under 4 Torr for 30 min. At the hot wire temperature of 1430 °C, the silicon nitride nanoparticles were mostly amorphous, although some of them have the crystalline lattice as shown by the magnified TEM image in Fig. 14.6a. At the wire temperature of 1730 °C, the silicon nitride nanoparticles were mostly crystalline, as shown in Fig. 14.6b, c.

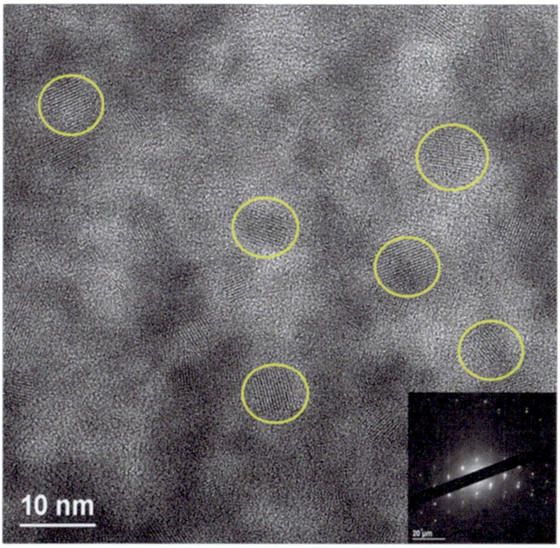

Fig. 14.5 TEM images of silicon nanoparticles captured for 10 s on the amorphous carbon membrane at room temperature during HWCVD with 20 % SiH_4–80 % H_2 at the wire temperature of 1560 °C (Lee et al. 2008)

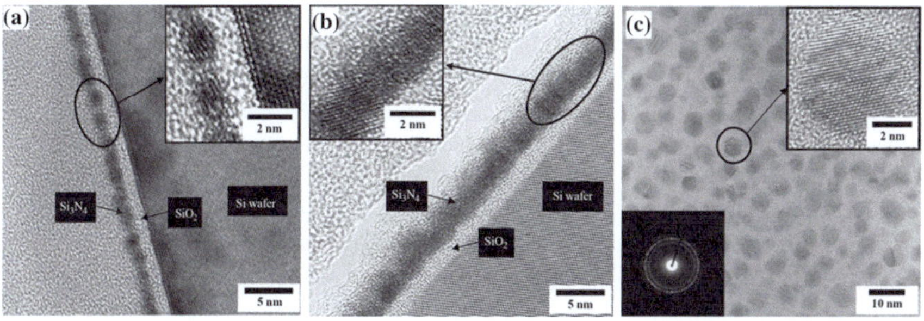

Fig. 14.6 HRTEM images of nanocrystalline silicon nitride deposited on a Si (100) wafer with a natural silicon oxide layer at the substrate temperature of 700 °C and at the wire temperatures of **a** 1430 °C and **b** and **c** 1730 °C under the reactor pressure of 4 Torr (Kim et al. 2009)

14.3.3 Amorphous Incubation Layer

When the substrate is amorphous as in the case of glass, CNPs of crystalline silicon will change to an amorphous structure upon landing on the substrate because CNPs are liquid-like and tend to accommodate their structure to that of the substrate. For this reason, even if crystalline CNPs are formed and contribute to deposition in the

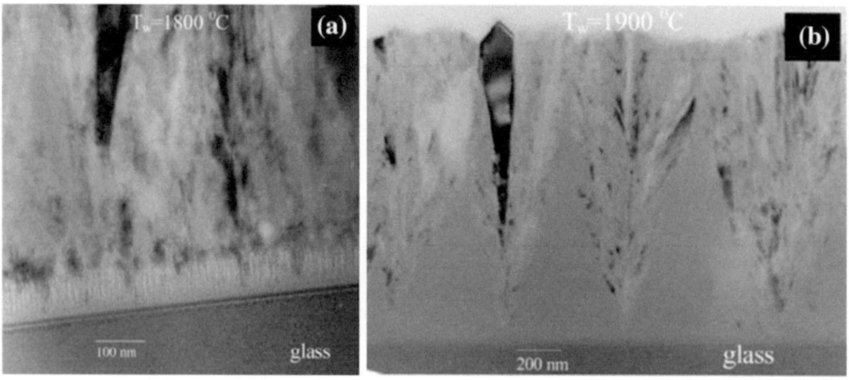

Fig. 14.7 Cross-section TEM images of HWCVD poly-Si films fabricated at wire temperatures of **a** 1800 °C and **b** 1900 °C (Rath et al. 2001)

hot wire or plasma CVD process, the initial layer of silicon on the glass substrate tends to have an amorphous structure, which is called 'amorphous incubation layer (AIL).' Figure 14.7a, b shows a typical AIL on the glass substrate at wire temperatures of 1800 and 1900 °C respectively.

In Fig. 14.7a, the AIL as thick as ~80 nm is grown on the glass substrate and in Fig. 14.7b, the gray triangle shaped AIL is formed on the glass substrate. Note that the AIL is more pronounced at the wire temperature of 1900 °C than at 1800 °C. With increasing wire temperature, the size of CNPs decreases, which has been confirmed reproducibly. This result would be attributed to increasing amount of negative charges with increasing wire temperature. Since the negative charges are the sites for charge-induced nucleation, the size of CNPs decreases with increasing wire temperature. The effect of the wire temperature on the size of CNPs was studied by Kim et al. (2006) as shown in Fig. 14.8.

The negatively charged silicon nanoparticles were captured on the carbon membrane at the temperatures of 1500 and 1800 °C. Under 0.16 Torr of a flowing gas mixture of SiH_4 (1 sccm) and H_2 (9 sccm), the clusters were captured for 30 s. In an attempt to maintain a substrate temperature of 450 °C, the distances between the substrate and the wire were 13 and 40 mm respectively for wire temperatures of 1500 and 1800 °C.

The nanoparticles captured at 1500 °C are larger (~2.5 nm) than those at 1800 °C (~1 nm). Since small CNPs are more liquid-like, they will be more easily transformed to the amorphous structure on glass than large CNPs. In other words, the formation of the AIL is favoured by small CNPs. This explains why the AIL is more pronounced at the wire temperature of 1900 °C than at 1800 °C in Fig. 14.7. The formation of the AIL is also favoured by the high substrate temperature in the range where silicon is not crystallized because the high substrate temperature makes the CNPs liquid-like.

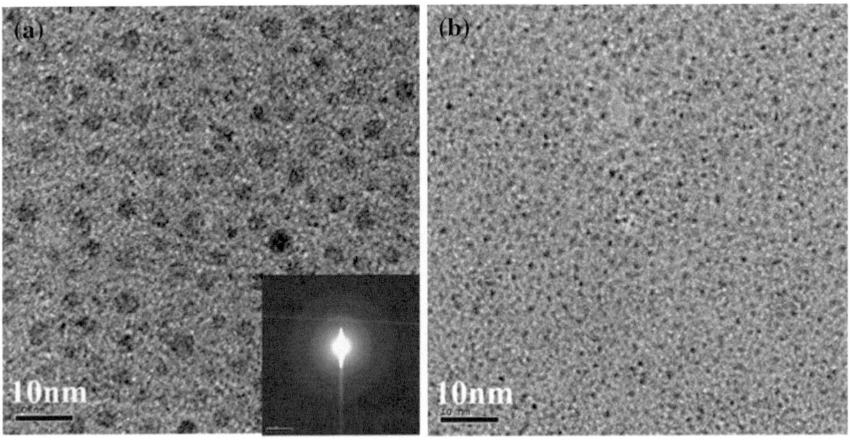

Fig. 14.8 TEM images of Si nanoparticles captured at the wire temperatures of **a** 1500 °C and **b** 1800 °C in the HWCVD process. The diffraction pattern in the inset indicates that the clusters are amorphous (Kim et al. 2006)

The silicon nanoparticles in Fig. 14.8 are amorphous, whereas those in Fig. 14.5 are crystalline. This difference is attributed to the fact that the SiH_4 concentration in Fig. 14.8 is smaller and as a result the size of silicon nanoparticles in Fig. 14.8 is smaller (1–2.5 nm) than that in Fig. 14.5 (7–8 nm). Besides, the substrate temperature in Fig. 14.8 is 450 °C whereas that in Fig. 14.5 is the room temperature. These results confirm again that the large size of CNPs and the low substrate temperature, which decreases the liquid-like tendency of CNPs, is favourable for maintaining the crystallinity of CNPs after landing on the amorphous substrate.

On the other hand, CNPs are small in the initial stage of HWCVD and continue to increase their size until 10–15 min, where the steady state is reached and the size does not increase anymore. In order to examine the size change of CNPs with processing time, the shutter should be installed above the substrate. It should be noted that in the experiments of Figs. 14.5 and 14.8, the shutter was not installed. Therefore, the observed nanoparticles in Figs. 14.5 and 14.8 have a characteristic of nanoparticles in the initial stage of the process.

The size increase of CNPs with the processing time is shown in Fig. 14.9. Silicon nanoparticles were captured for 10 s for TEM observation at the hot wire temperature of 1800 °C, the substrate temperature of 300 °C, the reactor pressure of 0.1 Torr and the flow rates of SiH_4, H_2 and PH_3 being 5, 12, 0.375 sccm, respectively. The distance between the wire and the substrate was 4 cm. Using the shutter above the substrate, the delay time for deposition was controlled. Without the delay (the delay time of 0 min), the size of silicon nanoparticles is 5–10 nm (Fig. 14.9a, d). With the delay time of 5 min, the size increases to 10–25 nm (Fig. 14.9b, e). With the delay time of 15 min, the size increases to 30–50 nm (Fig. 14.9c, f). Figure 14.9d–f are the higher magnification images respectively of

Fig. 14.9a–c. The silicon nanoparticle in Fig. 14.9d shows the lattice fringe, indicating the crystalline structure. However, many other nanoparticles were amorphous accommodating to the amorphous carbon membrane when the delay time was 0 min.

In this initial stage of deposition, the liquid-like property of small CNPs in Fig. 14.9a has a bad effect on crystallinity of the film because CNPs tend to transform to the amorphous phase. Large CNPs in Fig. 14.9c would produce a high crystallinity. This aspect is revealed by the Raman spectra and resistivity of the films deposited with different delay times as shown respectively in Fig. 14.10a, b. As the delay time increases up to 10 min, the crystallinity of the film continues to increase in Raman spectra (Fig. 14.10a). The resistivity of the film without the delay time (0 min) is $\sim 10^4$ Ωcm whereas the resistivity with the delay times of 3, 5 and 10 min is ~ 10 Ωcm, which is 3 orders of magnitudes lower. Note that the data of high and low resistivity have a different scale in the y-axis of Fig. 14.10b. The effect of the delay time on the electrical property of crystalline silicon films deposited on glass was also studied under some other conditions of HWCVD by Chung et al. (2012). It is clear from Figs. 14.9 and 14.10 that if the delay time is controlled by using a shutter, the AIL can be decreased significantly.

Another way to minimize the AIL is to add HCl, which makes the retrograde solubility of silicon as shown in Fig. 7.2 (Hwang 1999). Then, the gas phase nucleation would change the driving force at the substrate temperature from that for

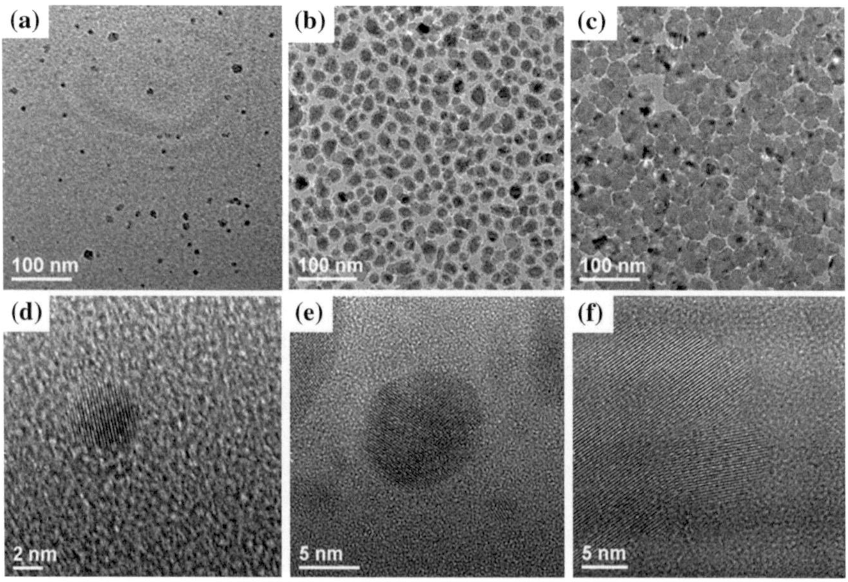

Fig. 14.9 TEM images showing the effect of the delay times of **a** 0 min, **b** 5 min and **c** 15 min on the size of nanoparticles captured on the carbon membrane for 10 s in the silicon HWCVD process. **d**, **e** and **f** are the higher magnification images respectively of (**a**), (**b**) and (**c**)

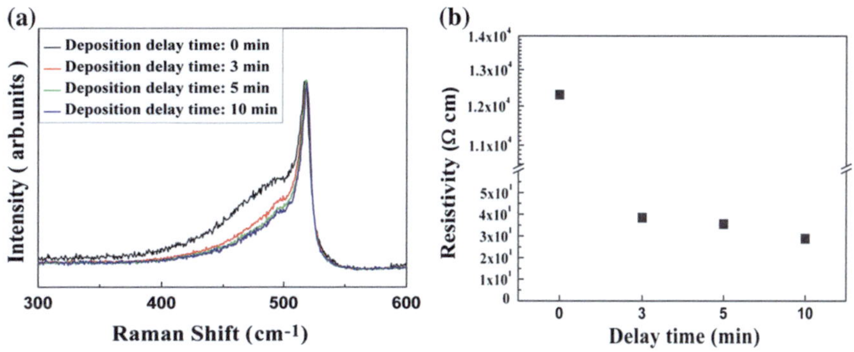

Fig. 14.10 **a** Raman spectra and **b** resistivity data showing the effect of the delay time. The silicon film was deposited for 4 min under the conditions of Fig. 14.9

deposition to that for etching. Increasing HCl increases the equilibrium amount of silicon to be etched and also the etching kinetics. From this fact, how HCl can be used to minimize the AIL and to increase the crystalline fraction was deduced by Chung et al. (2011) as follows.

CNPs formed in the gas phase can be either crystalline or amorphous, depending on the temperature they are formed. If they are formed in a sufficiently high temperature region near the hot wire, they will be crystalline. However, if they are formed in a low temperature region away from the hot wire, they will be amorphous. Crystalline CNPs contribute to the deposition of a crystalline film whereas amorphous CNPs contribute to the deposition of an amorphous film. Besides, the amorphous glass substrate contributes to the deposition of an amorphous film by transforming crystalline nanoparticles to amorphous nanoparticles upon contact.

HCl etches amorphous silicon much faster than crystalline silicon. Therefore, HCl etches not only amorphous silicon CNPs in the gas phase but also amorphous silicon formed on the glass substrate. This effect would minimize the formation of the AIL and maximize the crystallinity of the film at the low substrate temperature. The HCl concentration was controlled by the ratio of the flow rate between HCl and SiH_4, which is designated as $R_{HCl} = [HCl]/[SiH_4]$, where the bracket represents the flow rate. Similarly, the doping concentration was controlled by the ratio of the flow rate between PH_3 and SiH_4, which is designated as $R_{PH3} = [PH_3]/[SiH_4]$.

The effect of HCl on the film crystallinity is revealed by Raman spectra shown in Fig. 14.11. With increasing R_{HCl}, the crystallinity of the film continues to increase at the expense of the growth rate. It should be noted that for $R_{HCl} = 6.9$, the shape of Raman spectra is very close to that of a single crystal silicon wafer. The decrease of the film growth rate with R_{HCl} is revealed by the film thickness, which was 6290, 1660, 860 and 200 nm respectively for $R_{HCl} = 0$, 2.8, 4.2 and 6.9 for the same deposition time of 13 min. Films were deposited under conditions where the substrate-to-filament distance, substrate temperature, wire temperature and pressure were respectively 3 cm, 200 °C, 1800 °C and 0.1 Torr.

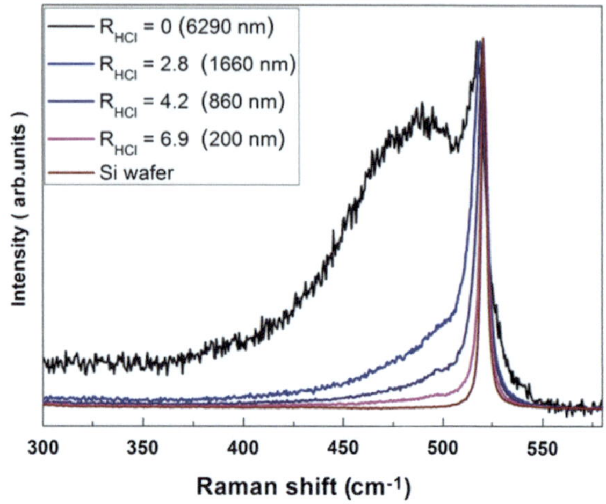

Fig. 14.11 Raman spectra of intrinsic silicon films with varying HCl concentration. *Brown line* is for a single crystal silicon wafer (Chung et al., 2011)

Figure 14.12 shows the HRTEM images of the silicon film deposited with $R_{HCl} = 6.9$. The thickness of the film was 200 nm. The HRTEM images in Fig. 14.12c, d as well as SADP in Fig. 14.12b confirm that the film consists of the crystalline silicon free of AIL. The high resolution image between the glass and the silicon film in Fig. 14.12d shows that AIL is practically absent with its thickness of 0 nm in comparison with ~5 μm for $R_{HCl} = 0$. High resolution image in the middle of the film in Fig. 14.12c shows that the structure is of crystalline silicon with grain boundaries, being free of an amorphous phase. The planes are (111) and (220). The film prepared with $R_{HCl} = 6.9$ has much better crystallinity than that prepared without HCl.

The high quality of the crystalline silicon films prepared by adding HCl was revealed by the electrical properties as shown in Fig. 14.13. A temperature dependent dark conductivity of intrinsic silicon films is shown in Fig. 14.13a. It is seen that the dark conductivity is enhanced with increasing HCl concentration. Note that the dark conductivity of the crystalline silicon prepared with $R_{HCl} = 6.9$ is as high as 3.74×10^{-5} Scm^{-1} at room temperature, which is comparable to that of crystalline silicon prepared by excimer laser annealing (ELA) techniques (Adikaari and Silva 2005).

Figure 14.13b shows the Hall mobility of n-type crystalline silicon films prepared with $R_{HCl} = 7.5$ as a function of doping concentration. A comparison is made with the values from n-type silicon film prepared with $R_{HCl} = 0$. It is again noted that Hall mobilities for all doping concentrations are comparable to those reported on the films prepared by RTA processing (Morimoto et al. 2002). For example, one of the best Hall mobility values reported is 17.5 $cm^2\ V^{-1}\ s^{-1}$ and the value obtained in Fig. 14.13b is 15.8 $cm^2V^{-1}\ s^{-1}$.

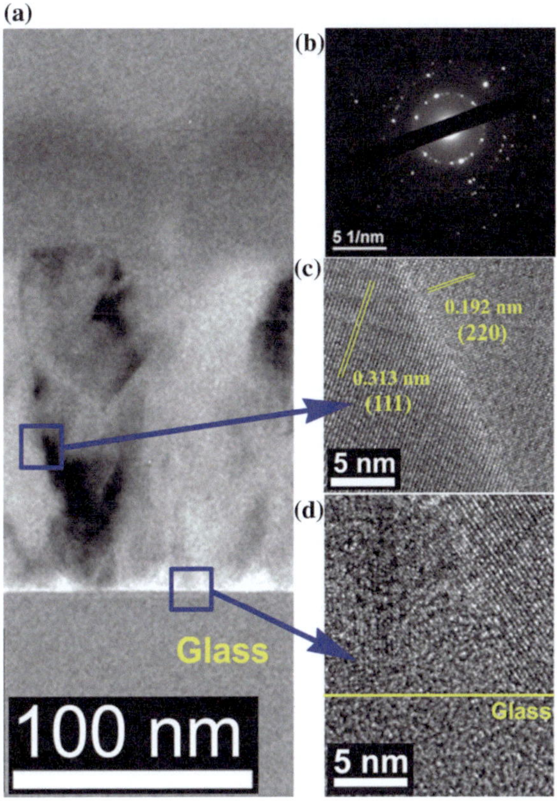

Fig. 14.12 TEM image of intrinsic crystalline silicon at $R_{HCl} = 6.9$ for **a** the bright-field image showing an overall view of the deposited film and **b** the corresponding selected area diffraction pattern (SADP). High resolution images of **c** and **d** are for the locations of the film designated in (**a**) (Chung et al. 2011)

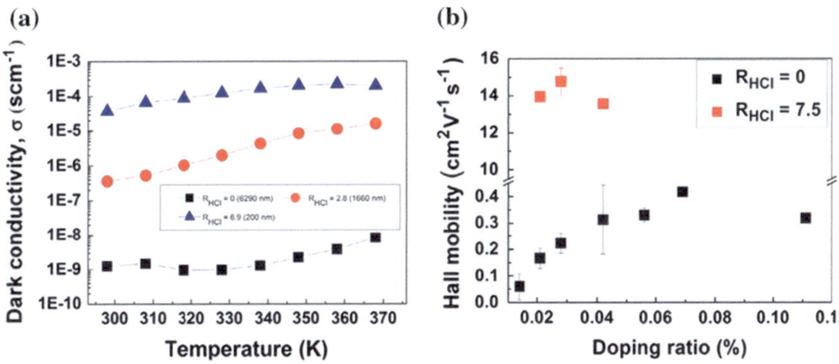

Fig. 14.13 Electrical properties of crystalline silicon films showing **a** the dark conductivity of intrinsic silicon films and **b** the Hall mobility of n-type crystalline silicon films prepared at $R_{HCl} = 7.5$ in comparison with those of n-type silicon films prepared at $R_{HCl} = 0$ (Chung et al. 2011)

14.4 Phenomena that Might Be Related with Charge-Enhanced Diffusion

Among the phenomena of charge-related enhanced atomic diffusion, there are electropulsing, electromigration and spark plasma sintering. In all these phenomena, the atomic diffusion is enhanced when the high current density typically in the range of 10^2–10^5 A cm^{-2} is applied by pulse. Electropulsing was shown to decrease the effective activation energy, 0.8 eV/(V/cm) in the structural relaxation process of amorphous $Cu_{50}Ti_{50}$ (Mizubayashi and Okuda 1989), enhance the crystallization of metallic glass (Holland et al. 2004) and enhance the atomic diffusion (Yang and Conrad 2001; Zhou et al. 2004; Hui et al. 2007; Jiang et al. 2009; Samuel et al. 2010; Qin et al. 2011). Electromigration (Orchard and Greer 2005; Bertolino et al. 2001, 2002; Garay et al. 2003; Zhao et al. 2007) also enhances atomic diffusion significantly regardless of the direction of electron flow.

In spark plasma sintering (SPS) (Shen et al. 2003; Munir et al. 2006, 2011), the electric current is used to activate sintering. SPS is called in various names such as pulsed electric current sintering (PECS), field-activated sintering technique, and current-activated pressure-assisted densification. In the process, the densification is enhanced anomalously by applying electric current with a typical density higher than ~ 100 A cm^{-1} by pulse.

Considering that the charge-enhanced diffusion of CNPs comes from weakening of the bond strength by charge, it might be possible that the enhanced diffusion might be related with weakening of the bond strength caused by the high current density. Since the charge in CNPs is static and the charge in current is dynamic, the analogy is not clearly justified. Systematic works need to be done to confirm the possibility for dynamic charge in high density to weaken the bond strength.

References

Adikaari A, Silva S (2005) Thickness dependence of properties of excimer laser crystallized nano-polycrystalline silicon. J Appl Phys 97(11):114305. doi:http://dx.doi.org/10.1063/1.1898444

Akedo J Aerosol deposition method for fabrication of nano crystal ceramic layer. In: Materials Science Forum, 2004. Trans Tech Publ, pp 43–48. doi:10.4028/www.scientific.net/MSF.449-452.43

Akedo J (2006) Aerosol deposition of ceramic thick films at room temperature: Densification mechanism of ceramic layers. J Am Ceram Soc 89(6):1834–1839. doi:10.1111/j.1551-2916.2006.01030.x

Akedo J (2008) Room temperature impact consolidation (RTIC) of fine ceramic powder by aerosol deposition method and applications to microdevices. J Therm Spray Tech 17(2):181–198. doi:10.1007/s11666-008-9163-7

Bardotti L, Jensen P, Hoareau A, Treilleux M, Cabaud B (1995) Experimental observation of fast diffusion of large antimony clusters on graphite surfaces. Phys Rev Lett 74(23):4694–4697

Barnes MC, Jeon I-D, Kim D-Y, Hwang NM (2002) Generation of charged clusters during thermal evaporation of gold. J Cryst Growth 242(3–4):455–462. doi:http://dx.doi.org/10.1016/S0022-0248(02)01417-3

Baski A, Fuchs H (1994) Epitaxial growth of silver on mica as studied by AFM and STM. Surf Sci 313(3):275–288

Bertolino N, Garay J, Anselmi-Tamburini U, Munir Z (2001) Electromigration effects in Al-Au multilayers. Scripta Mater 44(5):737–742

Bertolino N, Garay J, Anselmi-Tamburini U, Munir Z (2002) High-flux current effects in interfacial reactions in Au–Al multilayers. Philos Mag B 82(8):969–985

Buchholz S, Fuchs H, Rabe JP (1991) Surface structure of thin metallic films on mica as seen by scanning tunneling microscopy, scanning electron microscopy, and low-energy electron diffraction. J Vac Sci Technol B 9(2):857–861

Bunshah RF (1983) Processes of the activated reactive evaporation type and their tribological applications. Thin Solid Films 107(1):21–38. doi:http://dx.doi.org/10.1016/0040-6090(83)90004-4

Bustin WM, Dukek WG (1983) Electrostatic hazards in the petroleum industry. John Wiley and Sons, New York, USA

Chapman B (1980) Sputtering Chapter 6. Glow discharge process:201

Chidsey CE, Loiacono DN, Sleator T, Nakahara S (1988) STM study of the surface morphology of gold on mica. Surf Sci 200(1):45–66

Chung Y-B, Park H-K, Lee D-K, Jo W, Song J-H, Lee S-H, Hwang N-M (2011) Low temperature deposition of crystalline silicon on glass by hot wire chemical vapor deposition. J Cryst Growth 327(1):57–62. doi:http://dx.doi.org/10.1016/j.jcrysgro.2011.05.004

Chung Y-B, Lee S-H, Bae S-H, Park H-K, Jung J-S, Hwang N-M (2012) Effect of the initial structure on the electrical property of crystalline silicon films deposited on glass by hot-wire chemical vapor deposition. J Nanosci Nanotechno 12(7):5947–5951. doi:http://dx.doi.org/10.1166/jnn.2012.6415

Crane J, Warwick M, Smith R, Furlan N, Binions R (2011) The application of electric fields to aerosol assisted chemical vapor deposition reactions. J Electrochem Soc 158(2):D62–D67

DeRose J, Thundat T, Nagahara L, Lindsay S (1991) Gold grown epitaxially on mica: conditions for large area flat faces. Surf Sci 256(1):102–108

Fuchita E, Tokizaki E, Ozawa E, Inoue H, Sakka Y, Kita E (2013) High-temperature phase in zirconia film fabricated by aerosol gas deposition and its change upon subsequent heat treatment. J Ceram Soc Jpn 121(1412):333–337. doi:10.2109/jcersj2.121.333

Garay J, Anselmi-Tamburini U, Munir ZA (2003) Enhanced growth of intermetallic phases in the Ni–Ti system by current effects. Acta Mater 51(15):4487–4495

Holland TB, Löffler JF, Munir ZA (2004) Crystallization of metallic glasses under the influence of high density dc currents. J Appl Phys 95(5):2896–2899

Hui S, Wang Z-J, Gao T-J (2007) Effect of high density electropulsing treatment on formability of TC_4 titanium alloy sheet. T Nonferr Metal Soc 17(1):87–92

Hwang NM (1999) Deposition and simultaneous etching of Si in the chemical vapor deposition (CVD) process: approach by the charged cluster model. J Cryst Growth 205(1–2):59–63. doi:http://dx.doi.org/10.1016/S0022-0248(99)00247-X

Hwang J, Dubson M (1992) Atomically flat gold films grown on hot glass. J Appl Phys 72(5):1852–1857

Hayashi C Gas deposition. In: Materials Science Forum, 1997. Trans Tech Publ, pp 153–180. doi:10.4028/www.scientific.net/MSF.246.153

Inukai J, Mizutani W, Saito K, Shimizu H, Iwasawa Y (1991) Gold substrates for scanning tunneling microscopy of adsorbed species. Jpn J Appl Phys 30(12R):3496

Jaworek A (2007) Electrospray droplet sources for thin film deposition. J Mater Sci 42(1):266–297. doi:10.1007/s10853-006-0842-9

Jaworek A, Sobczyk A, Krupa A, Lackowski M, Czech T (2009) Electrostatic deposition of nanothin films on metal substrate. Bull Pol Acad Sci-te 57(1):63–70. doi:10.2478/v10175-010-0106-3

Jeon I-D, Barnes MC, Kim D-Y, Hwang NM (2003) Origin of positive charging of nanometer-sized clusters generated during thermal evaporation of copper. J Cryst Growth 247(3–4):623–630. doi:10.1016/S0022-0248(02)02058-4

Jeon I-D, Kim D-Y, Hwang N-M (2005) Spontaneous generation of charged atoms or clusters during thermal evaporation of silver: Dedicated to Professor Dr. Duk Yong Yoon on the occasion of his 65th birthday. Z Metallkd 96(2):186–190

Jiang Y, Tang G, Shek C, Zhu Y (2009) Effect of electropulsing treatment on microstructure and tensile fracture behavior of aged Mg–9Al–1Zn alloy strip. Appl Phys A 97(3):607–615

Jones AC, Hitchman ML (eds) (2009) Chemical vapour deposition: precursors, processes and applications. Royal Society of Chemistry, London

Kashu S, Mihara Y (1995) Preparation of PZT Deposited films by using an aerosol jet printing system and their electrical properties. J Jpn Soc Powder Metall 42:314–317

Kashu S, Fuchita E, Manabe T, Hayashi C (1984) Deposition of ultra fine particles using a gas jet. Jpn J Appl Phys 23(12A):L910–L912

Katakuse I, Ichihara T, Fujita Y, Matsuo T, Sakurai T, Matsuda H (1985) Mass distributions of copper, silver and gold clusters and electronic shell structure. Int J Mass Spectrom 67(2):229–236. doi:http://dx.doi.org/10.1016/0168-1176(85)80021-5

Katakuse I, Ichihara T, Fujita Y, Matsuo T, Sakurai T, Matsuda H (1986) Mass distributions of negative cluster ions of copper, silver, and gold. Int J Mass Spectrom 74(1):33–41. doi:http://dx.doi.org/10.1016/0168-1176(86)85021-2

Kawakami Y, Seto T, Ozawa E (1999) Characteristics of ultrafine tungsten particles produced by Nd:YAG laser irradiation. Appl Phys A 69(1):S249–S252. doi:10.1007/s003399900286

Kim J-Y, Kim D-Y, Hwang N-M (2006) Spontaneous generation of negatively charged clusters and their deposition as crystalline films during hot-wire silicon chemical vapor deposition. Pure Appl Chem 78(9):1715–1722. doi:http://dx.doi.org/10.1351/pac199870040789

Kim C-S, Youn W-K, Lee D-K, Seol K-S, Hwang N-M (2009) Low-temperature deposition of crystalline silicon nitride nanoparticles by hot-wire chemical vapor deposition. J Cryst Growth 311(15):3938–3942. doi:http://dx.doi.org/10.1016/j.jcrysgro.2009.05.035

Koch R, Winau D, Führmann A, Rieder K (1992) Intrinsic stress of polycrystalline and epitaxial Ag, Cu and Au films on mica (001). Vacuum 43(5):521–523

Lee S-H (2014) Microstructure control of silicon thin films through controlling charged nanoparticles generated during hot wire chemical vapor deposition. Ph.D Thesis Seoul National University, Seoul

Lee BS, Barnes MC, Kim D-Y, Hwang NM (2002) Spontaneous generation of charged clusters of a few nanometers during thermal evaporation of copper. J Cryst Growth 234(2–3):599–602. doi:http://dx.doi.org/10.1016/S0022-0248(01)01747-X

Lee S-S, Ko M-S, Kim C-S, Hwang N-M (2008) Gas phase nucleation of crystalline silicon and their role in low-temperature deposition of microcrystalline films during hot-wire chemical vapor deposition. J Cryst Growth 310(15):3659–3662. doi:http://dx.doi.org/10.1016/j.jcrysgro.2008.05.009

Levlin M, Laakso A, Niemi H-M, Hautojärvi P (1997) Evaporation of gold thin films on mica: effect of evaporation parameters. Appl Surf Sci 115(1):31–38

Li Y, Chen J, Ma J (2015) Properties of Cu_2ZnSnS_4 (CZTS) thin films prepared by plasma assisted co-evaporation. J Mater Sci: Mater Electron 26(9):6546–6551. doi:10.1007/s10854-015-3251-5

Marchand P, Hassan IA, Parkin IP, Carmalt CJ (2013) Aerosol-assisted delivery of precursors for chemical vapour deposition: expanding the scope of CVD for materials fabrication. Dalton Trans 42(26):9406–9422

Mizubayashi H, Okuda S (1989) Structural relaxation induced by passing electric current in amorphous Cu_{50} Ti_{50} at low temperatures. Phys Rev B 40(11):8057

Morimoto R, Izumi A, Masuda A, Matsumura H (2002) Low-resistivity phosphorus-doped polycrystalline silicon thin films formed by catalytic chemical vapor deposition and successive rapid thermal annealing. Jpn J Appl Phys 41(2R):501

Munir Z, Anselmi-Tamburini U, Ohyanagi M (2006) The effect of electric field and pressure on the synthesis and consolidation of materials: a review of the spark plasma sintering method. J Mater Sci 41(3):763–777

Munir ZA, Quach DV, Ohyanagi M (2011) Electric current activation of sintering: a review of the pulsed electric current sintering process. J Am Ceram Soc 94(1):1–19

Orchard H, Greer A (2005) Electromigration effects on compound growth at interfaces. Appl Phys Lett 86(23):231906

Panjawi N, Naik A, Warwick ME, Hyett G, Binions R (2012) The preparation of titanium dioxide gas sensors by the electric field assisted aerosol CVD reaction of titanium isopropoxide in toluene. Chem Vapor Depos 18(4-6):102–106

Peineke C, Schmidt-Ott A (2008) Explanation of charged nanoparticle production from hot surfaces. J Aerosol Sci 39(3):244–252. doi:http://dx.doi.org/10.1016/j.jaerosci.2007.12.004

Peineke C, Attoui MB, Schmidt-Ott A (2006) Using a glowing wire generator for production of charged, uniformly sized nanoparticles at high concentrations. J Aerosol Sci 37(12):1651–1661. doi:http://dx.doi.org/10.1016/j.jaerosci.2006.06.006

Poppa H, Heinemann K, Elliot AG (1971) Epitaxial orientation studies of gold on UHV-cleaved mica during early stages of nucleation and growth. J Vac Sci Technol 8(3):471–480

Qin R, Samuel E, Bhowmik A (2011) Electropulse-induced cementite nanoparticle formation in deformed pearlitic steels. J Mater Sci 46(9):2838–2842

Rath JK, Stannowski B, van Veenendaal PATT, van Veen MK, Schropp REI (2001) Application of hot-wire chemical vapor-deposited Si:H films in thin film transistors and solar cells. Thin Solid Films 395(1–2):320–329. doi:http://dx.doi.org/10.1016/S0040-6090(01)01288-3

Samuel EI, Bhowmik A, Qin R (2010) Accelerated spheroidization induced by high intensity electric pulse in a severely deformed eutectoid steel. J Mater Res 25(06):1020–1024

SchÜTz R, Hartnagel HL (1989) Brief communication. New molecular beam epitaxy technique: radiofrequency plasma-assisted evaporation epitaxy for low temperature InP growth. Int J Electron 67(2):233–234. doi:10.1080/00207218908921074

Shaw G, Parkin IP, Pratt KF, Williams DE (2005) Control of semiconducting oxide gas-sensor microstructure by application of an electric field during aerosol-assisted chemical vapour deposition. J Mater Chem 15(1):149–154

Shen Z, Peng H, Nygren M (2003) Formidable increase in the superplasticity of ceramics in the presence of an electric field. Adv Mater 15(12):1006–1009

Spear KE, Dismukes JP (eds) (1994) Synthetic diamond: emerging CVD science and technology, vol 25. John Wiley & Sons, Hoboken, New Jersey

Takagi T (1986) Ionized cluster beam technique for thin film deposition. Z Phys D Atom Mol Cl 3(2):271–278. doi:10.1007/BF01384816

Takagi T, Yamada I, Yanagawa K (1975) Ion Implantation in Semiconductors. Springer, US. doi:10.1007/978-1-4684-2151-4

Takagi T, Yamada I, Sasaki A (1976) An evaluation of metal and semiconductor films formed by ionized-cluster beam deposition. Thin Solid Films 39:207–217. doi:http://dx.doi.org/10.1016/0040-6090(76)90638-6

Yang D, Conrad H (2001) Exploratory study into the effects of an electric field and of high current density electropulsing on the plastic deformation of TiAl. Intermetallics 9(10):943–947

Zhao J, Garay JE, Anselmi-Tamburini U, Munir ZA (2007) Directional electromigration-enhanced interdiffususion in the Cu–Ni system. J Appl Phys 102(11):114902

Zhou YZ, Zhang W, Guo J, He G (2004) Diffusive phase transformation in a Cu–Zn alloy under rapid heating by electropulsing. Phil Mag Lett 84(5):341–348

Index

© Springer Science+Business Media Dordrecht 2016
N.M. Hwang, *Non-Classical Crystallization of Thin Films and Nanostructures in CVD and PVD Processes*, Springer Series in Surface Sciences 60,
DOI 10.1007/978-94-017-7616-5

MIX
Papier aus verantwortungsvollen Quellen
Paper from responsible sources
FSC® C105338
www.fsc.org

If you have any concerns about our products,
you can contact us on
ProductSafety@springernature.com

In case Publisher is established outside the EU,
the EU authorized representative is:
Springer Nature Customer Service Center GmbH
Europaplatz 3, 69115 Heidelberg, Germany

Printed by Libri Plureos GmbH
in Hamburg, Germany